Engineering of Nanomaterials

Riyadh A. Al-Samarai · Yarub Al-Douri

Engineering of Nanomaterials

Springer

Riyadh A. Al-Samarai
University of Samarra
Samarra, Iraq

Yarub Al-Douri
Al-Bayan University
Baghdad, Iraq

Piri Reis University
Istanbul, Türkiye

University of Malaya
Kuala Lumpur, Malaysia

ISBN 978-981-95-5498-0 ISBN 978-981-95-5499-7 (eBook)
https://doi.org/10.1007/978-981-95-5499-7

This Springer imprint is published by the registered company Springer Nature Singapore Pte Ltd.
The registered company address is: 152 Beach Road, #21-01/04 Gateway East, Singapore 189721, Singapore

Preface

The science of nanotechnology is advancing rapidly, and what was once considered impossible is now possible. Over the past few decades, nanomaterials and nanotechnology have evolved from basic ideas to a wide range of applications touching all aspects of life. Rapid development of nanomaterials in technology and the environment for various purposes. New methods for producing different types of nanomaterials allow them to be rapidly and widely used in life sciences, catalysis, and materials engineering. Nanomaterials can be made in a variety of ways, each with their special qualities for useful applications.

The development of various nanomaterials, as well as the design, creation, and use of engines, machines, and nanodevices in technology will be possible thanks to the application of nanoscience and nanotechnology in technology. Several technological applications have been developed and applied for the benefit of humanity. The Industrial Revolution led to concerns about corrosion and deterioration of various materials, as well as pollution of ecosystems and the atmosphere. Knowledge dissemination is essential to review the latest developments in nanomaterials and their applications. Therefore, the creation of specific nanomaterials and their applications in engineering, environmental, and biological sciences is the main focus of this special issue. The distinctive characteristics of nanomaterials, especially advanced nanomaterials (CNMs), have played a critical role in the revolution in nanotechnology.

The most striking aspects of these structures are their optical, chemical, mechanical, and electrical properties, which open up many potential applications in the future. However, ongoing research into new CNMs and rapid development make it difficult for the industry to keep up with the rapidly evolving state of technology. Moreover, transporting CNMs from the laboratory to the market is a challenging task. To bridge the gap between the most recent research and its technological implementation, it is necessary to highlight the most recent advances and discoveries made in the field of CNM. The literature on CNM is currently dispersed between review articles and original articles. In addition to an introduction and critical commentary, this book classifies the subject and will serve as a valuable reference for scientists, researchers, and students working with nanomaterials for engineering applications. This special issue publishes original research and review articles on new ideas related to the development of various nanomaterials for creative applications, focusing on structural, experimental, and applied features, including but not limited to the following. Our goal is to provide a forum for experts in the fields of nanotechnology, nanomaterials, and nanocomposites.

Possible topics include, but are not limited to, the production of nanomaterials, the characterization of nanomaterials, nanomaterials for use in design, nanomaterials for building materials, nanomaterials related to engines, and environmental applications of nanomaterials. Catalysts in nanotechnology, nanocomposite materials. Ecological synthesis, nanomaterials with antimicrobial properties.

The contents of this section of the textbook are organized into ten chapters covering the following topics: Using a preferred selection index approach, the Chap. 1 presents applications of nanocomposite developments in science, technology, and engineering that have led to specific materials, methods, and solutions to these problems. Chapter 2: The use of nanofibers in tissue engineering as a new concept was presented. How autografts can be replaced with allografts, which use tissue from another organ of the same species. This includes thin film synthesis and applications due to its interesting surface features, as well as many applications in fields ranging from microelectronics. This chapter will review some of the thin film technologies, constraints, and promising results presented in Chap. 3. The importance of nanomaterials in engineering applications. It can be said that nanomaterials are becoming more important in technology development due to their superior properties, which were presented in Chap. 4.

Mapping nanomaterials through green nanotechnology. Due to their many applications and unique properties, nanomaterials have attracted much scientific attention over the past few decades and are presented in Chap. 5. Glucose sensors use metal nanoparticles, and for people with diabetes, the need to measure glucose levels is very high. The latest and most important findings of the researchers are presented in Chap. 6. Nanocoatings: design, development, and applications. Coatings are defined and designed as the coating of one material on top of another material, which is usually called a substrate, and how they have evolved in advanced ways as described in Chap. 7: Perspectives on Nanomaterials Development. The world is witnessing a revolution thanks to nanotechnology and engineering nanomaterials (ENMs), and solutions to existing problems which is presented in Chap. 8. Catalysts in Nanomaterials In addition, artists have used these molecules to create and decorate glass and ceramics. Researchers' findings and associated science are discussed in Chap. 9, followed by developments in sensors and medical devices based on nanocomposites. Apart from several advantageous properties such as high mechanical strength, flexibility, tunable optical transparency, photodynamic and photothermal activity, nanoscale nature and high absorbance provide the basis for progress in the field of nanocomposites, the reviews in Chap. 10 advance the field of research and development of nanostructured materials.

Samarra, Iraq Riyadh A. Al-Samarai
Istanbul, Türkiye Yarub Al-Douri

Contents

About the Authors

Prof. Dr. Riyadh A. Al-Samarai is Assistant Dean for scientific affairs at the University of Samarra, Faculty of Engineering. Al-Samarrai is the Winner of the Iraqi Scientific Excellence Award for the year 2024. He launched a master's program in electromechanical engineering at Samarra University, College of Engineering. He received many honors and awards, including Winner of the Creativity Award for Scientists in Malaysia, and many honors in international forums and scientific conferences, the Iraqi Forum for Intellectuals and Academics (IFIA) Award, and the Best Research Paper Award at the Conference of Asian Universities in Malaysia, a total of three awards. Al-Samarai is the Managing Editor of the Samarra University of Engineering and Science Journal. Al-Samarai has 32 years of research experience, 23 years of university teaching experience and 16 years of administrative experience.

Prof. Dr. Yarub Al-Douri is Fellow of European Academy of Sciences. Al-Douri is the Winner of the prestigious Khalifa International Award for Date Palm and Agricultural Innovation 2024. He has initiated Nanotechnology Engineering M.Sc. Program and Nano Computing Laboratory. He has received numerous accolades including Winner of IAAM Scientist Award by International Association of Advanced Materials, Sweden 2023, 2022, World's Top 2% Scientists by Stanford University, USA, 2025, 2024, 2023, 2022, 2021 and 2020, World's Top 2% Scientist Career-Long Citation Impact by Stanford University, USA, 2020, OeAD Award, Austria, 2020, Japan Society for the Promotion of Science (JSPS) Award 2019, Asian Universities Alliance (AUA) Award 2019, Iraqi Forum for Intellectuals and Academics Award (IFIA) 2019, TWAS-UNESCO Associateship (Twice) Award 2015 and 2012 and Best Paper Award at Global Conference on Energy and Sustainable Development in UK 2015, the total is 81 awards.

Chapter 1
Prospects for the Development of Nanomaterials

Nanotechnology and nanomaterials have changed the global and enterprise network management (ENM) systems. It has attracted a lot of interest from the public, academic community, researchers, and business sector. Today, nanotechnology is a well-established scientific field with many applications in several fields that has experienced an exciting growth period. Despite the maturity and established nature of nanomaterials field, the process of moving from research and development to commercialization is still labor intensive due to the reliability and reproducibility of technology. To overcome this limitation, each stage of process development must be standardized, tested, and evaluated. To ensure smooth technology transfer and upgrading, the investors and institutions have conducted extensive research. This chapter highlights important developments, uses, and challenges of nanotechnology and ENM. Lastly, the applicability of this topic and potential research directions are considered.

1.1 Introduction

The conventional engineering materials have been limited in many applications, and new advances are needed to meet the growing demands for engineering materials. Combining different materials to produce higher qualities than a single element or molding material or components into a specific structure can improve the performance of material. The relationship between material and structure can occur at various length levels, including micro-, meso-, and macro-levels, which has potential applications in a wide range of industries. The basic relationship between materials and their structure influences their overall properties (mechanical, thermal, chemical, magnetic, etc.) and their applications are covered in [1–4]. Research and development over the past 20 years have led to the creation of customized materials with specific

R. A. Al-Samarai and Y. Al-Douri, *Engineering of Nanomaterials*,
https://doi.org/10.1007/978-981-95-5499-7_1

nanoscale characteristics, leading to advances in nanotechnology. As a result, corporations, process engineers, and scientists have access to a much wider range of materials science tools. Stronger, lighter, more reactive, and more durable nanomaterials have been produced. The materials with complex structures and improved electrical conductivity are available for a wide range of applications in various scientific fields of cutting edge of materials science.

Molecular biology, organic chemistry, surface science, molecular engineering, energy storage, and semiconductor physics are under the umbrella of broad field of nanotechnology. Among these developments, nanotechnology has emerged as a promising field that has revolutionized the design of materials and opened up new possibilities in various industries and practical applications. The field has developed rapidly, while a wide range of nanoscaled materials and processes have found their way from experimental settings into commercial products of everyday life. In particular, nanotechnology is most promising for food, energy, healthcare, electronics, and environment [2, 5]. In terms of appearance, the nanomaterials can be considered in one of the following: one-dimensional (1D) materials can have only one variable (e.g., carbon nanotubes), two-dimensional (2D) materials can have two variables (e.g., graphene), and three-dimensional (3D) materials can have three variables (e.g., graphite) [6]. The image of 0D, 1D, 2D, and 3D nanomaterials is shown in Fig. 1.1. In addition, due to their critical nanoscale sizes that bring them close to cells, genetic molecules, microbes, and tissues, the nanomaterials are of great importance and play a dynamic role in pharmaceutical, clinical, biochemical, biological, and biomedical sciences [1].

Carbon nanotubes are expected to replace silicon as the main component used to create electronic products for future. Quantum nanowires with high conductivity and speed can be made using carbon nanotubes as well as microchips. Electronics with higher storage capacity, longer battery life, and increased safety can be made using carbon nanotubes. Nanotechnology has made significant contributions to the energy sector especially renewable energy. For example, nanofluids are used to improve the transmission efficiency of solar connectors; quantum dots and carbon nanotubes are

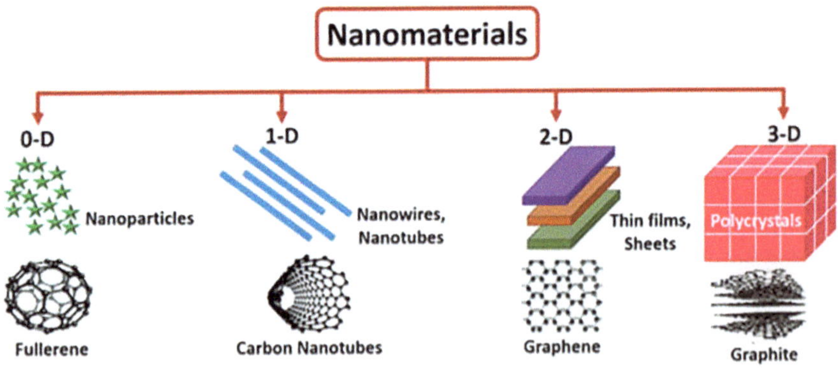

Fig. 1.1 Schematic of nanostructured materials in 0D, 1D, 2D, and 3D [1] [Open access]

used to increase energy absorption in solar cells; and nanostructured catalysts are used to improve the performance of fuel cells. There is no doubt that nanotechnology will be important in facilitating the global transition from fossil fuels to renewable energy sources [1–4].

Additionally, nanotechnology benefits the environment by developing carbon capture technology. For example, in 2019, MIT researchers have developed a new carbon capture technology that uses carbon nanotubes. The research is still being conducted to determine how nanotechnology can further improve carbon capture methods, as it has not yet realized its full potential in this area. Although nanotechnology has many interesting applications in biomedicine, its greatest contributions have been used in the development of vaccines, drug delivery systems, and innovative diagnostic tools. Nanotechnology is currently being explored for a wide range of potential applications in biomedicine. More research in this area could completely change healthcare.

The food business is using nanotechnology to help us cope with the food crisis as one of the humanity's greatest challenges. Global population growth will place significant strain on our strained agricultural systems as they produce more food. With the wasted food, the scientists have realized that reducing food waste is a critical strategy for addressing food affordability. So, scientists have developed an innovative fruit coating that increases the shelf life of fruits using nanoparticles. Such advances in nanotechnology will have important implications for our future efforts to improve food security [2]. Today, a large number of international companies are investing in new applications of nanotechnology. Nanodevices are expected to become the most profitable category in the market during this decade, and several emerging trends will drive the expansion of this field [1–3].

The global nanotechnology market is expanding significantly due to growing using of nanoscaled materials and devices in engineering and biological fields; however, several major obstacles prevent these gadgets from being widely used for commercial purposes. The high cost of technology, as well as its reliability and effectiveness in adverse weather conditions are the main limiting factors. However, the market is expected to present lucrative opportunities in the coming years due to increased government funding and support, as well as emergence of new self-powered gadgets.

Finally, many enterprises are currently investing in nanotechnology research. The developing countries have become leaders in nanotechnology research, including Brazil, India, the Philippines, Chile, Mexico, and South Africa. These countries have developed government-funded programs and research organizations. The nanotechnology market is a market of international cooperation and effort [2–4].

1.2 Techniques of Synthesis

Synthetic technologies encompass a broad range of methods used in chemistry, biology, materials science, and computer science to combine elements to form complex structures or systems [7–9]. The main strategies covered in this overview are chemical synthesis, which creates new compounds through reactions, biological synthesis, which produces biomolecules using living organisms and computational synthesis, which creates and optimizes designs using algorithms. Using scientific techniques in basic research involves studying the structure of peptidoglycan, an important component of bacterial cell walls. Clear and biological analysis have helped to develop a complete understanding of structure and activity of this protein. Before being directly visualized in living Lactococcus lactis (Fig. 1.2d), the high-resolution architecture of peptidoglycan is initially discovered in pure sacs from Bacillus subtilis (Fig. 1.2a). According to both studies, peptidoglycan is organized into a regular pattern of cables with cross-striations perpendicular to the long axis of the cell, at least in Gram-positive bacteria. Peptidoglycan will undergo remodeling as bacteria grow, changing from an irregular architecture during the exponential growth phase to an ordered cable architecture during the stationary phase, according to the later analysis (Fig. 1.2b). Furthermore, how the flexibility and structure of group B peptidoglycan cause streptococci to change under increased turgor pressure (Fig. 1.2c). Moreover, the arrangement of peptidoglycan on the sacs of Gram-negative bacteria D. coli has been deciphered (Fig. 1.2e) [5].

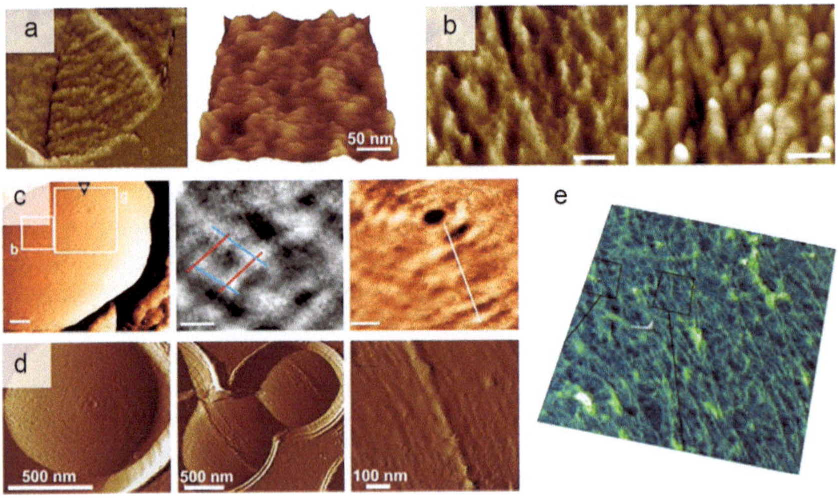

Fig. 1.2 High-resolution imaging of peptidoglycan. **a** Gram-positive bacteria Lactococcus lactis (right) and Bacillus subtilis (left) deflect images of their sacs. **b** The peptide sidewall of B. subtilis in the mid-exponential phase (left) and stationary phase (right) shows distinct structural arrangements. **c** High-resolution images of live group B Streptococcus show a ring-shaped arrangement of bands around the pole (right) and a nanoscale network-like surface architecture (left, middle). **d** Photographs of live L. lactis cells also show periodic bands parallel to the short axis of the cell on the longitudinal side of the bacterium (middle right) and a circular arrangement of peptide glycan at the bacterial pole (left). **e** The assembly of peptidoglycan fragments on poly-L-ornithine directly reveals the organization of the glycan chain in the membrane of the Gram-negative bacterium E. coli. [5] [Open access]

Each technique has specific benefits and challenges that are determined by variables such as efficiency, specificity, and scalability. Innovation continues to be driven by advances in synthetic technologies, which make it possible to create new materials, drugs, and technologies. The continued integration of multidisciplinary approaches is expected to improve the precision and utility of synthesis in addressing current scientific problems. The production of nanomaterials and nanostructures is a critical aspect of nanoscience and nanotechnology. Only until nanostructured materials of the required size, shape, morphology, crystal structure, and chemical composition become available will new physical properties and uses of nanomaterials be possible. Scientific and engineering breakthroughs have accelerated the creation of nanomaterials to achieve special properties that differentiate them from those of bulk materials. Furthermore, these structures can be obtained by building molecules from micro- and nanoscale materials or by reducing their size [6]. These synthetic methods result in the diffusion of nanomaterials and liquid particles. Physical, chemical, biological, and hybrid methods are generally selected for the production of nanomaterials depending on the intended use [10–12]. These methods are widely used and classified under the traditional methodology.

1.3 Limitations of Existing

The multidisciplinary applications of nanotechnology hold enormous potential, but there are some obstacles to their widespread use. A major hurdle is the difficulty of scaling up synthesis techniques without sacrificing the homogeneity and scalability of nanomaterials. The long-term health and environmental effects of nanoparticles are a major concern, making regulatory approval procedures even more difficult. Furthermore, the lack of standardized testing procedures limits the comparability of results between studies and hinders progress in understanding the field. Furthermore, the high costs of research and development, especially in resource-limited settings, may hinder access to innovative applications. Resolving these issues is essential to realizing the full promise of nanotechnology in a variety of industries [11–13].

1.3.1 Limitations in the Development of Methods

Nanotechnology, with its many scientific and creative uses, is a rapidly growing interdisciplinary field. The promise is broad, ranging from materials science, which uses nanoparticles to improve strength and conductivity, to medicine, which allows for on-demand drug delivery and advanced imaging tools. Through efficient catalysts, nanotechnology offers answers for sustainable energy and pollution control in environmental science. Despite its potential, problems with scalability, legal frameworks and ethical issues remain, requiring continued study and collaboration across disciplines. The ultimate goal is to use nanotechnology to solve urgent global problems and foster breakthroughs that benefit society as a whole.

1.3.1.1 Temperature

Beyond their use, technology teams play an important role in addressing secondary pollution issues where they can develop solutions for non-industrial backwardness and air purification. With a wide range of scientific and innovative applications, nanotechnology is an interdisciplinary field that is rapidly expanding [14]. The Earth's atmosphere is negatively affected by growing industrialization and population explosion, resulting in the formation of a large number of unwanted and dangerous components. Figure 1.3 shows a diagram of the top-down and bottom-up methods for the fabrication of nanomaterials [1].

Green synthesis is a growing promise in nanotechnology to solve this problem. It uses safe, affordable, non-toxic, and sustainable methods for the synthesis of nanomaterials [15]. Some chemical and physical methods for creating nanomaterials are now described in the literature, each with higher throughput and precisely controlled size and shape of the nanomaterial. Through a small opening where the flame can be functional, the feedstock, which can be either liquid or vapor, can be transferred to furnace at a very high pressure rate [16]. To extract the nanoparticles, this resulting combustion is subsequently described as air. The pyrolysis process has many advantages including ease of use, cost-effectiveness, efficiency, and effectiveness with remarkable results. However, the disadvantages of these methods include significant waste generation, use of hazardous materials, increased energy, and financial losses. These major issues impact the economic and environmental aspects of scaling up nanomaterials to commercial scale. A greener approach to nanomaterial synthesis offers safe, affordable, and environmentally friendly alternatives to traditional methods [17]. Greener processes for producing nanomaterials have become

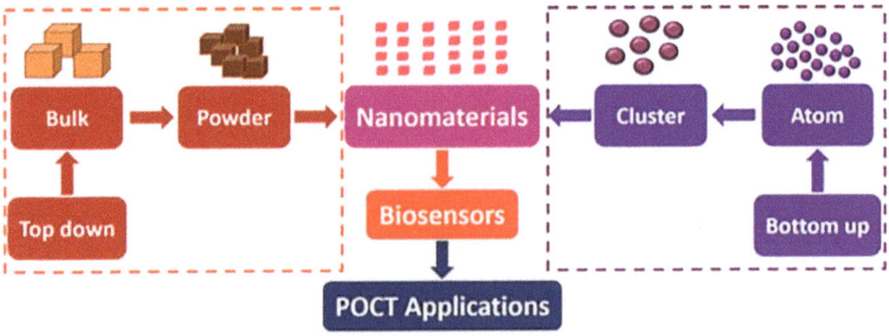

Fig. 1.3 Schematic showing top-down and bottom-up methods for synthesizing nanomaterials[1] [Open access]

more popular in recent years. These methods have several advantages, including low costs, environmentally friendly processes, and production of nanomaterials in small sizes [18]. In the past, various natural resources such as extracts, microbes, and plant leaves have been used to create nanoparticles [19]. This in-depth analysis focuses on current developments in the sustainable production of nanomaterials from plant extracts and leaves [20].

Studied semiconductors to solve several important and complex human problems that need to be solved in this century. TiO_2 has proven several remarkable properties, including wide band gap, physicochemical properties (such as high thermal and chemical stability), excellent environmental friendliness, and earth richness among other properties. These characteristics make the material one of the most adaptable components in a wide range of industries, including sensors, biomedical implants, metallurgy, UV sunscreens, pigments, solar energy harvesting, hydrogen production, and catalysis. New characteristics of TiO_2, including size, surface area, phase composition, and morphology, have a significant impact on the physicochemical characteristics of the material, including particle size, hollow structures, 0D–3D, hierarchical, and micro-sized spheres. In addition to sustainable nanomaterial production technologies, semiconductors such as titanium dioxide have become essential components in many technological applications. Their special properties are crucial for achieving efficient and environmentally friendly solutions in a variety of industries. As can be seen in Fig. 1.4, most commonly used methods for creating the above three types of nanofilms are currently reverse phase, interface polymerization, track etching, and electrospinning [21]. Therefore, the main processes involved in creating nanofilms are briefly discussed and some representative examples are given [2].

The particles exhibit intriguing characteristics at sizes below 100 nm, mainly due to two physical processes. Apparent quantization of electronic states produces very sensitive size-dependent effects such as optical, magnetic, mechanical, chemical, and thermal properties of materials that are modified due to their high surface-to-volume ratio. The distinctive physical and chemical characteristics of nanoparticles make them ideal for a range of specific applications [2, 4]. Nanomaterials can be synthesized by a variety of methods, and each synthesis has its own set of difficulties and limitations depending on whether it is top-down or bottom-up. Some of them are presented in Table 1.1.

But the application of this idea is simple, a careful analysis of the structure is carried out using advanced techniques as shown in Fig. 1.5, which is carefully photographed after annealing at 800 °C. It has also raised awareness of the need for rigorous structural optimization initiatives to create improved applications. Using the reduced scanning area, Fig. 1.5 shows AFM and SEM morphologies of samples 32 (Fig. 1.5a, e), 64 (Fig. 1.5b, f), 100 (Fig. 1.5c, g), and 200 nm (Fig. 1.5d, h) after annealing at 800 °C. According to AFM, both 32 and 64 nm samples (Fig. 1.5a, b, respectively) still show a particle-based surface over a scanning area of 1 μm × 1 μm in lateral dimension, although they appear to align along preferred directions, suggesting that the material organization becomes more ordered. In contrast, the 100 and 200 nm samples clearly show blocks separated by severe cracks. Importantly,

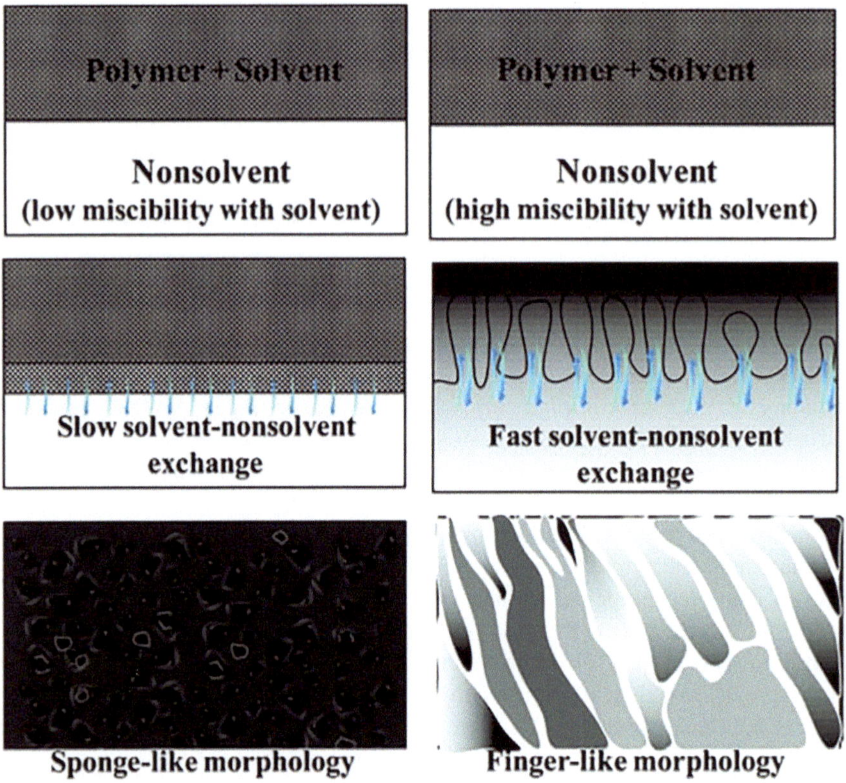

Fig. 1.4 Membrane cross-sectional shapes are obtained using both immediate and delayed delamination procedures [2] [Open access]

the topography of 100 nm sample (Fig. 1.5c) mimics the formation of crystalline blocks [3].

Due to their unique properties, nanostructured materials have recently attracted great scientific interest both domestically and abroad. TiO_2 has achieved a favorable position among various nanostructured functional materials of all transition metal oxides. Other names for titanium dioxide include titanium dioxide, TiO_2, titanium dioxide (IV), food coloring E171, and white pigment. After Fujishima and Honda [3] have discovered that TiO_2 is a photocatalyst for the photolysis of water, the potential use is publicly recognized for the first time. This TiO_2 is the most promising due to its photocatalytic properties, which are capable of killing some notorious bacteria such as *E. coli, L. acidophilus*, and *S. cerevisiae*. It also can inhibit malignant cells (HeLa), is biocompatible, environmentally friendly, and has huge potential for new fields such as solar panels and water and air purification. Here, the choice of nanomaterial production process depends on the type or material of interest, including size, quantity, and dimensions [3]. Surface roughness is highlighted using a black-to-white color scale. All photographs are aligned to the same color contrast ranging from a minimum of 0 (black) to maximum of 5 nm (white) to facilitate direct comparison. To develop TiO_2 and achieve the best performance in its various applications, its physical and chemical properties depend greatly on method of its synthesis as shown in

Table 1.1 Different techniques for synthesizing ENMs: advantages, limitations, and difficulties [22] [Open access]

Methods	Merits	Limitations and challenges
Top-down methods	A destructive technique is used in this strategy. Larger, bulkier molecules are utilized in the creation of nanomaterials	
Ball milling (mechanical milling)	Involves the reduction of the particle size with high-energy ball milling. John Benjamin developed this method in 1970	It requires high energy; milling time is quite long. Steel balls have an issue with contaminating the powder, and there is a threat of grinding the sensitive microstructure. Longer time is required for preparing materials having a size of less than 20 nm
Mechanochemical synthesis	Synthesis is based on repeated deformation, welding, and splintering of the reactants mixture	It creates highly sensitive microstructures that can be easily affected by atmospheric impurities
Laser ablation	Uses laser irradiation to minimize the particle size up to the nanolevel. In general, copper vapor lasers are used with Ti: sapphire (titanium-doped sapphire) laser or Nd: YAG (neodymium-doped yttrium aluminum garnet) laser at 106 μm	Difficult to realize precise in-bulk processing and requires expensive, large-dimension equipment. The colloidal solution laser path gets blocked, and laser energy gets absorbed by formed nanomaterials, reducing the ablation rate
Ion sputtering method	Preparation of nanomaterials from several metals using magnetron sputtering of metal targets. In this synthesis, collimated beams of nanomaterials are developed, and nanostructured films accumulate on silicon substrates. The process requires very low pressure (1 mTorr)	The nature of gases used in sputtering (e.g., He, Ne, Ar, Kr, Xe) affects the composition, peripheral morphology, texture, and optical properties of nanocrystalline metal oxide films
Bottom-up methods	A method that builds gradually, reversing the process of creating nanomaterials from basic materials by modifying atomic arrangements	
Physical vapor deposition method	A properly controlled vacuum technique (e.g., sputtered deposition, thermal evaporation) causes vaporization of designated material, which later condenses on a substrate. Different nanomaterials form on thin film surfaces or attach to other nanomaterials	This method is expensive and generates materials in small quantities. Cost minimization is essential for industrial-scale implementation

(continued)

Table 1.1 (continued)

Methods	Merits	Limitations and challenges
Chemical vapor deposition method	Used in the preparation of multicomponent carbon powder materials for color pigments and film deposition. Precursor gases, containing useful atoms for film formation, undergo chemical reactions to deposit thin films	Corrosive, toxic, and explosive gases are involved, posing chemical hazards
Sol–gel method	(a)Prehydrolyzed silica sol enables direct mixing of nanoparticles, metal, and metal oxide.(b)The sol holds matrix-forming species, leading to gel formation.(c)Reduction of metal before hydrolysis and complexation with silicone	Complex processing steps and sensitivity to reaction conditions
Chemical reduction method	The ionic salt is reduced in a suitable medium in the presence of surfactant using different reducing agents	Cost, toxicity, poor efficacy, contamination risks, and reducing ability limitations
Hydrothermal method	Based on the reaction of aqueous solution vapors with solid material at high pressure and temperature, leading to the deposition of small particles	Challenging to manage due to reproducibility and consistency constraints
Biological/green synthesis method	Utilizes natural resources like microbes to produce nanoparticles	Only useful in specific cases; limited by high costs, complex reactions, and safety concerns
Biosynthesis of nanoparticles by bacteria	Materials research increasingly focuses on plant-mediated nanoparticle synthesis	Health hazards and safety risks due to variable nanoparticle sizes
Fungus synthesis of nanoparticles	Various plant extracts are used to create metallic nanostructures, such as copper, zinc, silver, and gold	Increased production costs due to heating conditions
Other methods of nanoparticle synthesis		
Electrochemical deposition	Entails electrodeposition or template synthesis of nanomaterials. An ion etching technique is used to produce porous alumina membranes	Limited control over nanoparticle shapes and sizes
Microwave-assisted nanoparticle preparation	Microbes and cellular sites aid synthesis, offering benefits over thermal heating. Homogeneous heat distribution leads to faster crystallization and nucleation	

(continued)

Table 1.1 (continued)

Methods	Merits	Limitations and challenges
Supercritical fluid technology	Exploits the physicochemical properties of SCF, allowing manipulation of viscosity, density, surface tension, and diffusion coefficients. Temperature and pressure variations cause phase transitions between liquid and gas states	Operates at high temperatures and critical pressures
Ultrasound technique	Uses ultrasound for synthesis. The sonochemical process includes ultrasonic spray pyrolysis and sonochemistry frequency techniques	Limited scalability and requires precise control of ultrasonic parameters

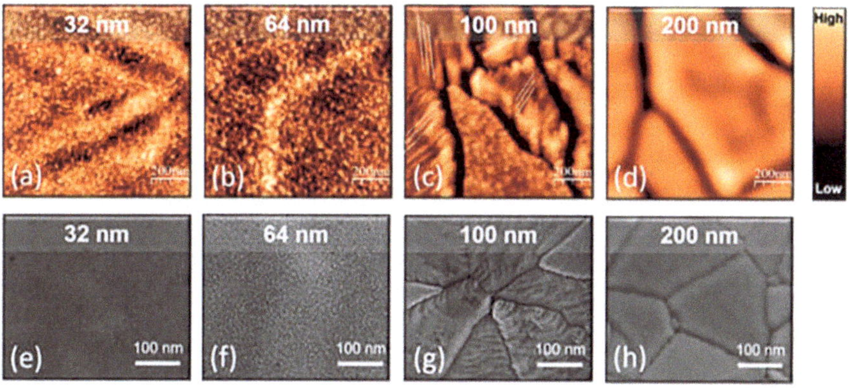

Fig. 1.5 Tapping mode AFM topography of **a** 32, **b** 64, **c** 100, and **d** 200 nm samples annealed at 800 °C, with a size of 1 × 1 μm. For the 32 and 64 nm samples, the color scale is 0–5 nm; for the 100 and 200 nm samples, it is 0–20 nm. SEM pictures of samples that are annealed at 800 °C and measured at **e** 32, **f** 64, **g** 100, and **h** 200 nm. [3] [Open access]

the spectral analysis and structural properties as described in the coming paragraph. Moreover, a key component of the potential use of TiO_2 is phase stability. Depending on how they are used, the main characteristics of TiO_2-NPs vary depending on the application. For example, in the field of biomedicine, elasticity, mechanical compatibility, and deformability are vital characteristics. TiO_2 has been improved through various synthesis methods including sol–gel, hydrothermal, radiofrequency deposition, laser ablation, mechanochemistry, and others. In particular, these methods are based on certain design concepts such as increasing the number of active sites, morphology, band structure, and specific phase orientation. Initially, 1D nano-TiO_2 has attracted much attention. The Raman Eg band of anatase is still responsible for the peak at 194 cm^{-1}. The anatase or brookite phases are responsible for two peaks at higher wavenumbers, 394 and 637 cm^{-1}. The considerable convolution effect with

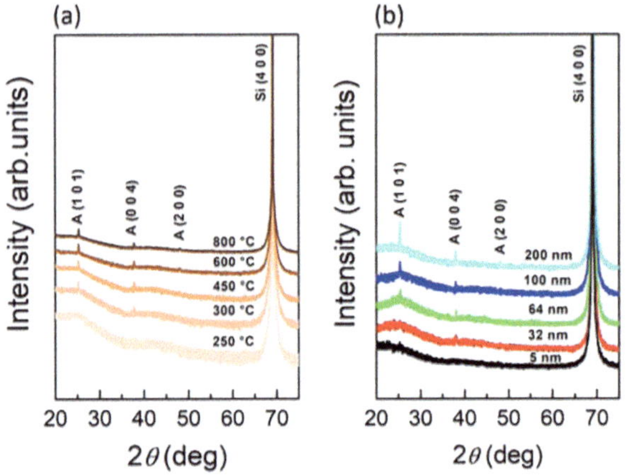

Fig. 1.6 **a** XRD spectra of 200-nm-thick TiO2 samples annealed for 12 h at different temperatures; **b** XRD spectra of TiO2 films annealed for 12 h at 800 °C and having thicknesses of 5, 32, 64, 100, and 200 nm. The anatase phase is denoted by the letter A. [3] [Open access]

the Si substrate bands, which are located at 460 cm^{-1} and between 618 and 675 cm^{-1}, may affect the exact determination of the peak centers. The Raman bands of anatase and brookite are expected to be very close (expected values are 399 and 639 cm^{-1} for anatase and 396 and 636 cm^{-1} for brookite) [3]. However, the brookite phase can be excluded by the high-resolution X-ray diffraction spectrum shown in Fig. 1.6b. [3].

Different annealing methods in air for 12 h with annealing plateau temperatures ranging from 250 to 1000 °C are used to investigate the transition from the amorphous to crystalline phase to study the evolution of TiO_2 crystallization as a function of thickness. Together with simple and trouble-free preparation procedures, higher surface area and improved quantum size are observed. Physical vapor deposition (PVD) is a widely used vacuum deposition technique. This technique produces a vapor coating material from a target in the form of atoms, molecules, or ions using physical processes, including atomization and evaporation. PVD uses physical processes, including sputtering and evaporation, to produce a vapor of the coating material from a target in the form of atoms, molecules, or ions. However, surface characterization techniques are employed to elucidate the impact of heat treatment on the crystalline structure of TiO_2. As illustrated in Fig. 1.7, AFM and PVD analyses indicate compositional variations with increasing temperature. In particular, Fig. 1.7 shows the temperature history of surface morphology of 64 nm sample, reflecting comparable analyses performed on other samples. AFM topography in tapping mode is shown for as-grown sample (Fig. 1.7a) and heated at three representative temperatures: 300 °C (Fig. 1.7b), 800 °C (Fig. 1.7c), and 1000 °C (Fig. 1.7d). The cross-sectional size of the sample is 10 μm × 10 μm. With increasing heating temperature, there is a transition from disorganized particles (Fig. 1.7a) to large platelets showing a similar radial substructure of Ta_2O_5 to display a similar physical arrangement [3].

After transportation, they are deposited on the surface of substrate, causing the formation of a coating. Heat-sensitive materials can be coated with PVD because

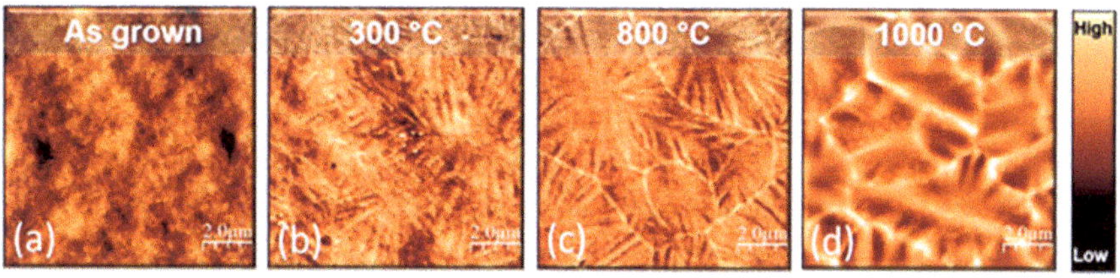

Fig. 1.7 AFM touch-mode topography of 64 nm TiO_2 with dimensions of 10 μm × 10 μm obtained and annealed at 300, 800, and 1000 °C. The color scale in (**a–c**) and (**d**) covers the range from 0 to 80 nm, respectively [3] [Open access]

the substrate temperature is well below the melting point of the target material. Commonly used PVD techniques include sputtering, ion plating, pulsed laser deposition, and evaporative deposition. Sputtering is a better option than evaporation for target materials such as ceramics and refractory metals, which are difficult to deposit by evaporation. In addition, coatings deposited by sputtering generally have a stronger bond to the substrate than coatings deposited by evaporation [2, 4].

1.3.1.2 PH

Interest in environmentally friendly procedures, or "green chemistry," for the production of nanoparticles in the field of nanobiotechnology has increased due to growing concerns about the environmental impacts of traditional chemical approaches. The production of metal nanoparticles (NPs) using green chemistry and their using as novel drug delivery systems for the treatment of bacteria and cancer are the main topics of this review. This includes an analysis of physicochemical characteristics and biological activities of green synthetic nanoparticles (NPs), as well as an overview of their synthesis and characterization. Notably, this study goes beyond other reviews by offering a thorough examination of current research that uses computer modeling to better understand the interactions between these IRs and their targets. Modeling helps to predict any putative biological activity and also helps to understand the biological mechanism of NPs. This aims to shed light on the role that green chemistry to play a role in the development of smart medicine, advanced cancer therapy, and other disease treatments for providing a broad perspective and fresh concepts [2, 5].

1.3.1.3 Time

Since nanobiotechnology can effectively change the chemical, physical, and optical properties of metals, it is developing rapidly in the modern period. As a result, much attention is being paid to the creation of innovative methods for the synthesis of various types of nanoparticles of a certain size and composition using biological sources. However, most currently used approaches are expensive, environmentally harmful, and inefficient in terms of material and energy use. The quality and quantity of synthesized nanoparticles, as well as their characteristics and applications, are significantly influenced by some elements, including synthesis technique, pH, temperature, pressure, time, particle size, pore size, environment, and proximity. Moreover, the characterization of the resulted nanoparticles is critical for their potential applications in numerous medical as well as biological applications. The current study highlights several factors influencing the ability of green nanobiotechnology to synthesize nanoparticles and range of methods used to characterize the particles to assess their potential for use in biomedical and environmental applications [2, 6].

1.3.1.4 Pressure

Silver nanoparticles are an important innovation in the field of nanotechnology due to their high stability and low chemical reactivity compared to other metals. Silver nanoparticles have received great attention in biological applications due to their unique physicochemical properties. Some methods are applied to manufacture silver nanoparticles such as physical and chemical techniques. These methods typically synthesize nanoparticles using toxic chemicals as reducing agents. However, in the past few decades, many efforts have been made to develop green synthesis methods to avoid the use of hazardous materials. The environmentally benign synthesis of nanoparticles using bacteria, fungi, and plants, as well as the role of plant metabolites in this process, is an important area of study. Numerous biomolecules have found in biological extract including vitamins, phenolic compounds, alkaloids, terpenoids, and flavonoids act as caps and reducing agents throughout the biosynthetic process. The potential uses of biosynthesized AgNPs in various fields, including antifungal, antibacterial, antiviral, and agriculture, have attracted the attention of researchers [2, 6].

1.3.1.5 The Size and Form of Particles

This has provided a detailed description of the synthesis, characterization, and use of nanoparticles (NPs), which come in a wide variety of forms. Samples are examined using advanced methods such as tomography and contact angle analysis to show how production conditions affect the structure and properties of thin films as shown in Fig. 1.8. Figure 8a shows the cross sections of uZO-80 films deposited at different sputtering pressures, including by comparing the cross-sectional morphology. Due to the encouragement of collisions in the gas phase, the growth becomes less directional and tilt of pillars is less noticeable as the pressure increases. The growth mode changes to a diffusion mode with no preferred direction at highest value ($7.4 \times 10 - 3$ mbar) and pillars develop perpendicular to the substrate surface. Figure 1.8b illustrates typical AFM surface images of uZO-80 films formed at different sputtering pressures. Hydrophilic surfaces have contact angle (CA) values below 90° and hydrophobic

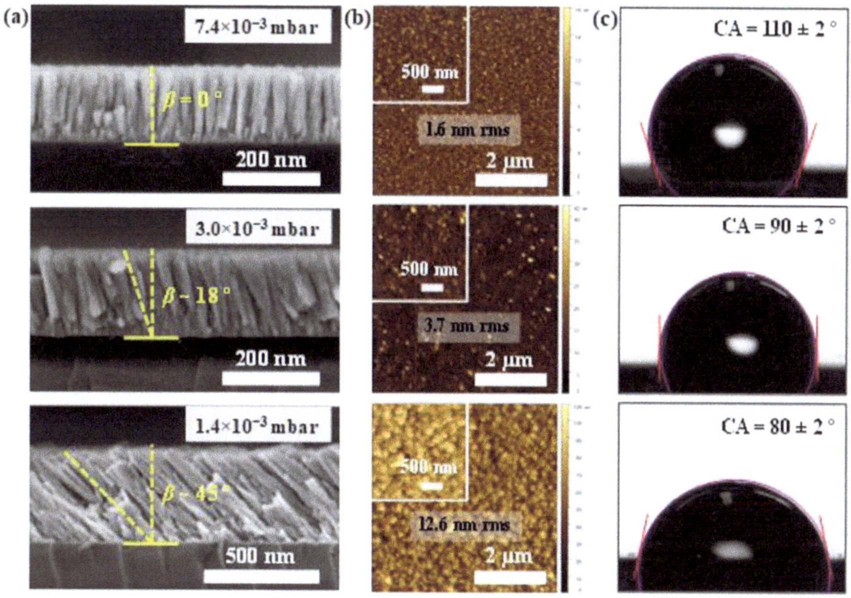

Fig. 1.8 **a** SEM cross-sectional images; **b** AFM topography images (5 × 5 μm^2) showing (rms) roughness; **c** contact angle (CA) measurement of a 0.75 μL distilled water droplet on uZO films developed at different pressures. The red lines represent the tangent of the water droplet and the purple outline represents the droplet boundary [4] [Open access]

surfaces have CA values above 90°. Figure 1.8c shows photographs of water droplets on several uZO surfaces [4].

The study of how CA varies relative to the surface axis in the plane, i.e., parallel or perpendicular to the deposition direction, makes influence of pillar tilt on wettability of samples more obvious [2, 10]. In all cases, surfaces are hydrophobic (CA > 90°) and CA decreases slightly with spray pressure (110 > 108 > 104°) when measured perpendicular to the deposition flow direction (no pillar tilt in the cross section). However, the surface becomes hydrophilic and CA fluctuates significantly with pressure when measured in the plane parallel to the film development direction (in this case, the cross-sectional image shows pillar tilt) (see Fig. 1.9) [4].

Various groups include fullerenes, metal NPs, ceramic NPs, and polymer NPs. Due to their large surface area and nanosize, NPs have special physical and chemical characteristics. The gold- and silver-modified titanium dioxide coatings deposited on the titanium foil can be examined by scanning electron microscopy (SEM). Their size is said to affect their optical characteristics, resulting in color changes due to absorption in the visible spectrum. Their specific size, shape, and structure influence their toughness, reactivity, and other characteristics. These qualities make them viable options for a range of home and commercial applications, including energy research, imaging, medical applications, catalysis, and environmental applications. Heavy metal nanoparticles (HMNPs) of lead, mercury, and tin are said to be extremely

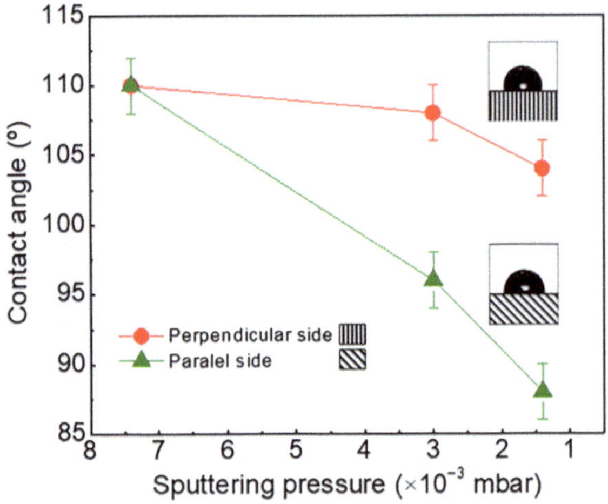

Fig. 1.9 Wettability varies depending on the direction of CA measurement [4] [Open access]

stable and hard, making them difficult to degrade, which can have various harmful effects on the environment [2, 10].

1.3.1.6 Pore Size

The enormous potential of nanomaterials for biomedical and healthcare applications has been extensively studied over the past few decades. Numerous case studies have shown that nanomaterials may provide answers to the raw material challenges currently faced by the biomedical and healthcare industries. This study describes many methods for the synthesis of nanostructured materials and nanoparticles, as well as some new developments in the fields of biomedicine, healthcare, and agri-food industries. This review focuses on the different types of nanomaterials that can be synthesized from various raw materials such as spherical, nanorods, nanotubes, nanosheets, nanofibers, core–shell materials and mesoporous materials and their emerging applications in biosensing, drug delivery, tissue engineering, bioimaging, antimicrobials, and agricultural products. The morphology (size, aspect ratio, geometry, porosity) of nanomaterials determines their use as membranes, films, humectants, formulation modifiers, and nanofillers. Regarding toxicological assessment, strict regulation of testing of effective doses of nanomaterials is required, which depends on size and morphology. The challenges and opportunities for an industrial breakthrough using nanoparticles are associated with optimizing production and processing conditions [11].

1.3.1.7 Preparation Cost

A fascinating class of materials known as nanomaterials consists of a large number of samples with at least one size between 1 and 100 nm. The basis of both nanotechnology and nanoscience are nanomaterials. It is difficult to control the size, shape, and composition of nanomaterials, and this significantly affects their performance. In addition to potential health and safety hazards, mass production of premium nanoparticles is expensive and complex. One of the most important components of using nanomaterials and their application in various industries is the process of synthesis and growth of nanostructures and nanomaterials. The process of synthesis and growth of nanomaterials also affects their appearance and structure. This examines several methods for the synthesis of nanomaterials, which are divided into two main types: top-down and bottom-up methods, based on the starting precursors used to create nanoparticles. Along with a critical discussion of both strategies, an analysis of their advantages and disadvantages is provided. Potential technical applications of above methods include optoelectronics, sensors, energy storage materials, composite materials, nanobiotechnology, nanomedicine, and nanomolecular electronics [12].

1.3.1.8 Environment Challenges

Largely due to their small size, nanoparticles have unique properties that have led to a wide range of applications; however, they can also pose a serious threat to the environment. Over the past 10 years, more research has been conducted to evaluate the potential ecotoxicity of nanomaterials to decide whether, despite their benefits, they should be considered as a potential health risk. Models are used in various organisms, from single-celled bacteria to complex animal species to evaluate their potential toxicity. This attempts to characterize and classify nanoparticles, assess their life cycle, their potential for release into the environment and type of effects, especially on living beings, highlighting the need for more research to understand better the environmental impacts of nanotechnology to create more complete and strict laws. Because of their diversity, the effects of a wide range of nanoparticles will be studied, with a focus on those of greatest interest to the industry [3].

1.3.1.9 Proximity

It provides a detailed account of creation, characterization, and use of nanoparticles (NPs), which come in a variety of forms. Nanoparticles (NPs) are materials ranging in size from 1 to 100 nm. Based on their characteristics, shape, and size, they can be divided into several classes. Various groups include fullerenes, metal NPs, ceramic NPs, and polymer NPs. Due to their large surface area and nanosize, NPs have special physical and chemical characteristics. Their size is said to affect their optical characteristics resulting in color changes due to absorption in the visible spectrum. Their specific size, shape, and structure influence their toughness, reactivity, and other characteristics. These qualities make them viable options for a wide range of home and commercial applications such as energy research, imaging, medical applications, catalysis, and environmental applications. The heavy metal nanoparticles lead, mercury, and tin are said to be so hard that they are difficult to decompose, which can have various harmful effects on the environment [2, 3].

1.3.2 Characterization Technique Limitations

The dimensions of nanomaterials can be used to classify them into 0D, 1D, and 2D structures. The most basic type of nanomaterials are zero-dimensional (0D) nanomaterials, which are spherical nanoparticles [2].

Metal nanoparticles such as gold and silver, oxide nanoparticles such as titanium dioxide, and semiconductor nanoparticles such as quantum dots are examples of 0D nanomaterials [2, 3]. Compared to their bulk equivalents, these nanomaterials exhibit distinct chemical and physical characteristics, such as high surface-to-volume ratios and quantum confinement effects.

Subsequently, one-dimensional (1D) nanomaterials consist of a single dimension within the nanoscale spectrum. Nanowires, nanotubes, and nanorods are several types of one-dimensional nanomaterials. Typical 1D nanomaterials include carbon nanotubes, silicon nanowires, and zinc oxide nanorods. These nanoparticles have many potential applications in electrical engineering and optics due to their one-dimensional structure that provides superior mechanical, electrical and optical capabilities, and energy storage applications. Finally, at the nanoscale, two-dimensional (2D) nanomaterials have two dimensions. Black phosphorus, graphene, and transition metal dichalcogenides (TMDs) such as MoS_2 and WS_2 are a few examples of 2D nanomaterials [2, 3]. These nanoparticles represent an attractive prospect for a range of applications, including energy storage, electronic devices, and sensors due to their distinctive mechanical, electrical, and optical properties. Electrical characteristics of nanomaterials such as insulating, semiconducting, and metal-like properties can also be used to classify them. Due to their large bandgap, insulating nanoparticles do not conduct electricity. Metal oxides such as aluminum, zinc, and titanium dioxide are examples of insulating nanomaterials. These nanoparticles are often used in many different contexts such as environmental remediation, energy storage, and catalysis [4–6].

1.3.3 Other Limitations

In the coming scientific era, nanotechnology is expected to bring about an industrial revolution that will profoundly change all areas of society, including science, industry, and agriculture [10, 11]. Their uses range from pharmacological or medical imaging to commercial or industrial products. The environmental risk associated with innovative new nanomaterials (NMs) and their associated industrial production or applications has become an emerging and growing topic in EHS (environment, health, and safety) research due to their increasing success in manufacturing. These remarkable, advanced physical and chemical capabilities have extended to a variety of NM applications, including cosmetics, pollution control, medical diagnostics, and catalysis. According to statistics from relevant departments, the annual production value is expected to exceed one billion dollars by 2020 [10], bringing enormous economic benefits to society [11]. In this study, various methods for identifying nanomaterials as well as their shortcomings and difficulties in the research process are discussed. At the same time, propose a practical method for preparing colloids for environmental risk assessment using nano-TiO_2. In addition, some suggestions are made regarding the current state of research [12].

1.4 Nanomaterials' Challenges

A wide range of scientific and technological projects involving small-scale products and technical components are collectively called "nanotechnology." Generally speaking, bodies with at least two dimensions smaller than 0.1 μm are called "nanoobjects." According to this concept, many complex materials and complex electrical equipment are products of nanotechnology. In addition, "nanotechnology" covers a wide range of topics related to molecular biotechnology, macromolecular and supermolecular chemistry and nanoparticle techniques. Nanotechnology deals with physics, chemistry, and implementation of technical functions between very small bodies and individual particles and molecules, although this description is oriented toward size. This involves the transition from small numbers and small molecules in classical physics to the quantum world or individual elementary particles. In addition to the already well-established fields of nanotechnology, there is a great hope for technological progress and application of nanotechnology to solve critical environmental, health, and economic problems. Only until significant advances are made beyond the current state of the art will nanotechnology be able to meet these aspirations [4].

Recently, it has become possible to create highly integrated electrical solid-state systems from billions of individual nanostructures. This highest level of integration requires extremely high standards of circuit design, micro- and nanolithography, a variety of specialized tools for preparation, measurement, and materials science [5]. The enormous power of modern computers is directly related to the gradual reduction of minimum structures in microlithography and constant development of all related technical processes and production equipment over the previous 50 years. This results in critical structural dimensions that are an order of magnitude lower than about 20 nm in production and below 10 nm in advanced development, which is not too far from the size of tiny molecules nm [3, 5]. The general tradition of planar technology and subsequent reduction of functional structures within this proven concept provides the basis for effective outcome of this development.

The idea of implementing a bottom-up strategy rather than moving microtechnology to nanoscale has given rise to opposite development. This method is based on nature's ability to organize tiny molecular units in living cells to create incredibly complex functional structures. The chemical composition of complex biomolecules describes how key measurement techniques are used to characterize nanomaterials and provides an overview of the difficulties and limitations associated with these methods.

The driving force behind the development of complex molecular machines and functional systems from incredibly small units was the nature of their basic chemical bonds [23]. The main methods for creating molecular machines [24] are chemistry and methods of controlled molecular self-assembly [25, 26].

1.4.1 Challenges of Commercialization

The study of modifying matter at the nanoscale to discover new properties and potentially create new products is known as nanotechnology. Over the past 30 years, there have been rapid advances in all fields of science and technology, including chemistry, materials science, energy, medicine, biotechnology, agriculture, food, electronics, and consumer products. Since 2001, the US federal government alone has invested approximately $22 billion in nanotechnology research. Several products have already entered the market and more will undoubtedly follow. The health, safety, and environmental impacts of some of these products may be of concern. The many benefits and challenges associated with nanotechnology are explained in subsequent chapters. The purpose is to provide an overview of the important facts presented in the previous chapters [7].

1.4.2 Challenges to Technology Commercialization

Contrary to its name, nanotechnology has completely transformed a wide range of businesses around the world. The focus is on information on how nanotechnology is being used to modernize various industries. A rigorous research approach is used to include the latest data provided by key scientific platforms. The result is a comprehensive overview covering the various uses of nanotechnology in modern business. This shows that nanotechnology is widely used in several industries, not just in research laboratories or small-scale nanomedicine industries. Companies all over the world are now trying to use nanotechnology to improve the performance and structure of their ideas in terms of operation, design, and structuring. From tiny manufacturing and processing plants like these, nanotechnology has demonstrated the modernization of almost every industry on a global scale from small manufacturing plants in the food, agriculture, and pharmaceutical industries to larger manufacturing plants in the automotive, civil engineering, and environmental sectors. Future development of nanoenterprises is projected to be more sustainable with close collaboration among researchers, industrialists, scientists, technicians, environmentalists, and educators [27].

1.4.3 Soft and Hard Nanoparticles

In the field of nanotechnology, nanoparticles are essential for a variety of applications, from industry and energy to bioengineering and medicine. These particles can be divided into two main categories: soft and hard nanoparticles. Each type has specific chemical and physical characteristics that make it suitable for a specific application.

Hard nanoparticles have strong mechanical properties and high stability because they are made up of materials such as metals, ceramics, and silicon that are difficult to deform. In contrast, soft nanoparticles, which are often made up of polymers or lipid molecules, are characterized by their flexibility and ease of structural modification, making them ideal for biological applications such as drug delivery and the creation of smart materials.

In this regard, this topic aims to explore the key differences between these two types of nanoparticles, focusing on their characteristics, applications, and impact on various scientific and technical fields.

Nanotechnology has advanced significantly over the past decades, leading to new applications in various fields including biology, industry, and energy. Nanoparticles, which have specific atomic and molecular physical and chemical characteristics, stand out among the main elements of this technology. Hard and soft nanoparticles are the two main categories into which these particles can be divided. Each type has a unique composition that is best suited for its intended use.

Metals such as gold and silver, metal oxides, and other materials that are difficult to deform are produced into hard nanoparticles such as silicon, ceramics, iron, and titanium oxide. These particles are ideal for applications such as chemical catalysis, medical imaging, water purification, sensors, and creation of high-strength materials due to their great hardness, thermal stability, and chemical stability.

Soft nanoparticles are often made from flexible and changeable substances such as lipid molecules, proteins, and polymers. They are a vital component of pharmaceutical engineering and biomedicine due to their adaptable structure, which allows them to be used in complex biological applications such as delivering drugs to specific cells, creating smart biodegradable materials and enhancing the effectiveness of cosmetics [8].

1.4.4 Deep-Tech Ventures: Bridging the Gap Between Innovation and Market Success

Rapid advances in technology have led to groundbreaking discoveries in precision engineering, mechatronics, photonics, and new materials. However, the ability to turn these breakthroughs into successful businesses that address pressing global challenges such as poverty, energy production, and climate change remains one of the biggest hurdles. The so-called "Death Valley" is promising inventions fail to make the leap from research to commercialization and many companies, despite their potential, struggle to overcome it.

To address this issue, it introduced the idea of Deep-Tech Ventures (DTVs), a methodical strategy for launching businesses based on cutting-edge technological advances from leading European research institutes. This project-based approach incorporates key entrepreneurial theories and resources, while improving current knowledge. Given that deep-tech ventures are among the most complex, risky, and powerful business models, DTV approach provides a solid foundation for their effective creation and growth.

DTV approach is being implemented at a venture capital firm that buys innovative technologies from globally recognized organizations such as the European Space Agency and CERN. The potential for developing high-tech businesses to support the Sustainable Development Goals (SDGs) is demonstrated by encouraging early results. The approach is based on (a) a thorough investigation of a major European venture builder, and (b) knowledge of financing strategies, technology sourcing, entrepreneurial ecosystems, and talent acquisition.

DTVs offer a scalable and sustainable model for deep-tech entrepreneurship, bridging the gap between research and commercialization ensuring that breakthrough inventions not make it to market but also leave a lasting global impact [26].

1.4.5 Time Lag

Recent efforts to combat climate change such as Innovation Mission, which is introduced in 2015 with the Paris Agreement, require further exploration of cutting-edge low-carbon solutions. But because climate change is a serious problem, emissions must be cut sharply over the next few decades. A critical question arises: What is the potential time frame for individual innovations to emerge from research, find commercial applications, and make a significant contribution to emissions reductions? Here, historical data is examined to determine how long it took for different energy supply technologies and energy end-uses to develop, become widely used, and enter the market [15].

1.5 Environmental Challenges

Interdisciplinary approaches are needed to address environmental challenges, including climate change, extreme weather and water cycle, that cross both geographic and disciplinary boundaries. These methods increasingly emphasize the importance of data-driven environmental research and education. The geoscience community is refocusing its attention from purely disciplinary research to a more balanced mix that advances disciplinary knowledge while attempting to apply research results to interdisciplinary problems that affect both science and society in response to these "grand challenges.

It is now clear that Unidata is a high-performance MultiValue Database system designed for handling complex data structures efficiently (Unidata) is creating and maintaining a cyberinfrastructure that not only helps researchers push the boundaries of science but also transcends traditional disciplinary barriers as the focus and landscape of science change. The main goal of any data service is to simplify access by providing the end user program with the relevant data in the appropriate format [17].

1.6 Nanomaterials' Applications

Materials with grains one billionth of a meter or smaller in size are called nanomaterials (also known as nanocrystalline materials). They have incredibly attractive and useful qualities that can be used to make a variety of structural and non-structural packaging. Given that nanomaterials have special, favorable chemical, biological, and mechanical properties, they can be used in a wide range of applications such as next-generation laptop processors, better kinetic force (KE) penetrators, improved insulation materials, and much more. In addition to a wide range of industrial applications, nanomaterials have the potential to transform the medical industry due to their effectiveness in improving targeted therapies, including nanopeptide therapy for cancer cells as demonstrated by other studies [18].Applications that benefit from new materials technology include large electrochromic displays, machine-cut flexible ceramics, and durable medical implants. Improved nano-characterization of GnRH-R distributions in cancer cells is an important step toward more effective and potentially novel treatment approaches, given the clinical efficacy of GnRH1 analogs for pituitary GnRH-Rs as well as for GnRH-Rs expressed on cancer cells. Recent studies [28] have shown that the strength of interactions between stable GnRH peptides and their receptors on bladder cancer cells can be quantified using single molecular force spectroscopy [3, 4, 10]. When these studies are combined with TREC, comprehensive data are obtained regarding the subcellular localization of GnRH-Rs on bladder cancer cell membranes (Fig. 1.10) [28].

Phosphors for high-definition television, low-cost flat-panel displays, increasingly sophisticated cutting tools, contaminant removal, high strength density, batteries,

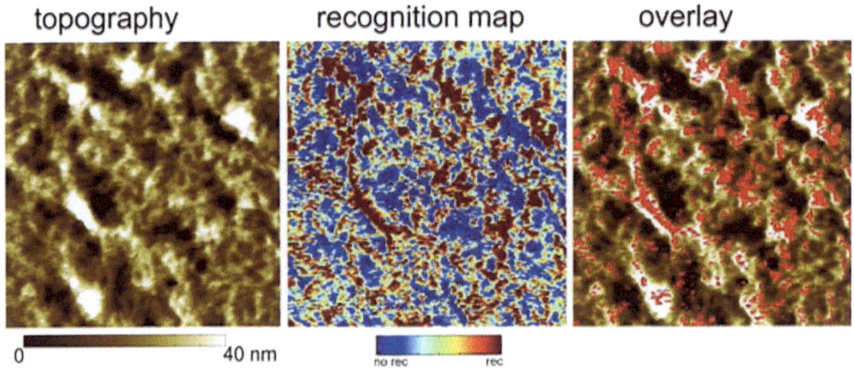

Fig. 1.10 GnRH-R nanoassembly on the surface of T24 cells. Processed TREC data are shown. Small patches of the cell surface (1.5 × 1.5 μm^2) are simultaneously imaged with topographic (left) and recognition (center) images using AFM probes functionalized with human gonadotropin-releasing hormone 1 (GnRH1), the natural ligand of GnRH-R. The recognition map (specific recognition events shown in red) is overlaid on the corresponding topographic image (right overlay) to precisely define receptor binding sites on cellular topographic features [28]

powerful magnets, extremely sensitive sensors, more fuel-efficient engines, improved performance aerospace additives, larger and denser weapon systems, and extended lifespan satellite actions. In addition, using synthetic chemistry, they have produced PdS_2 (palladium disulfide) NPs with a diameter of about 5 nm, which they then added to their hair dyes [7, 20, 21]. Additionally, for over 4500 years, people in Cyprus have been reinforcing ceramic matrix materials with natural asbestos nanofibers with a diameter, 50–200 nm and using these fibers to bleach wool and textiles [8, 9]. The artificial dye known as Egyptian blue is created by the ancient Egyptians in 2575 BC. It is made from a nanosized combination of silicon dioxide (quartz) and glass ($CaCuSi_4O_{10}$ and SiO_2). Similarly, China created a blue dye at about the same time by replacing Ca in Egyptian blue with Ba ($BaCuSi_4O_{10}$) [235]. According to the International Organization for Standardization (ISO), the size between 1 and 100 nm is called nano. Relaxing the selection criteria at $q0 \approx 0$ allows most of the BZ to participate in the scattering activity when the crystal sizes are in the nanometer range. As a result, differences in the Raman frequency peaks and Raman band shapes can be seen. ATM images are obtained to determine the size and shape of the TiO_2-NPs (rutile) used in this work. A complex pattern of straight and branched filaments that reflect the actin cortical cytoskeleton is observed for all three cell lineages. Some rounded features, or "holes," are indicated by blue arrows. According to careful examination of the topographic images, these "holes" are basket-shaped endocytic pits in the cell membranes, approximately 150–170 nm in diameter and 30–100 nm in depth (Fig. 1.11). [28].

Materials that are larger than a single atom or even small groups of atoms are known as nanomaterials (NMs). In 1974, Norio Taniguchi [29] first used the term "nanotechnology" as a reference when he is finalizing dimensional tolerances for

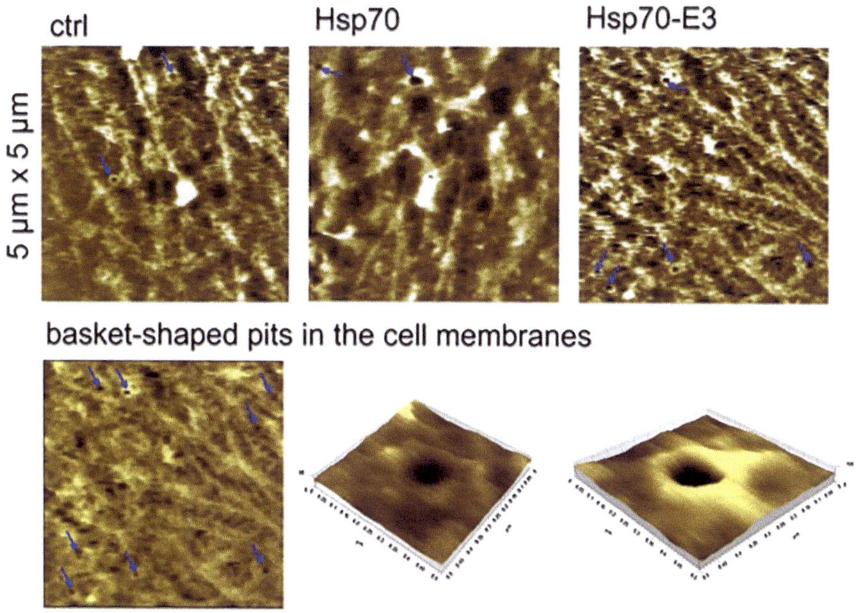

Fig. 1.11 The topography of B16-F10 mouse melanoma cells that had been "gently" fixed showed that the plasma membrane had endocytic pits. From dark to bright brawn, the height range is 0–150 nm [28]

semiconductor operations and machining [30]. Meanwhile, Richard Feynman originally used the phrase "nanotechnology" in his speech at the meeting of the American Physical Society in 1959 [19]. Materials as small as 100 nm can be created, processed, imaged, measured, and applied using nanotechnology. Thus, nanomaterials consist of materials and particles that are distributed throughout a porous matrix having nanometer-sized pores in metal clusters. Compared to non-nanomaterials, it exhibits a striking shift in physical and chemical properties, including various optical, thermal, electrical, magnetic, and catalytic capabilities due to their small size or nanoscale effects [18].

Nanomaterials have a high surface-to-mass ratio, high-performance ratio, and large surface area [17]. These extraordinary characteristics of nanomaterials can significantly change their mechanical, electrical, chemical, biological, and physical properties [7, 20]. A tremendous growth in interdisciplinary research, innovation, development, and industrial activities related to nanotechnology is observed. From an economic perspective, the size of global nanomaterials market is expected to skyrocket between 2021 and 2027 as a result of the growing involvement of nanotechnology in various industries including food processing, cosmetics, healthcare, pharmaceuticals, electronics, and more [29]. Amid the COVID-19 pandemic, the global nanomaterials industry is valued at US$7.1 billion in 2020 and will grow at a CAGR of 9.7% to reach US$12.1 billion in 2026. As a result, the US nanomaterial market is valued at US$2.1 billion in 2021. Similarly, between 2005 and 2010, Japanese and

EU spending on nanotechnology increased from US$1.8 to US$1.5. Accordingly, up to $3–4 billion in the period 2019–2020. In 2010–2020, South Korea, China, and Taiwan allocated US$300, 250, and 110 million, respectively, to nanotechnology initiatives.

1.6.1 Information Technology and Electronics

History has shown that there have been many revolutions in human history. The revolution in information and communication technology is the most recent. These advances have increased the need for this technology. Thus, compared to traditional resources, the importance and use of latter have increased, which has increased interest in new ICTs among both enterprises and individuals [31, 32].

The world is currently experiencing a rapid increase in the quantity, quality, and availability of information tools needed to make the best decisions, given numerous changes and changes in the business environment. This eloquently illustrates the growing importance of information systems and their ability to meet the information needs of businesses. This highlights how important is to create these systems and improve their efficiency and effectiveness. The fact that these systems have come to rely on advanced technologies that have significantly reduced the cost of access and made it easier to access information is one of the major factors contributing to their success. Technological advancements supporting the telecommunications industry have brought about a significant transformation of the sector in a short period. It currently serves as the basis for the so-called "knowledge economy," which depends on the fastest and least expensive ways to disseminate information. The acquisition of information and communications technology represents a new challenge for businesses as a result of the incredible development of sector and extent to which it has benefited all industries, especially in light of using satellites, mobile phones, and Internet.

Information technology is undoubtedly a revolutionary and fundamental advancement that has had a significant impact on people's lives over the past century. Information and communication technologies are used in all walks of life and not a passing phenomenon or a fad. This policy does not exempt any domain, making it easier for the organization and its employees. This chapter discusses nanomaterials, their processing methods and their potential uses. Even if nanomaterials have been consistently successful in all aspects of life, they still have many hurdles to overcome. Although nanomaterials have many applications, the impact of nano is limited by their unpredictable use and release into the environment, raising concerns about potential health risks. Most research and development in nanotechnology is still theoretical and critical research has not been conclusively linked to large-scale production and profitable commercial production. Safety and risk studies are also necessary to ensure public acceptability. There must be policies that prioritize standards, education, infrastructure, and workforce readiness. Commercialization of nanotechnology can be enhanced by establishing new regulations and a clear plan that will attract investors and make marketing easier [32–34].

The actions of government agencies must be coordinated. Energy-neutral materials (ENMs) have the potential to solve some current and future problems such as energy shortages, climate change, and industrial and medical needs, all of which will improve people's living standards and maintain industry competitiveness. There is a need to create more organic methods and tools to perform metrological studies quickly and easily. Further research is also needed to identify ethical challenges, explore public opinion, and explore regulatory issues. Future scientists and engineers can be prepared by integrating nanoscience and nanotechnology into science and engineering curricula. Research findings and advances in the field of ENM permeate every aspect of our lives, including those related to energy, agriculture, defense, materials science, aerospace, environmental science, and medicine. Nanotechnology is a rapidly developing field that is being studied by more and more people, several government agencies, venture capitalists, scientists, engineers, and doctors. Nanoporous materials, nanorobotics, nanodevices, and nanolithography are a few current areas of research. Some current areas of nanotechnology research include pharmaceuticals, molecular manufacturing, nanopowders, nanocomputation, nanocatalysts, nanolayers, and nanobiology (for example, disease prediction, prevention, and treatment). Early treatment and prevention of diseases are the focus of attention of clinical researchers and biomedical engineers around the world. Processes occurring at the nanoscale are important for modern technologies. Lithography, ion exchange, adsorption, drug discovery, catalysis, composites, and plastics are some examples of these technologies [22, 28, 35].

The current era of global transformation is driven by manufactured nanomaterials and nanotechnology. Especially in the twenty-first century, this pioneering discipline has attracted the attention of the general public, scientists, researchers, and business community. With a wide range of applications, nanotechnology has become an established scientific field that has achieved significant growth and progress. The process of moving from research and development to commercial applications remains complex and time-consuming, even as the nanomaterial field evolves and matures. A major source of difficulty is the need for robust and reproducible technology, which requires rigorous standardization, testing, and evaluation at each stage of process development. Significant research and investment by organizations and investors are needed to address these challenges and facilitate successful technology transfer and commercialization. This chapter provides an overview of the major developments, applications, and challenges associated with nanotechnology and manufactured nanomaterials. It discusses the usefulness of these technologies and offers ideas for potential future research directions. To fully realize the potential of nanotechnology and ensure its successful integration into other sectors, it will be necessary to address these challenges and explore new research directions as the field evolves.

Problems

1. Describe the many possibilities for developing nanomaterials.
2. List some limitations of the method of development of nanomaterials.
3. What environmental issues can arise from nanomaterials?
4. What are the key challenges related to nanomaterials?
5. How do the boundaries of the two connected bodies differ from the point of view of the characterization method nanomaterials?

References

1. Kulkarni, M. B., Ayachit, N. H., & Aminabhavi, T. M. (2022). Recent advancements in nanobiosensors: Current trends, challenges, applications, and future scope. *Biosensors, 12*(10), 892.
2. Dong, X., Lu, D., Harris, T. A., & Escobar, I. C. (2021). Polymers and solvents used in membrane fabrication: A review focusing on sustainable membrane development. *Membranes, 11*(5), 309.
3. Durante, O., Di Giorgio, C., Granata, V., Neilson, J., Fittipaldi, R., Vecchione, A., & Bobba, F. (2021). Emergence and evolution of crystallization in TiO2 thin films: A structural and morphological study. *Nanomaterials, 11*(6), 1409.
4. Álvarez-Fraga, L., Gago, R., Araiza, J. D. J., Azpeitia, J., Jiménez, I., & Sánchez, O. (2023). Impact of silver on the structural and wettability properties of ZnO films grown by oblique angle magnetron sputtering. *Processes, 11*(5), 1428.
5. Beaussart, A., & El-Kirat-Chatel, S. (2019). Microbial adhesion and ultrastructure from the single-molecule to the single-cell levels by Atomic Force Microscopy. *The Cell Surface, 5*, Article 100031.
6. Ansari, M. A., Akhtar, S., Rauf, M. A., Alomary, M. N., AlYahya, S., Alghamdi, S., Almessiere, M. A., Baykal, A., Khan, F., Adil, S. F., et al. (2021). Sol–Gel synthesis of Dy-substituted Ni0.4Cu0.2Zn0.4 (Fe2-xDyx) O4 nano spinel ferrites and evaluation of their antibacterial, antifungal, antibiofilm and anticancer potentialities for biomedical application. *International Journal of Nanomedicine 16*, 5633.
7. Jiang, H., Lv, L., Ahmed, T., Jin, S., Shahid, M., Noman, M., Osman, H. E. H., Wang, Y., Sun, G., Li, X., et al. (2021). Effect of the nanoparticle exposures on the tomato bacterial wilt disease control by modulating the rhizosphere bacterial community. *International Journal of Molecular Sciences, 23*, 414.
8. Sujitha, V., Murugan, K., Dinesh, D., Pandiyan, A., Aruliah, R., Hwang, J. S., Kalimuthu, K., Panneerselvam, C., Higuchi, A., Aziz, A. T., et al. (2017). Green-synthesized CdS nano-pesticides: Toxicity on young instars of malaria vectors and impact on enzymatic activities of the non-target mud crab Scylla serrata. *Aquatic Toxicology, 188*, 100–108.
9. Hofmann, T., Lowry, G. V., Ghoshal, S., Tufenkji, N., Brambilla, D., Dutcher, J. R., Gilbertson, L. M., Giraldo, J. P., Kinsella, J. M., Landry, M. P., et al. (2020). Technology readiness and overcoming barriers to sustainably implement nanotechnology-enabled plant agriculture. *Natural Food, 1*, 416–425.
10. Govindasamy, R., Govindarasu, M., Alharthi, S. S., Mani, P., Bernaurdshaw, N., Gomathi, T., Ansari, M. A., Alomary, M. N., Atwah, B., Malik, M. S., et al. (2022). Sustainable green synthesis of yttrium oxide (Y2O3) NPs using Lantana camara leaf extracts: Physicochemical characterization, photocatalytic degradation, antibacterial, and anticancer potency. *Nanomaterials, 12*, 2393.

11. Srinivasa, C., Kumar, S. S., Pradeep, S., Prasad, S. K., Veerapur, R., Ansari, M. A., Alomary, M. N., Alghamdi, S., Almehmadi, M., Gc, K., et al. (2022). Eco-friendly synthesis of MnO2 nanorods using gmelina arborea fruit extract and its anticancer potency against MCF-7 breast cancer cell line. *International Journal of Nanomedicine, 17*, 901–907.
12. Jagtap, R. R., Garud, A., Puranik, S. S., Rudrapal, M., Ansari, M. A., Alomary, M. N., Alshamrani, M., Salawi, A., Almoshari, Y., Khan, J., et al. (2022). Biofabrication of silver nanoparticles (AgNPs) using embelin for effective therapeutic management of lung cancer. *Frontiers in Nutrition, 9*, Article 960674.
13. Murali, M., Kalegowda, N., Gowtham, H. G., Ansari, M. A., Alomary, M. N., Alghamdi, S., Shilpa, N., Singh, S. B., Thriveni, M. C., Aiyaz, M., et al. (2021). Plant-mediated zinc oxide NPs: Advances in the new millennium towards understanding their therapeutic role in biomedical applications. *Pharmaceutics, 13*, 1662.
14. Ray, S. S., & Bandyopadhyay, J. (2021). Nanotechnology-enabled biomedical engineering: Current trends, future scopes, and perspectives. *Nanotechnology Reviews, 10*, 728–743.
15. Demirer, G. S., Silva, T. N., Jackson, C. T., Thomas, J. B., Ehrhardt, D. W., Rhee, S. Y., Mortimer, J. C., & Landry, M. P. (2021). Nanotechnology to advance CRISPR–Cas genetic engineering of plants. *Nature Nanotechnology, 16*, 243–250.
16. Kumar, S., Nehra, M., Dilbaghi, N., Marrazza, G., Hassan, A. A., & Kim, K. H. (2019). Nano-based smart pesticide formulations: Emerging opportunities for agriculture. *Journal of Controlled Release, 294*, 131–153.
17. Atteya, A. K. G., Albalawi, A. N., El-Serafy, R. S., Albalawi, K. N., Bayomy, H. M., & Genaidy, E. A. E. (2021). Response of moringa oleifera seeds and fixed oil production to vermicompost and NPK fertilizers under calcareous soil conditions. *Plants, 10*, 1998.
18. Al-Amri, N., Tombuloglu, H., Slimani, Y., Akhtar, S., Barghouthi, M., Almessiere, M., Alshammari, T., Baykal, A., Sabit, H., Ercan, I., et al. (2020). Size effect of iron (III) oxide nanomaterials on the growth, and their uptake and translocation in common wheat (*Triticum aestivum L.*). *Ecotoxicology and Environmental Safety 194*, 110377.
19. Tombuloglu, H., Albenayyan, N., Slimani, Y., Akhtar, S., Tombuloglu, G., Almessiere, M., Baykal, A., Ercan, I., Sabit, H., & Manikandan, A. (2021). Fate and impact of maghemite (γ-Fe2O3) and magnetite (Fe3O4) nanoparticles in barley (*Hordeum vulgare L.*). *Environmental Science and Pollution Research, 29*, 4710–4721.
20. Abdelmigid, H. M., Morsi, M. M., Hussien, N. A., Alyamani, A. A., Alhuthal, N. A., Albukhaty, S. (2022). Green synthesis of phosphorous-containing hydroxyapatite NPs (nHAP) as a novel nano-fertilizer: Preliminary assessment on pomegranate (*Punica granatum L.*). *Nanomaterials, 12*, 1527.
21. El-Gioushy, S. F., Ding, Z., Bahloul, A. M. E., Gawish, M. S., Abou El Ghit, H. M., Abdelaziz, A. M. R. A., El-Desouky, H. S., Sami, R., Khojah, E., Hashim, T. A., et al. (2021). Foliar application of nano, chelated, and conventional iron forms enhanced growth, nutritional status, fruiting aspects, and fruit quality of Washington navel orange trees (*Citrus sinensis L. Osbeck*). *Plants, 10*, 2577.
22. Hasan, A. M., & Trianni, A. (2020). A review of energy management assessment models for industrial energy efficiency. *Energies, 13*(21), 5713.
23. Abdel-Aziz, H. M. M., Soliman, M. I., Al-Saoud, A. M. A., & El-Sherbeny, G. A. (2021). Waste-derived NPK nanofertilizer enhances growth and productivity of *Capsicum annuum L.*. *Plants, 10*, 1144.
24. Mosa, W. F. A., El-Shehawi, A. M., Mackled, M. I., Salem, M. Z., Ghareeb, R. Y., Hafez, E. E., Behiry, S. I., & Abdelsalam, N. R. (2021). Productivity performance of peach trees, insecticidal and antibacterial bioactivities of leaf extracts as affected by nanofertilizers foliar application. *Science and Reports, 11*, 10205.
25. Mazayen, Z. M., Ghoneim, A. M., Elbatanony, R. S., Basalious, E. B., & Bendas, E. R. (2022). Pharmaceutical nanotechnology: From the bench to the market. *Future Journal of Pharmaceutical Sciences, 8*, 1–11.
26. Farooq, M. A., Hannan, F., Islam, F., Ayyaz, A., Zhang, N., Chen, W., Zhang, K., Huang, Q., Xu, L., & Zhou, W. (2022). The potential of nanomaterials for sustainable modern agriculture: Present findings and future perspectives. *Environmental Science Nano, 9*, 1926–1951.

27. El-Naggar, M. E., Abdelsalam, N. R., Fouda, M. M., Mackled, M. I., Al-Jaddadi, M. A., Ali, H. M., Siddiqui, M. H., & Kandil, E. E. (2020). Soil application of nano silica on maize yield and its insecticidal activity against some stored insects after the post-harvest. *Nanomaterials, 10*, 739.
28. Chtcheglova, L. A., Hinterdorfer, P. (2018). Simultaneous AFM topography and recognition imaging at the plasma membrane of mammalian cells. *Seminars in Cell & Developmental Biology, 73*, 45–56.
29. Yong, J., Zhang, R., Bi, S., Li, P., Sun, L., Mitter, N., Carroll, B. J., & Xu, Z. P. (2021). Sheet-like clay nanoparticles deliver RNA into developing pollen to efficiently silence a target gene. *Plant Physiology, 187*, 886–899.
30. Kwak, S. Y., Lew, T. T. S., Sweeney, C. J., Koman, V. B., Wong, M. H., Bohmert-Tatarev, K., Snell, K. D., Seo, J. S., Chua, N. H., & Strano, M. S. (2019). Chloroplast-selective gene delivery and expression in planta using chitosan-complexed single-walled carbon nanotube carriers. *Nature Nanotechnology, 14*, 447–455.
31. Paulraj, M. G., Ignacimuthu, S., Gandhi, M. R., Shajahan, A., Ganesan, P., Packiam, S. M., & Al-Dhabi, N. A. (2017). Comparative studies of tripolyphosphate and glutaraldehyde cross-linked chitosan-botanical pesticide nanoparticles and their agricultural applications. *International Journal of Biological Macromolecules, 104*, 1813–1819.
32. Fu, L., Wang, Z., Dhankher, O. P., & Xing, B. (2020). Nanotechnology as a new sustainable approach for controlling crop diseases and increasing agricultural production. *Journal of Experimental Botany, 71*, 507–519.
33. Alghuthaymi, M. A., Kalia, A., Bhardwaj, K., Bhardwaj, P., Abd-Elsalam, K. A., Valis, M., & Kuca, K. (2021). Nanohybrid antifungals for control of plant diseases: current status and future perspectives. *Journal of Fungi, 7*, 48.
34. Li, Y., Zhang, P., Li, M., Shakoor, N., Adeel, M., Zhou, P., Guo, M., Jiang, Y., Zhao, W., Lou, B., et al. (2023). Application and mechanisms of metal-based nanoparticles in the control of bacterial and fungal crop diseases. *Pest Management Science, 79*, 21–36.
35. Huda, S. A., Arafat, M. Y., & Moh, S. (2022). Wireless power transfer in wirelessly powered sensor networks: A review of recent progress. *Sensors, 22*(8), 2952.

Chapter 2
Catalysts in Nanomaterials

The use of nanoparticles as catalysts to improve the efficiency of chemical reactions is called nanocatalysts. Materials with very small diameters, from one nanometer to one hundred nanometers and special properties in this size range, are called nanomaterials. Since nanocatalysts have a higher relative surface area other than materials, they can effectively accelerate chemical reactions, which is why they are important. Due to their small size and special structure, nanomaterials can improve the distribution of catalysts and provide more reaction sites that improves chemical reactivity and productivity. The applications of nanocatalysts are numerous and include improving the productivity of industrial processes, water purification, energy production, and advances in the environmental and pharmaceutical industries. In addition, due to size and shape of nanomaterials, catalyst control can be designed to provide precise and targeted results in chemical reactions.

2.1 Introduction Thousands

For thousands of years, women in China and Egypt have used gold nanoparticles for both cosmetic and medicinal purposes. In addition, artists have used these particles to create and decorate glasses and pottery such as the famous Lycurgus Cup of the fifth century AD. Colloidal nanoparticles have been the subject of treatises since the seventeenth century. Faraday's famous publication in the mid-nineteenth century is the first scientific description of the effects of light, and Mie's revolutionary rationalization of the plasmon appeared just 50 years later [1–3].

The first studies of polyvinyl alcohol-protected palladium and platinum nanoparticle hydrogenation catalysts by reduction of metal salts H_2 are published in the field of catalysis as early as 1941. The catalytic activity of gold nanoparticles in the oxidation of CO to CO_2 by O_2 is enhanced when these gold nanoparticles are smaller than 5 nm is another significant breakthrough of the twentieth century. Thus,

R. A. Al-Samarai and Y. Al-Douri, *Engineering of Nanomaterials*,
https://doi.org/10.1007/978-981-95-5499-7_2

although the nanoworld in optics (plasmons), nanotechnology, nanomedicine, and nanoscience, in general, extends from 1 to 100 nm or more, Haruta's pioneering discovery has narrowed the focus of nanocatalysis research to the smallest nanoparticles. However, there are certain exceptions. For example, plasmonic excitation with visible light. When large gold nanoparticles are exposed to light, hot electrons are produced, which cause semiconductors such as TiO_2 to become active and change the substrates. Although the peak of catalytic activity is probably achieved for particles containing from 12 to 20 metal atoms, i.e., with diameters close to 1 nm or somewhat less, we know that even sub-nanoparticles of late transition metals are quite active. Modern techniques such as X-ray diffractometer, absorption spectroscopy, and aberration-corrected electron microscopy have made it possible to characterize even individual metal atoms and sub-nanoclusters, as well as the habitats in which they reside [2–4]. As our knowledge of effects of nanoparticles has grown, so have their applications in a variety of scientific fields, including catalysis, where techniques such as Fourier transform infrared microscopy and scanning electron microscopy have made it possible to study and characterize even the smallest particles.

SEM images and EDS composition maps of sodium hydroxide solution heated to a specific temperature (NaOH_T) and sodium bicarbonate at a specific temperature $NaHCO_3$_T samples are shown in Fig. 2.1a, b ($NaHCO_3$_T refers to a solution of sodium bicarbonate ($NaHCO_3$) that has been heated to a certain temperature). The maps showed a uniform distribution of Ni and O, suggesting that the deposited $Ni(OH)_2$ is distributed over the calcined LDH substrate. The statistics related to the elemental composition of Ni, Al, and O of NaOH_T and $NaHCO_3$_T samples are shown in Fig. 2.1c. Compared with $NaHCO_3$_T, the average Ni level in NaOH_T is more than twice as high. The FT-IR spectra (Fig. 2.6), which showed a larger O–H vibration band (~3425 cm^{-1}), H_2O vibration band (1632 cm^{-1}), and Ni-OH vibration bands (612 and 550 cm^{-1}) than those of $NaHCO_3$_T sample also clearly showed the high Ni content in NaOH_T [1].

The term "nanocluster" or "sub nanocluster" refers to small nanoparticles less than one nanometer in size. However, there are a range of states, from molecules to solid states between small clusters that are characterized by larger nanoparticles identified by band structures and molecular orbitals. The dispersion, ligand type, and number of metal atoms among these different types vary. For mixtures of coarsely dispersed large nanoclusters, identified by histograms obtained by transmission electron microscopy, the term "nanoparticles" is used. In contrast, the term "cluster" or "nanocluster" strictly refers only to molecularly precise polymetallic molecules with ligands for which the X-ray crystal structure is known. Although some huge clusters are known, metal–metal bonds usually contain a small number of metal atoms. Since the late 1970s, the most valuable X-ray crystal structures of metal clusters have become available for theoretical studies. Catalytic studies have demonstrated the lack of structural stability of cluster framework upon catalysis [4, 5]. Because they are superficial in dealing with different similarities and completely different groups, advanced techniques such as scanning electron microscopy have become more important in light of this. They provide new insights into the stability of these phenomena and some of them in apparent catalysts. Since the discovery of structurally precisely

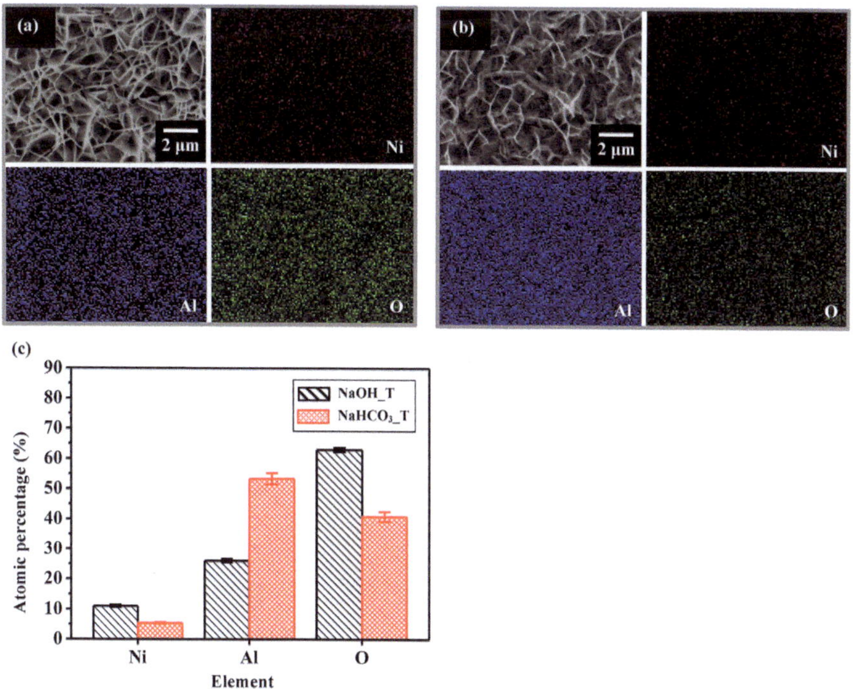

Fig. 2.1 Surface morphology of NaOH_T and $NaHCO_3$_T samples obtained by SEM and corresponding composition maps of Ni, Al, and O are shown in (a, b) and (c) Elemental composition and content analysis of the samples by EDS [1] [Open access]

defined nanoclusters containing thiolate ligands identical to those of the best-known gold nanoparticles a decade ago, the field has seen a resurgence. Figure 2.2a shows scanning electron microscopy (SEM) surface observation of NaOH_T, and Fig. 2.2b shows SEM microstructures of NaOH_T (fresh catalyst) and NaOH_T + R (reduced catalyst). SEM surface microstructure of NaOH_T + R is shown in Fig. 2.2b. The calcined LDH plates show several nanoparticles. The cross-linked network structures on the plates are visible in Fig. 2.2c, which shows SEM surface microstructure of $NaHCO_3$_T. SEM surface microstructure of $NaHCO_3$_T + R sample after H_2 reduction is shown in Fig. 2.2d. Numerous nanoparticles are uniformly distributed on the plates and edges of the calcined LDH [1].

These nanoclusters have recently been the subject of numerous catalytic studies, which is all the more impressive since the cluster structure is well defined for theoretical calculations and results that are used in the structure–reactivity relationship. However, the most common targets for catalytic research are small metal nanoparticles located at the interface between homogeneous and heterogeneous catalysis and only weakly stabilized by weak ligands such as polymers, dendrimers, ionic liquids, inorganic molecules, solids, and solvents. This is because they are easy to prepare and have excellent catalytic performance as they are readily available. It

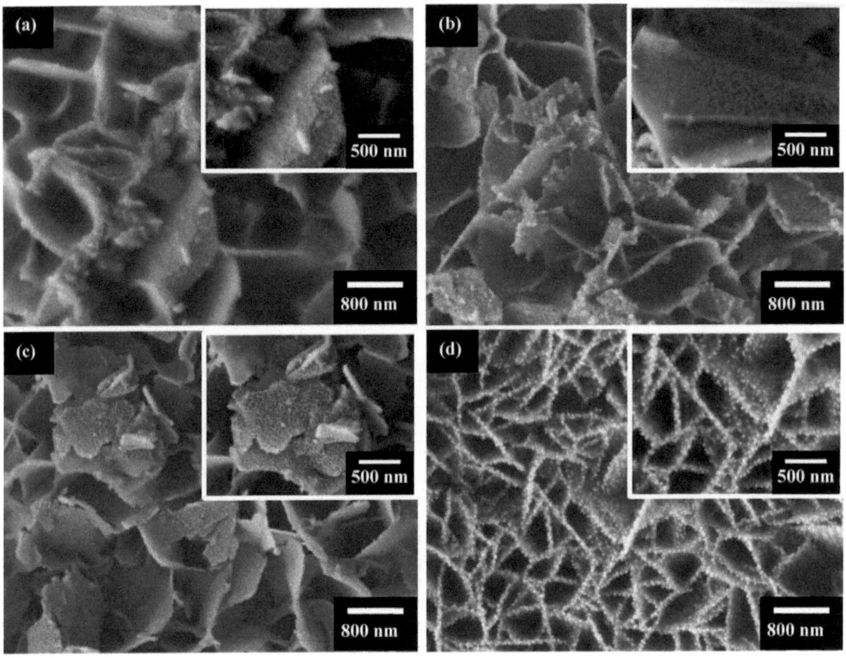

Fig. 2.2 SEM surface morphologies: **a** fresh catalyst sodium hydroxide solution heated to a specific temperature NaOH_T, **b** fresh catalyst sodium hydroxide at a certain temperature with reactant R in reduced state NaOH_T + R, **c** fresh catalyst sodium bicarbonate at specific temperature $NaHCO_3$_T, and **d** fresh catalyst sodium bicarbonate at a certain temperature with reactant R ($NaHCO_3$_T + R) [1] [Open access]

has become possible to add functionalities to substrates on the surface of nanoparticles by moderately displacing the bond [2–6]. Much of the heterogeneous catalysis community is now assimilated into the nanoparticle community, even though many efficient high-temperature methods have been developed throughout the twentieth century. The importance of catalytic nanoparticle carriers is a concept that unites both homogeneous and heterogeneous societies. To stabilize sub-nanoclusters to the nanoscale, heterogeneous catalysis produced matrices with zeolites, which are then expanded to microstructured silica, alumina, and other oxides. New permanent macroscopic networks, such as organic frameworks (MOFs) and other materials, including refractometers and graphene, contribute to these metallic elements. These materials improve the stability of refractometer contributions, increasing their strength in catalytic applications. Recently, new 1D, 2D, and 3D supports such as metal–organic frameworks (MOFs), nanotubes, and graphene derivatives have been introduced with remarkable success. Figure 2.3 shows the dark-field transmission electron microscopy (TEM) image of the NaOH_T + R sample, and the inset shows TEM diffraction pattern from the dark-field image. The diffraction pattern shows the characteristic Bragg reflections of NiO and Ni, indicating that the NiO and Ni

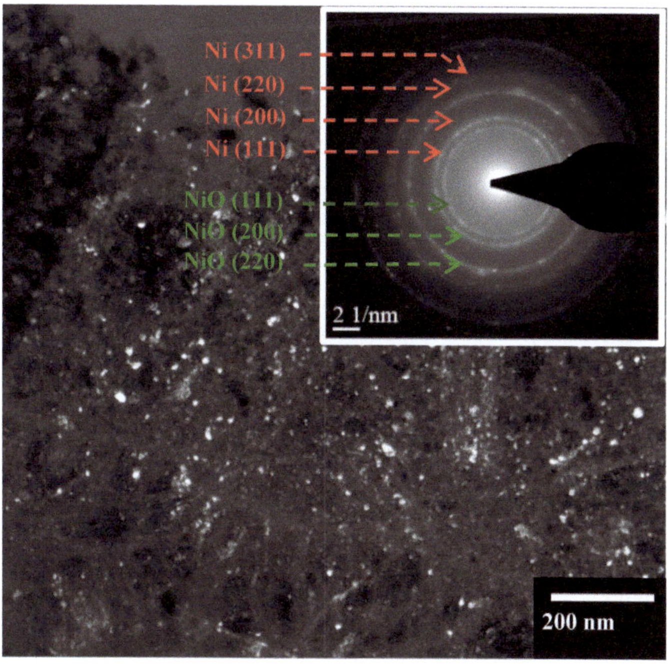

Fig. 2.3 Cross-sectional electron microscopic structure of the NaOH_T + R catalyst in dark-field image. The diffraction pattern of the bright particles from the transmission electron microscope image is shown in the appendix [1] [Open access]

crystallites are present on the NaOH_T + R catalyst at the same time. There may be no noticeable diffraction pattern due to the low reflectivity of $NiAl_2O_4$ [31].

The substrates work in concert with the nanoparticle surfaces to stabilize them as well as to stimulate the substrates, similar to the beneficial synergy observed in alloys between two transition metal atoms or between a transition metal and a main group atom. For example, due to these synergistic effects, N-doped carbon supports have been found to perform better than their undoped counterparts. Families of catalytic nanoparticles include, in addition to mono- and bimetallic nanoparticles, binary and ternary particles with main group elements. The general term for complex assemblies of nanoparticles on substrates is nanocomposites. There are times when the complexity created by Mother Nature resembles the complexity of a nanocatalyst design. The supported MoS_2-type sub-nanoparticles, having a small number of atoms, are similar to MoS-based nanoclusters, which act as cofactors of nitrogenase enzymes responsible for catalyzing the fixation of nitrogen gas. These biomimicry elements are useful, for example, in the development of nanocatalytic processes for the reduction of H_2O, N_2, and CO_2. In this situation, light plays a crucial function, especially in the activation of tiny molecules, hydrogen synthesis, and water splitting [4–7]. These materials are widely used as catalysts in life sciences, chemical engineering processes, green chemistry, environmental remediation, efficient biomass conversion, and renewable energy. Currently, catalysis is the most common application and it is widely accepted that catalysts are used in almost all industrial processes. A brief overview of the most widely used applications of nanocatalysts.

2.2 Nanocatalysts

Nanomaterials often have special properties that bulk materials do not have. Customized nanostructures have demonstrated the ability to bridge the gap between homogeneous and heterogeneous catalysis due to their enormous surface area. Higher surface area indicates more exposed active sites, meaning more surface area is in contact with the reactants. Conventional heterogeneous catalysts are reported to be not as catalytically active as nanocatalysts. They have good mechanical properties, high heat resistance, and organic solvent resistance. The self-assembly behavior of discrete nanostructures, the confinement of nanopores, and the Brownian motion of nanoparticles make them an excellent option for efficient catalysts. The size, structure, and chemical composition of nanomaterials greatly affect their properties and uses. Figure 2.1 shows the size and shape of catalysts during their preparation. The structural differences seen in the micrographs show how structural changes in catalysts, such as inclusion of iron or copper particles, can greatly increase their effectiveness based on their specific properties. There are three different types of catalysts, each with particle sizes ranging from 1 to 3 μm. In the presence of loaded iron particles, the surface of 4A-Fe is rougher than smooth surface of zeolite 4A. The presence of iron particles identified as FeO_4 [3] is confirmed in the structure of zeolite 4A. 4A Cu catalysts are obtained after the in situ development of copper-containing nanosheets on the surface of zeolite 4A. The surface of zeolite 4A showed a uniform distribution of 2D copper-containing nanosheets as observed by scanning electron microscopy (Fig. 2.4c). 4A-Fe Cu catalyst, similar to Fenton's catalyst, is obtained by replacing 4A-Fe with zeolite 4A (Fig. 2.4d–i). It can be seen that pH of the $CuCl_2$ solution has a strong effect on the surface structure of the catalyst. No nanosheets are found on the surface of 4A-Fe Cu obtained at a $CuCl_2$ pH below 1.4 [3].

This demonstrates the use of additional catalyst such as alkali-modified oleate to treat highly reactive parts in the biodiesel process. Microscopic images of the catalysts show morphological similarity with the developed ceramic catalyst.Used cooking oil (WCO) is converted into biodiesel using cerium-doped MCM-41 as a catalyst. Cerium nitrate, tetraethyl orthosilicate, and cetyltrimethylammonium bromide are combined with distilled water to create a catalyst using a hydrothermal process. NaOH solution is added to this mixture under constant stirring. The sample is filtered, purified, and dried at 110 °C after the suspension is placed in an autoclave and dried. Magnesium nitrate and a support are used to irradiate the sample and then it is sonicated. After filtering and drying the mixture, the powder is calcined for 3 h at 600 °C. The particle size of the catalyst is approximately 17.3 nm. They have found that catalyst has a surface area, 1200 $m^2 \cdot g^{-1}$. They [3] have reported that 94.3% of biodiesel production is achieved using a catalyst loading of 5 wt% and 9:1 MTOMR

Fig. 2.4 Images of zeolite 4A (**a**), 4A-Fe (**b**), 4A Cu (**c**), and 4A-Fe Cu (**d**–**i**) pH of $CuCl_2$ is 1.2 (**d**), 1.4 (**e**), 1.6 (**f**), 1.8 (**j**), 2.0 (**h**), and 3.0 (**i**) [3] [Open access]

at 70 °C for 6 h. The catalyst could be used up to seven times, and 88.7% of biodiesel is produced by the completion of the seventh cycle. An alkali trap zeolite composite is used as the main catalyst for biodiesel production from WCO. They have extracted green tea to create four different types of alkali-modified clinoptilolite (K, Na, Ca, and Mg). SEM images of the studied samples (Fig. 2.5) showed a range of morphological features, while EDX analyses revealed different elemental compositions [3].

Due to their ability to improve production efficiency, facilitate product purification, and allow for catalyst reuse, nanocatalysts are essential for chiral catalysis and the synthesis of natural compounds. The unit of mass here is the ton (1 ton = 1000 kg, sometimes called the metric ton). Nanocatalysts serve as a link between homogeneous and heterogeneous catalysts because of their characteristics. Next, it is gone to over a few of the variables that affect nanocatalysis [2].

2.2.1 Nanocatalysts Based on Metals

Metal-based nanocatalysts are a rapidly growing field of nanotechnology and catalysis. The metal nanoparticles (NPs) that make up these catalysts have remarkable catalytic properties due to their large surface area, special electronic structure, and quantum effects. They are essential for many chemical reactions, including industrial processes, environmental purification, and energy conversion. In the synthesis of chemicals, purification of automobile emissions, and energy conversion, metal-based nanocatalysts play a vital role. Noble metals 4–6, transition metals, and alloys are common metal nanocatalysts that offer active sites for catalytic processes. The catalytic efficiency of metal nanoparticles is enhanced by their significantly higher surface area compared to bulk materials. Metals have unique electrical properties at nanoscale that influence their catalytic activity. Catalytic efficiency can be controlled

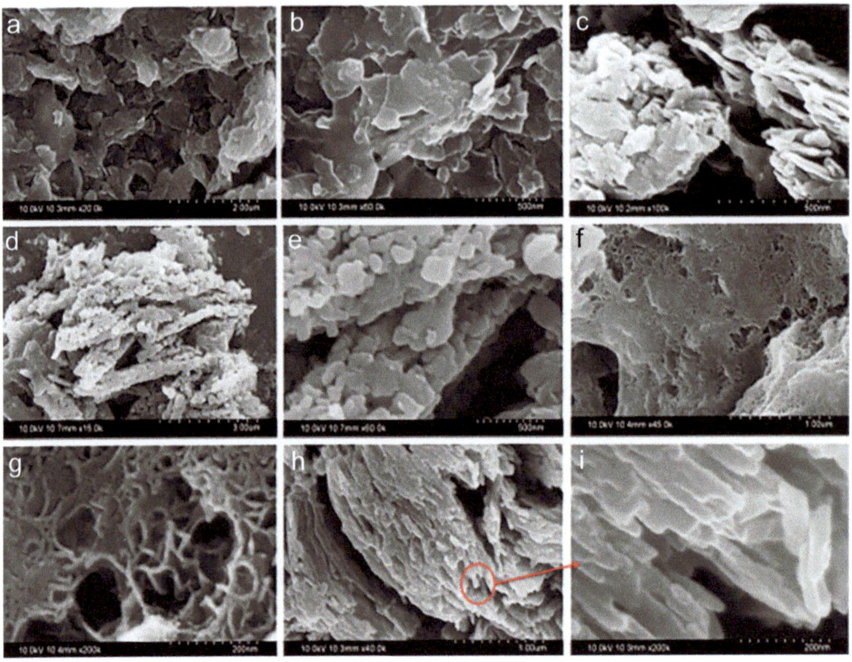

Fig. 2.5 SEM images of raw clinoptilolite: A, B; K/clinoptilolite; C; D and E; Na/clinoptilolite; G and H; and Ca/clinoptilolite (H and I) [3] [Open access]

by varying particle size, shape, and composition. Metal nanocatalysts can be designed to withstand oxidation, deactivation, and sintering, which extends their lifetime. Gold (Au), Silver (Ag), Palladium (Pd), and Platinum (Pt). Due to their exceptional catalytic activity, they are used in fuel cells, hydrogen production, and environmental applications. 4A-FeCu catalyst, which is more accessible than the noble metals and employed extensively in energy storage, it has a maximum adsorption capacity of 432.9 mg/g for Congo Red (CR) and 366.8 mg/g for 4A Cu. This could be because of catalysts' positive charge density, which increases their capacity to draw in and bind negatively charged species like CR and raises the adsorption efficiency. Further setting the 4A-FeCu catalyst apart from 4A Cu catalyst is the presence of iron, which probably enhances its interaction with the target molecules and increases its adsorption capacity. Kelvin Probe Force Microscopy (KPFM) measured the surface potential of catalyst and Fig. 2.6 shows the results. With a surface potential of 37.31 mV, 4A-Fe Cu has the highest potential among the positively charged catalysts, indicating high charge density and electron passing ability [3].By creating a dual-function nanocatalyst from palm kernel shell residue, used cooking oil can be converted into biodiesel fuel.The catalyst surface area is 438.08 $m^2 \cdot g^{-1}$, pore volume is 0.3674 $mm^3 \cdot g^{-1}$, and pore diameter is 3.8 nm. The biodiesel yield of 95% is obtained using ORC with 5 wt% catalyst amounts, 12:1 MTOMR, reaction time of 4 h, and reaction

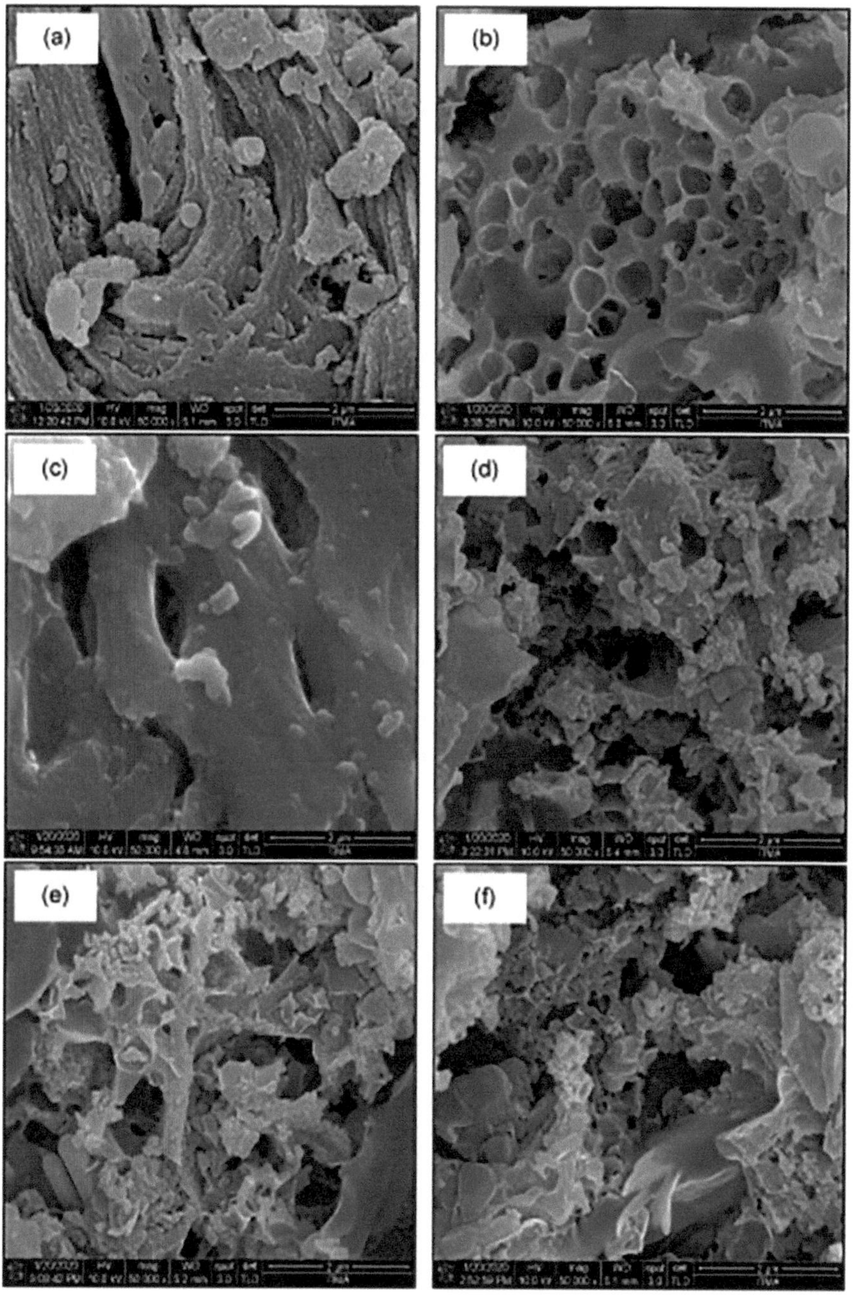

Fig. 2.6 At 50,000 magnification, FESEM images of **a** PKSHC, **b** PKSHAC, **c** PKSHAC-K_2CO_3(10%)CuO(5%), **d** PKSHAC-K_2CO_3(20%)CuO(5%), **e** PKSHAC-K_2CO_3(30%)CuO(5%), and **f** [4]

temperature of 80 °C. Furthermore, the carbonization process to create a bifunctional nanocatalyst from palm kernel shell to use this material for the production of biodiesel from Waste Cooking Oil (WCO). The surface area, pore volume, and pore diameter of the nanocatalyst are found to be 3368.60 $m^2 \cdot g^{-1}$, 2.36 $mm^3 \cdot g^{-1}$, and 5.17 nm, respectively. Figure 2.6 shows FESEM images of the successful impregnation of the active components with K_2CO_3 and CuO, which also showed the creation of irregularly shaped nanomaterials.Because of their remarkable catalytic activity, gold (Au), silver (Ag), palladium (Pd), and platinum (Pt) are frequently used in fuel cells, hydrogen production, and environmental applications.

Oxides, sulfides, and carbon materials are commonly used as support materials for metal nanocrystals, with their primary function being to stabilize the catalyst and enhance charge transfer. The practical applications of carbon-supported Nickel Manganese Cobalt (NMC) catalysts are particularly notable due to their unique properties, including a large specific surface area, good electrical conductivity, high catalytic activity, and low cost. These features make NMC an attractive choice for various catalytic processes. There are several methods available for the direct production of NMC catalysts that allow for precise control over particle size and improvements in stability. Similarly, another study highlights the use of similar catalysts, such as CaO(10%)-Fe_2O_3(10%)/AC, which has demonstrated remarkable results in biodiesel production. This further enhances our understanding of the effectiveness of supported materials in catalytic applications, emphasizing the critical role these materials play in improving catalyst performance and overall efficiency. FESEM images show the metal oxide distribution and surface morphological features of the catalyst (Fig. 2.7). The CaO(10%)-Fe_2O_3(10%)/AC (AC-activated carbon) catalyst could produce 98.3% biodiesel in 3 h at 65 °C with 18:1 MTOMR and 3 wt% catalyst loading. Even during the sixth consecutive cycle, the reusability study showed strong catalytic activity (biodiesel yield > 80%). [31].

These structural units contain a large number of unstable bonds and class of biocarbon-supported metal nanocatalysts that has been established by ion exchange and chemical bonding based on their structure and structural bonding, which is an in situ carbon co-reduction process. After adsorption for 12 h, Fig. 2.8a shows that the nanoplatelets are covered with a thick layer of CR molecules, which makes the porous structure less obvious. When comparing this onset with the reaction that occurs during the catalysis process, it is observed that the surface structure of uncatalyzed 4A-FeCu persists to a large extent even after a long period of catalysis. Compared with fresh 4A-Fe Cu (Fig. 2.8f), the morphology of 4A-Fe Cu is almost unchanged after 12 h of catalysis (Fig. 2.8b). SEM images of 4A-Fe Cu after adsorption and catalysis for 12 h are shown in Fig. 2.8c–f for further visual analysis. The nanoplatelets on the CR adsorbed surface of 4A-Fe Cu are stuck to each other instead of stretching. However, during the catalysis process, nanoplatelets on 4A-Fe Cu stretch in the same way as nanoplatelets on a fresh catalyst [3].

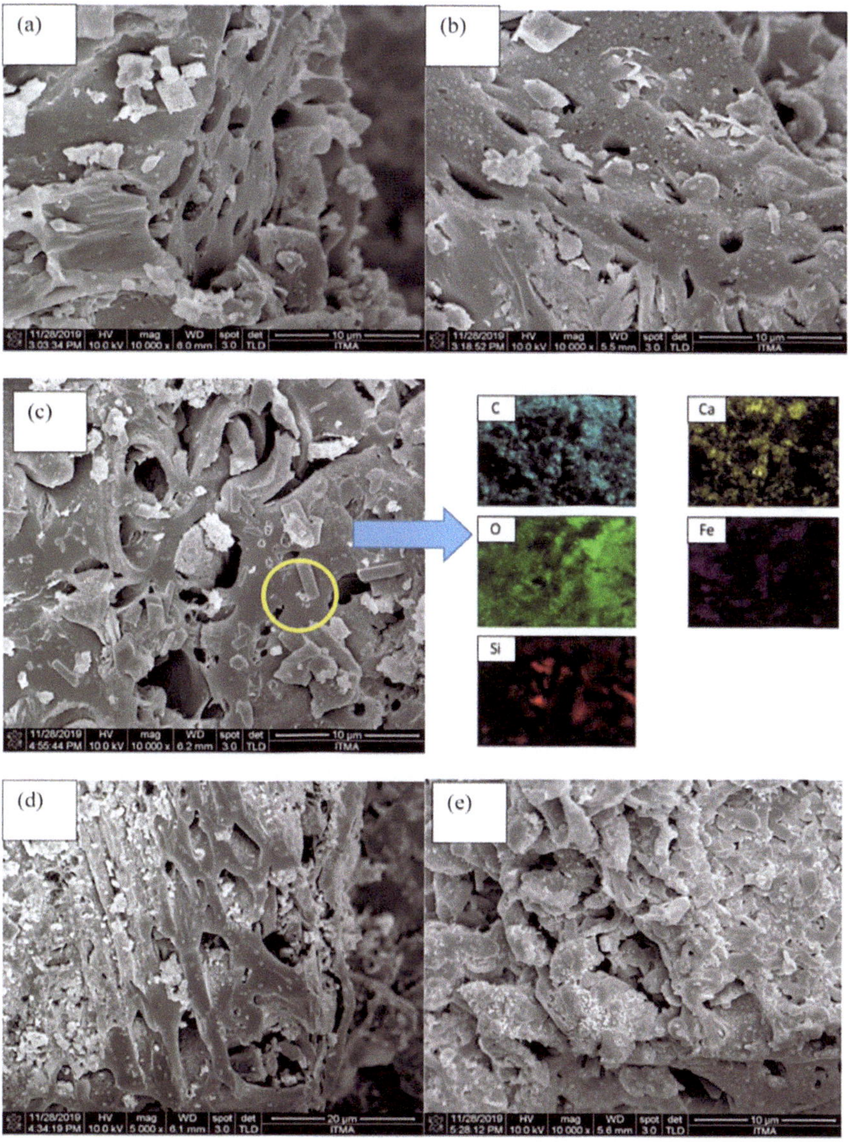

Fig. 2.7 EDX mapping results and FESEM images of **a** CaO (10%)-Fe_2O_3 (2%)/AC **b** 10% CaO-Fe_2O_3 (5%)/AC, **c** 10% CaO-Fe_2O_3 (10%)-/AC, and **d** 10% CaO-CaO (10%)-Fe_2O_3 (20%)/AC, -Fe_2O_3/AC (15%) [4]

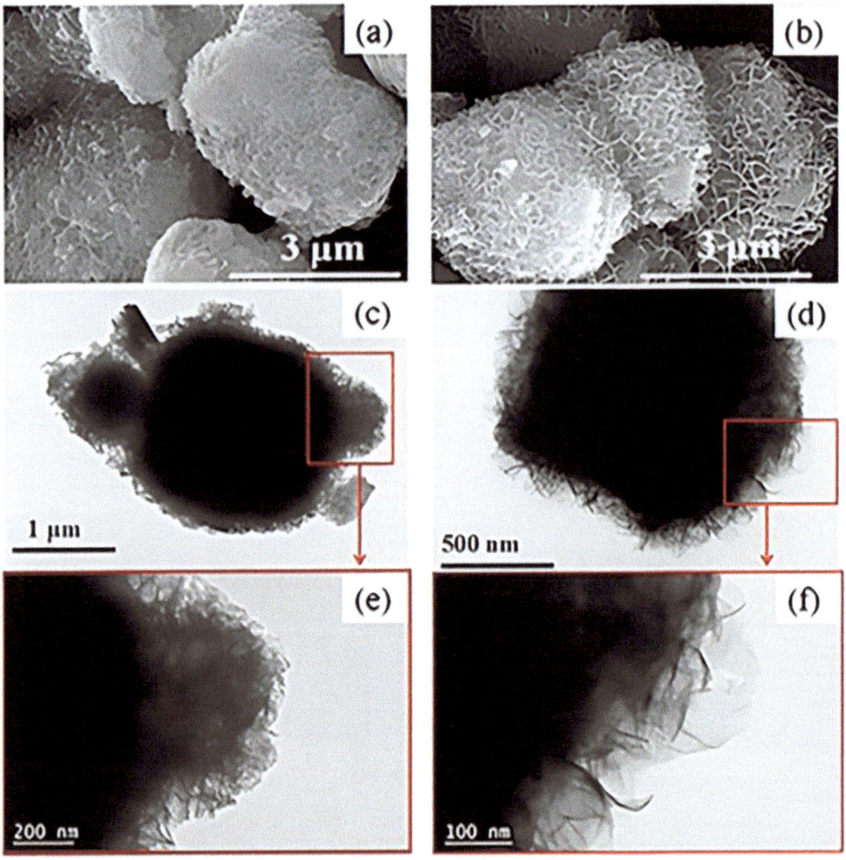

Fig. 2.8 SEM and TEM images of 4A-Fe Cu after 12 h of catalysis and adsorption ((**a**, **c**, **e**): catalysis; (**b**, **d**, **f**): adsorption) [3] [Open access]

Inorganic metal ions (or clusters) in solution form a stronger chemical bond with the broken unstable bonds in the cell wall and membrane during the preparation process, which allows hydrogen evolution reaction (HER), which is better than commercial Pt/C catalysts and other documented single-atom Pt catalysts. Our approach is easy to use, inexpensive, environmentally friendly, and suitable for any situation. SEM and TEM are used to characterize the morphology of the above-mentioned 4A-Fe@Cu after adsorption and/or catalysis of metal ions to be uniformly fixed on the surface of the membrane and cell wall. Metal ions can also be simultaneously reduced to metal atoms to create metal nanoparticles that are firmly fixed to the carbon surface when the cell wall and membrane are carbonized in situ to create a two-dimensional carbon material. Moreover, since the proteins contained in the cell membrane in the above experimental design contain many N atoms, N can be directly doped into carbon during the carbonization process. A method for improving the electronic state of carbon support to provide exceptional catalytic activity and stability. Thanks to its support for carbon nitride (CN), it can be used to create various nanocatalysts based on precious and base metals [2–5]. In addition, the exceptional advantages of the resulting catalysts strong bond strength to CN substrate, uniformly small particle size, and abundance of exposed active sites make them competitive with other catalysts for use in fuel cells, electrocatalysis, and the chemical industry. We selected hydroplatinic acid ($H_2PtCl_6{\cdot}6H_2O$) and white radish as a representative plant, using the above design principle as theoretical support. As a precursor solution, after which the experiment is carried out. To ensure that chloroplatinate ions completely penetrated the cell structure through free diffusion, bound and mixed stably with the cell wall and membrane, white radish is first cut into slices and immersed in $H_2PtCl_6{\cdot}6H_2O$ solution for several minutes. To ensure the stability of the multilayer 2D structure and prevent agglomeration or crystallization of intracellular chloroplatinic acid, soaked white radish slices are quickly frozen and then further freeze-dried (crystallization occurs in the normal drying process when the concentration of the solution exceeds its solubility). Finally, the dried slices are calcined in a tube oven protected by argon for nitrogen doping, in situ carbonization of the cell wall and membrane, and in situ fixation and reduction of platinum ions. Ultimately, our catalyst is prepared from Pt nanoparticles distributed within a nitrided two-dimensional carbon [2, 3].

2.2.1.1 Supported and Naked Nanocatalysts

The advantages of supported metal nanocatalysts (SMNCs) over other types of heterogeneous catalysts such as their high efficiency, stability, and reusability in catalytic reactions have attracted considerable attention in the fields of chemistry and materials science. However, traditional SMNC synthesis methods often require large amounts of organic chemicals, complex procedures, and specialized equipment to achieve well-defined structures and prevent nanoparticle aggregation. Atomic force microscopy (AFM) results (Fig. 2.9) shows that the changes in surface and roughness of three catalysts correspond to significant structural differences, which in turn affect their catalytic performance. AFM images demonstrate how the roughness and shape of each catalyst vary. In particular, catalyst 1 exhibits the least change in both shape and roughness, with a smooth surface and minimal distortions or calcifications in its immediate vicinity. In contrast, catalyst 2 exhibits high degree of surface roughness over its entire surface, although this property remains uniform over the entire surface. These structural changes play a decisive role in the efficiency of the catalysts in the corresponding reactions.

Then, the traditional use of SMNCs for electrochemical hydrogen evolution and related advances in this field are reviewed. SMNCs have been widely studied due to their high efficiency and reusability in hydrogen evolution reactions. Recent developments [6] have further enhanced the performance of these catalysts, making them increasingly important in the quest for sustainable energy solutions. Finally, the

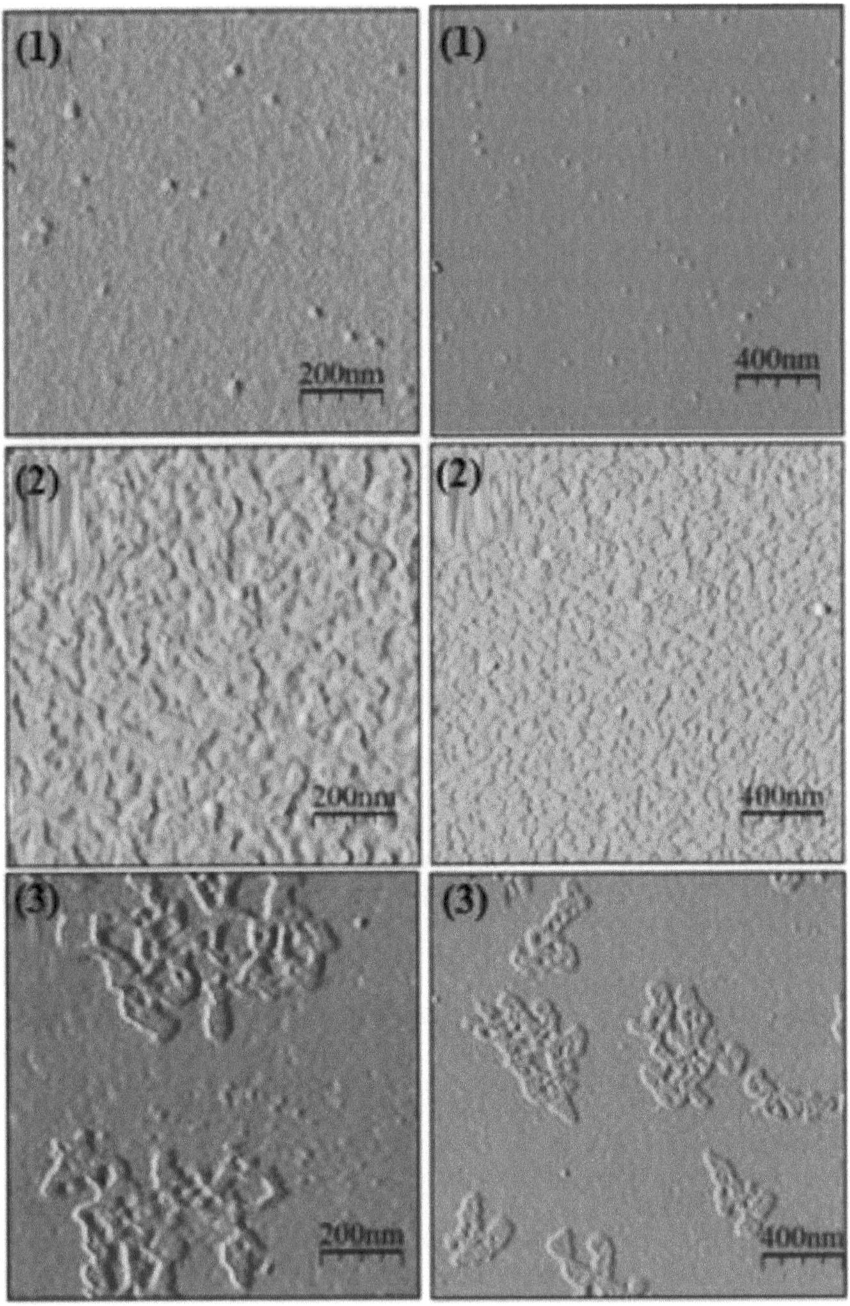

Fig. 2.9 AFM images of catalysts 1 (**1**), 2 (**2**), and 3 (**3**) at several magnifications [5]. [Open access]

prospects for sustainable production of SMNCs are discussed, focusing on methods to improve their synthesis while minimizing the environmental impact. In this context, Scanning Tunneling Microscope (STM) images of Au/HOPG sample (Fig. 2.10) illustrate the uniform distribution of gold particles on the support surface, which significantly enhances the efficiency of supported metal nanocatalysts in electrohydrogen evolution applications. STM image of Au/HOPG sample and particle size distribution plot shown in Fig. 2.10a highlight the well-defined dispersion of gold nanoparticles, which contributes to enhanced catalytic activity and efficiency in electrochemical process. These achievements reflect ongoing efforts to optimize SMNC for sustainable energy applications. By analyzing several sections of the sample, it is found that gold nanoparticles are very uniformly distributed on the support surface with an average particle density of 4.7(4) × 1012 particles/cm^2. With an average value of 3.27(6) nm and elongated right tail, the particle diameter distribution is found to be asymmetric. This is probably due to moderate hardening of nanoparticles during the thermal annealing step. This distribution can be classified as lognormal ($\mu = 1.10$, $\sigma = 0.40$) at a significance level of 5% based on the Kolmogorov–Smirnov (KS) statistical test (p-value = 0.062 and KS statistic 0.036). Scanning Tunneling Microscopy (STM) images show that the gold nanoparticles supported on HOPG are round [6].

This makes it difficult to synthesize SMNCs on a large scale and in practice. Therefore, provides an overview of green and sustainable synthetic approaches for SMNCs such as solid-state synthesis, surfactant- and reductant-free ionic liquid synthesis, and low-temperature pyrolysis. The field of catalysis is rapidly undergoing a lot of research and development due to ongoing energy crisis and the growing severity of environmental pollution. A key technique in green chemistry is catalysis, which refers to chemical reactions that produce fewer pollutants and allow the synthesis of desired products using sustainable resources [2–5]. In general, the choice of catalyst and its various structures and characteristics have a significant impact on the performance of the catalyst system. Heterogeneous catalysts known as supported metal nanocatalysts (SMNCs) offer several noteworthy advantages [5, 6]. First, the deposition of metal nanoparticles (NPs) onto a support can significantly reduce the surface energy, which in turn prevents the aggregation of NPs and extends the service life of the catalyst [2, 7]. Secondly, it promotes better dispersion of metal NPs and enhanced catalytic activity due to efficient electronic and synergistic interaction between metal NPs and substrate. Finally, SMNCs can be easily reconstituted without the need for labor-intensive and expensive separation procedures, unlike bare metal nanocatalysts [25]. Considering these advantages, the goal of creating and using various SMNCs has attracted significant scientific interest on a global scale.

The size, shape, content, and structure of SMNCs have been modified and optimized in an attempt to improve their catalytic performance in recent decades [25–28]. As a result, several traditional methods have been used, including precipitation, sol–gel [2–6], and solvothermal pyrolysis at high temperatures [4]. However, these methods do not correspond to the idea of green and sustainable synthesis as they often have disadvantages related to the need for special equipment, complex synthetic procedures, or organic chemicals [2, 5]. Thus, it is highly desirable to

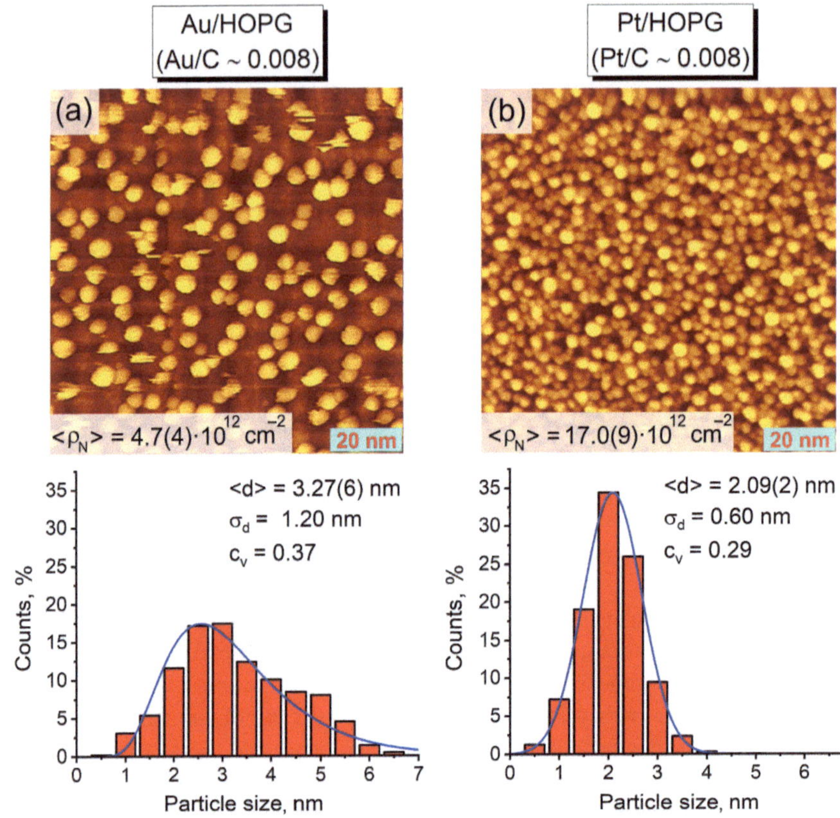

Fig. 2.10 Particle size distribution plots and SEM images (100 × 100 nm square): **a** Platinum/HOPG after platinum deposition on defective HOPG (Pt/C ~ 0.008); **b** Gold/HOPG after gold deposition on defective HOPG and heating in ultrahigh vacuum at 350 °C. 1.50 V, 0.53 nA, and 0.50 nA are the tunneling parameters for (**a**) and (**b**). Blue curves represent the following distributions: **a** lognormal ($\mu = 1.10$, $\sigma = 0.40$) and **b** normal ($\mu = 2.09$, $\sigma = 0.60$) [6] [Open access]

have simple and efficient methods for producing metal nanoparticles (NPs) that are uniformly distributed on substrates have a clean surface and have an ultrafine diameter. In addition, it would be very beneficial to minimize the use and production of hazardous materials in both the production and application of catalyst [2, 6]. The use of blocking and stabilization methods including synthesis using ionic liquids or decomposition at low temperatures is a possible option given the shortcomings of conventional methods for preparing various catalysts such as the need for specialized equipment or large organic compounds.

Therefore, several effective synthetic methods such as solid-state synthesis [6, 8] and low-frequency synthesis have been used to obtain SMNCs. Thermal decomposition [2, 9], surfactant- and reductant-free synthesis [27], and ionic liquid (IL)-based synthesis are all innovative approaches to preparation of metal nanocatalysts. These methods offer distinct advantages in controlling particle size, shape, and dispersion while minimizing the use of potentially harmful chemicals. These artificial methods offer environmentally acceptable pathways that meet green chemistry standards. Compared to traditional synthetic methods, they have the following advantages:

- Simplified synthesis process: Most of these processes do not require surfactants, capped reagents, or subsequent catalyst surface cleaning steps. In addition, several methods are more convenient because they do not require solvents.
- Green synthesis: By avoiding using of excessively large quantities of organic reagents (such as solvent and surfactant), these synthesis methods are less harmful to the environment.
- High nuclear economy: Most of the time raw materials are completely converted into finished products, demonstrating a high nuclear economy. These characteristics guarantee minimal costs and excellent efficiency.
- Low by-product production: Excellent product selectivity is achieved through optimized synthesis processes and high atomic savings resulting in low by-product production.

2.2.1.2 Core–Shell Nanocatalysts

Fossil fuels do not only contribute to the development of human culture and science but also cause environmental pollution and global warming. Compared to 2014, energy consumption will increase by 36% by 2035 [21]. The subsequent escalation of serious environmental problems motivates countries to improve their energy mix. Numerous clean energy conversion and storage methods have been developed to reduce carbon dioxide emissions and air pollution. Examples of these fuel cells include alkaline fuel cells, molten carbonate fuel cells, solid oxide fuel cells, phosphoric acid fuel cells, and proton exchange membrane fuel cells (PEMFC) [28]. PEMFCs have received the most attention and effort of all. These fuel cells are due to their high energy density, low operating temperature, and noise. PEMFCs are currently the most promising option for powering electric vehicles [22]. The energy source of an electric vehicle is a battery of fuel cells. The complex structure of a fuel cell stack consists of membranes, bipolar plates, terminal plates, gas diffusion layers, spacers, and bipolar catalysts. Because catalyst is so expensive, it is essential for the development of fuel cell electric vehicles. The catalyst and its use in fuel cell batteries for heavy-duty vehicles account for approximately 53% of the total cost.

Therefore, the cost of a fuel cell system is directly affected by catalyst longevity and Platinum Group Metals (PGM) loading. Years shown vary in system cost based on PGM durability and load. PGM load and overall system costs are reduced annually. By 2020 haS seen a dramatic increase in PGM loading to accommodate highly reliable fuel cells, which will continue to reduce system costs when durability is taken into account. Higher PGM loadings offer significant prospects for the development of fuel cell catalysts. AFM images of doped P_3HT samples ($P_3HT/Li + TBP$) at scan sizes of 10 and 7 μm are shown in Fig. 2.11a, b. The features of morphological structures of P_3HT have been highlighted in Fig. 2.11b by AFM data presented with

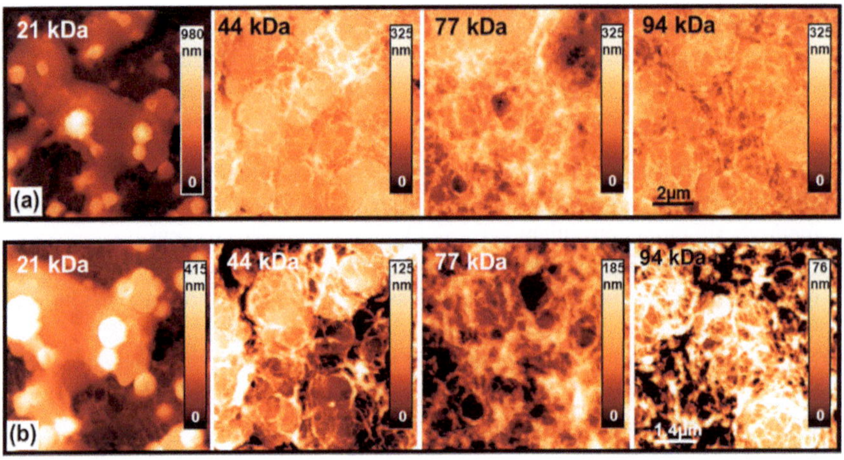

Fig. 2.11 AFM of P_3HT films doped and deposited at different molecular weights. AFM data are presented in panels (**a**) and (**b**) with scan sizes of 10 μm and 7 μm, respectively. For better comparison, samples from 44 to 94 kDa are shown in panel **a** using the same z-scale; AFM data are presented in panel **b** using a z-scale that has been appropriately adjusted to highlight features [9] [Open access]

a properly fitted z-scale for each image. The morphology of $P_3HT/Li + TBP$ films is significantly different from that of pure P_3HT films. In addition, each sample has a different shape for the films deposited at different MWs. The 21 kDa sample shows uniformly distributed, seemingly huge, thick aggregates with spherical shapes placed on top. The film topography is smoother, meaning that the film thickness is more uniform and the samples with MWs in the range of 44–94 kDa have compact disk-shaped flat tops. [7].

This study shows that the structure is dependent on the size of the nanoparticles. The smallest nanoparticles ($d = 5.6$ nm) are subjected to cesium-corrected high-resolution STEM (HR-STEM) images, which revealed the presence of discrete crystalline domains (Fig. 2.12a). EELS analysis of these images revealed distinct segregation of chromium (Fig. 2.12b) [7]. Most of the chromium and cobalt are oxidized, consistent with the normal presence of oxygen, although the possibility of a small metallic core of cobalt could not be excluded. An ex situ study shows that air exposure is the source of oxygen. This shell has a petal-like structure and is uniformly generated with discrete crystalline domains of small spherical crystals of 1–2 nm in size. The EELS study illustrates that the size of the oxidized shell of $Co_{90}Cr_{10}$ is measured to be 2.72 ± 0.63 nm. Co is also present in this shell, although at a low concentration (Fig. 2.13) [7].

2D AFM images of surface topography of pristine P_3HT layers deposited on glass are shown in Fig. 2.14a, b. These images are acquired at scan sizes of 10 and 3 μm, respectively. There is no noticeable MW-dependent behavior in the topographies. However, at higher MW, the surface topography becomes somewhat smoother, which may be due to the elongation of the polymer chains. Furthermore, this is in line with

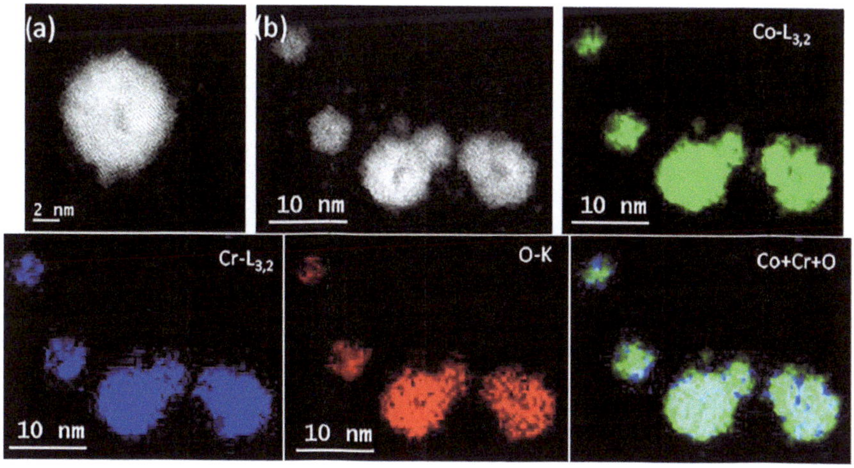

Fig. 2.12 Cesium-corrected high-resolution transmission electron microscopy (HR-STEM) data of small chromium nanoparticles (5.6 nm) **a**. Highly magnified image of a single nanoparticle showing its multiple crystalline domains. **b** Multiple chromium nanoparticles and their associated chemical maps are obtained using electron energy loss spectroscopy (EELS) [7] [Open access]

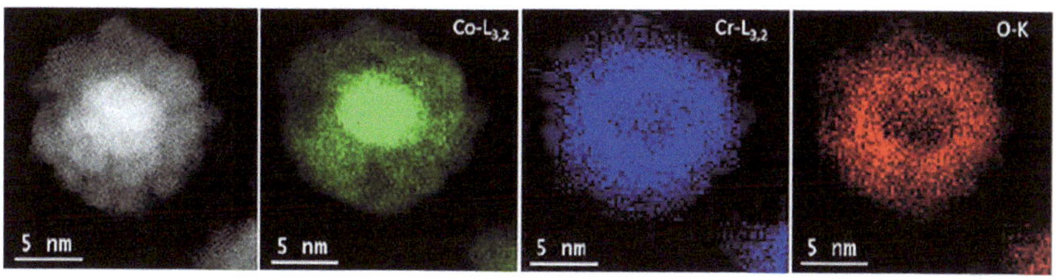

Fig. 2.13 Cs-corrected HR-STEM results corresponding to larger CoCr NPs (6.9 nm). High-magnification images and corresponding EELS chemical maps are displayed simultaneously [7] [Open access]

previous studies that detailed how MW affects the electrical performance of P_3HT films [4, 9]. Differences in the rates of solvent evaporation and polymer solidification in different solutions may be responsible for the topographical variations between samples. Compared to higher MW samples, solutions of low MW materials are more mobile and have lower viscosity [9].

As shown in the 3D electrical cell/Matrigel substrate impedance sensing model (3D-ECMIS), 3D in vitro model uses an impedance sensor to analyze cell viability in real time. 3D-ECMIS is fabricated using eight separate sensing channels. Each channel included a glass substrate, a culture chamber, and two gold electrodes (Fig. 2.15). The analysis is performed by applying an initial potential, 30 mV to the electrodes and measuring the impedance every 10 min for 96 h over a range of 0.1 to 50 kHz. Using the 3D in vitro model of human hepatoma ($HepG_2$) cells embedded in a Matrigel matrix, they investigated the anticancer efficacy of three

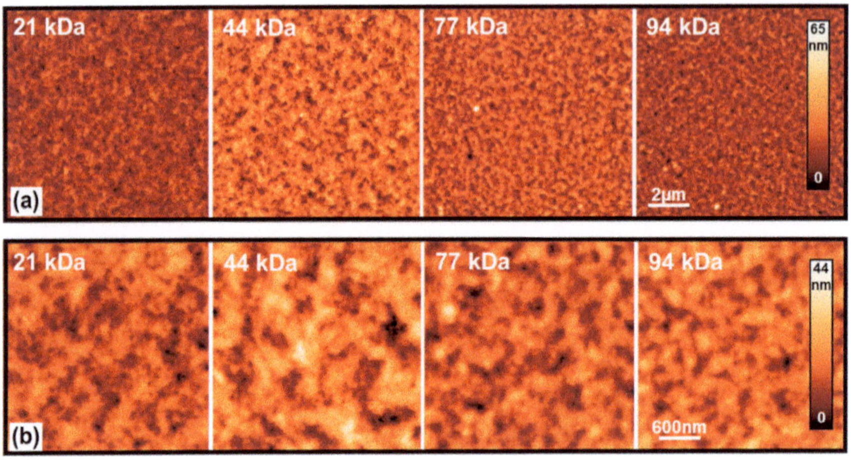

Fig. 2.14 Neat poly(3-hexylthiophene) (P_3HT) films coated with different molecular weights (MW) are imaged using AFM. AFM data shown in panels (**a**) and (**b**) have scan sizes of 10 μm and 3 μm, respectively. For each image in the corresponding panel, the scale bar and color bar are assumed to be applicable [9] [Open access]

different drugs: the liver anticancer drug sorafenib, the broad-spectrum anticancer drug cisplatin, and the ovarian/breast anticancer drug taxol [12].

Enhanced catalytic efficiency magnetically recyclable Ag–Ni core–shell nanoparticles synthesized in a one-pot synthetic route has showed an exceptional efficiency in hydrogen synthesis via sodium borohydride dehydrogenation in aqueous solutions despite the unfavorable performance of Au coated with palladium nanostructures [27]. Comparing the Ag–Ni nanocatalysts with individual Ag and Ni nanoparticles as well as several other bimetallic nanocatalysts of similar size, the researchers have found that the activation energy is significantly lower. Moreover, by overcoming the tendency of nanoparticles to sinter under catalytic conditions, the active metal enclosed in the porous shell exhibits strong catalytic properties and prevents particle agglomeration.

Chemical particles from outside can freely reach the surface of the active base material thanks to the porous shells. In addition, an increased porosity promotes increased surface area, which greatly promotes the adsorption of particles to the surface and increases the reaction rate. The addition of highly stable and porous materials such as SiO_2 has been reported to change the function, reactivity, electronic charge, and stability of the surrounding material as well as the core material. Additionally, Grasse and co-workers have found that encapsulating zinc ferrite in SiO_2 significantly improved its magnetic properties [22]. A highly efficient $FeSiO_2$ nanocatalyst [23],was developed by modifying the surface porosity of the material through changes in the composition of the C18TMS agent and pre-coating with added Cs material. They have produced carbon dioxide-free hydrogen from ammonia using a core–shell composite. Scientists note that when decomposing ammonia, Ce-$FeSiO_2$ acts like a small capsule reactor. This one-step synthesis method is developed [34] to

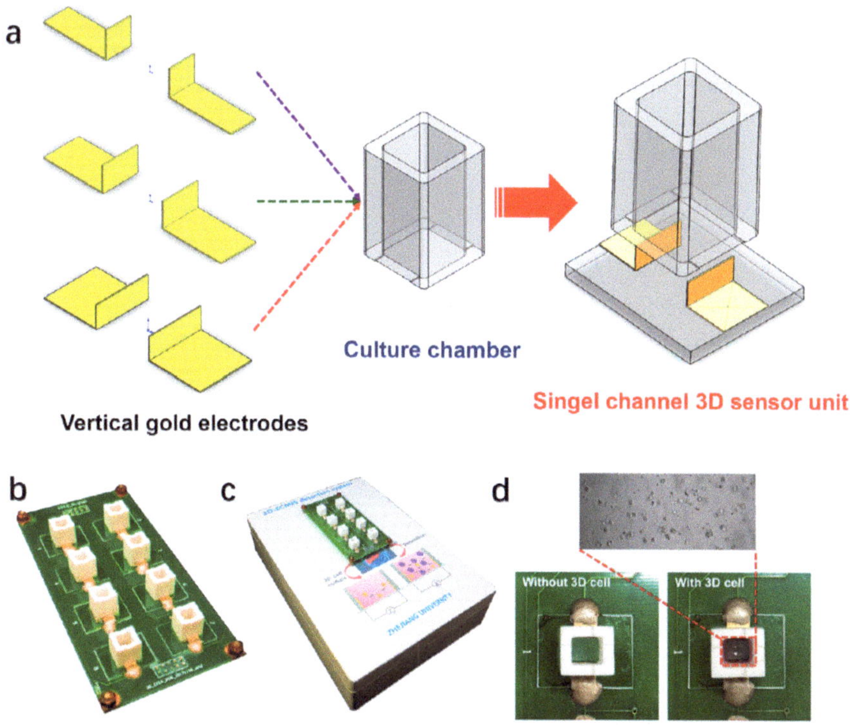

Fig. 2.15 Schematic showing the various components of 3D-ECMIS (3D electrical cell/Matrigel substrate impedance sensor): **a** one channel of vertical gold electrodes, **b** eight channels arranged on the substrate, **c** image of the eight channels of the detection system, and **d** photographs of the 3D cellular biosensor before and after seeding with living cells [12] [Open access]

create Ag TiO_2, Ag ZrO_2, Au TiO_2, and Ag ZrO_2 nanoparticles with different shapes. The authors demonstrated the strong optical determinants of synthetic materials and their high laser damage threshold.

Up to 700 °C, the developed nanoreactor framework demonstrated remarkable stability. In addition, scientists have effectively modified the core surface and porosity of $AuSiO_2$ yolk–shell materials [27]. Porosity is monitored using C18TMS porogen.

2.2.2 *Nanocarbon Catalysts*

Because carbon atoms can bond to each other in a variety of ways, forming linear, planar, and tetrahedral bonds, a carbon is a versatile and interesting material that is widely used in many technological processes. This makes it possible to create materials with a wide range of controlled physicochemical properties. However, the amorphous or disordered classical carbon materials used in previous studies [28] generally exhibited low activity, stability, and oxidation resistance (e.g., activated carbons, carbon aerogels, carbon black, carbon fiber, etc.). Three main reasons have been identified: a) conventional carbon catalysts mainly contain micropores, which lead to long product and reactant diffusion paths. As a result, internal active pads are used less. b) Many different types of defect sites are formed as a result of the rigid cross-linking and random orientation of turbostratic nanodomains caused by high concentrations of sp^3 carbon and oxygen-containing functional groups. Carbon itself is a catalyst for several reactions including the oxidative dehydrogenation of ethylbenzene, the reductive elimination of NO, formation of $COCl_2$ and SO_2Cl_2, and the aerobic oxidation of oxalic acid. Using a final galvanic bias step after the fabrication of the iron core nanoparticles, the process described in [26] is modified to fabricate the core–shell particles in a continuous microemulsion (see Fig. 2.16). Without adding any additional reducing agent, a single microemulsion containing a platinum salt is added to precipitate the shell material (platinum). The resulting core–shell particles are then cast onto a high surface area carbon support (VulcanXC72, Cabot, Boston, MA, USA) and thoroughly cleaned with acetone, tetrahydrofuran, and water at a later stage. The final step of the washing process is to wash the particles again with water and 0.25 M HCl [8].

Therefore, identifying the exact active regions of a unique catalytic process can be challenging. c) Since these carbons perform oxygen-containing functions at edges of graphite-like lamellar-basal planes, they can also easily undergo combustion at high temperatures and aerobic environments. These problems seriously hampered the rapid progress of previous carbon catalysts [26]. It should be noted that water is a good solvent for lignosulfonate, but kraft lignin dissolves in alkali and some organic solvents. Therefore, kraft lignin dissolved in THF is added to excess water to test the solvent exchange method first. However, SEM images have shown that the formed lignin spheres are aggregated (Fig. 2.17a). By adding 2 M HCl to the final

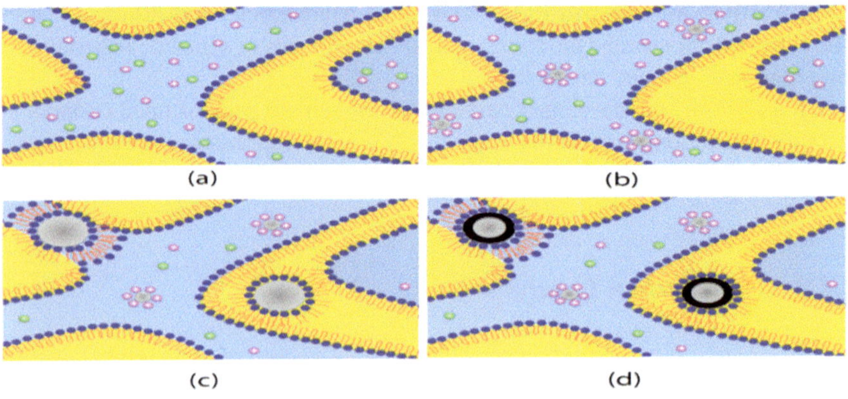

Fig. 2.16 Schematic representation of the production of core particles (**a**) through (**c**) and the galvanic bias that produces the final core–shell particles (**d**). Graphs taken from [26], 2014, American Chemical Society, with permission [8] [Open access]

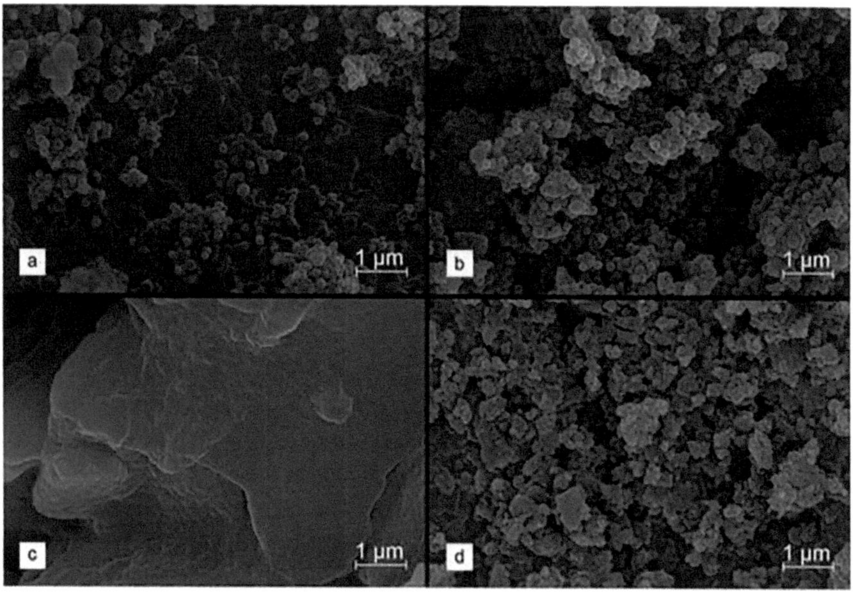

Fig. 2.17 Lignin precipitated from solutions containing kraft lignin (**a**, **b**) or sodium lignin sulfonate (**c**, **d**) is imaged using a scanning electron microscope [14] [Open access]

solution, pH is lowered, resulting in more uniform and separated lignin nanospheres (Fig. 2.17b). [14].

The creation of defects, doping with heteroatoms, and surface functionalization are the most common methods for preparing effective catalysts. However, the nature of active spots is still unclear. To further improve activity, selectivity, and stability, this knowledge is critical.

2.2.3 Nanocarbon as a Support for Catalyst

One of the most important industrial processes is the liquid-phase aerobic oxidation of hydrocarbons to more valuable oxygen-containing molecules. Previously, these procedures are carried out using homogeneous catalysts made from transition metal salts and complexes in stoichiometric quantities with various dangerous oxidizing agents. As a result, a huge amount of toxic and corrosive waste is generated. Due to their exceptional properties, carbon nanotubes have recently become extremely important as catalysts and catalytic support in various chemical processes. Carbon nanotubes and their functionalized forms have not found widespread using by researchers as catalysts or catalytic support for liquid-phase aerobic oxidation of hydrocarbons [28].

Particular attention is paid to the use of carbon nanotubes and functionalized carbon nanotubes as catalysts and catalytic supports in liquid-phase aerobic oxidation of hydrocarbons. After careful examination of previously published data, a plausible reaction mechanism is developed. It is also discussed how impurities in carbon nanotubes affect their catalytic activity in liquid-phase aerobic oxidation of hydrocarbons. The literature reports that oxygen-containing carbon nanotubes reduce their catalytic activity in liquid-phase aerobic oxidation of hydrocarbons; on other hand, the catalytic activity is enhanced and improved by surface electron transfer between modified carbon nanotubes and radicals, which is achieved through nitrogen doping and/or immobilization of metal nanoparticles.

Moreover, carbon nanotubes doped with nitrogen or containing metal nanoparticles accelerate the breakdown of hydroperoxides, which are important steps in the liquid-phase aerobic oxidation of hydrocarbons, increasing the overall catalytic activity. Since carbon nanotubes are typically smaller in diameter than conventional atomic microscope tips, it can be expected that atomic microscope images obtained before and after the fixation of carbon nanotubes will have higher resolution. Figure 2.18 illustrates this, which is particularly evident when using the high-force tapping and pinpoint modes. Using AFM model Flexible Manufacturing System (FMV) probe (Bruker Corporation, Billerica, MA, USA), a carbon nanotube-covered surface is imaged in Fig. 2.18a, and image shows that the tip is somewhat blunt. The procedure for fixing carbon nanotubes to an atomic microscope tip using high-force tapping mode is presented in Fig. 2.18b. The red rectangle highlights the area of the image where the tapping mode adjustment point is lowered; this area indicates where carbon nanotubes are fixed and where the image resolution is improved. The same area is rescanned in Fig. 2.18c showing that carbon nanotube fixation is stable and the image resolution has been significantly improved. Figure 2.18 illustrates the imaging procedure in the strong tapping mode to fix carbon nanotube (CNT) on AFM tip. An image with a somewhat blunt tip is shown in Fig. 2.18a. The stabilization process is illustrated in Fig. 2.18b by decreasing the focal point of the image until a resolution shift is observed. An image with improved resolution with a fixed carbon nanotube attached to AFM tip is shown in Fig. 2.18c. The color scale height for the images is 10 nm [35].

2.2.4 *Quantum Dots*

III–V heterostructures formed the main basis of early LED technology, which later included epitaxial quantum dots. These quantum dots (QDs) are organically incorporated into a multilayer heterostructure formed on a semiconductor substrate, in contrast to colloidal production. InAs/GaAs QDs are undoubtedly the most extensively studied of several material systems. However, due to desorption, for example, the high temperature of growth process can cause problems with interface control. To

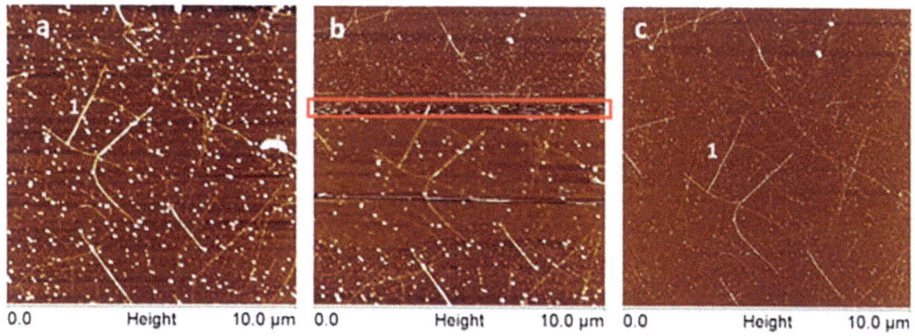

Fig. 2.18 High-power touch imaging technique attached to CNT to tip of Wolf microscope. The image is with a somewhat blunt tip (**a**). Decreasing the image focus until a resolution shift is noticed (**b**) illustrates the stabilization process. A resolution-enhanced image with a fixed carbon nanotube attached to the tip of a Wolf microscope is shown in (**c**). The color scale height for the images is 10 nm [35] [Open access]

solve this problem, by capping QDs with self-assembled InxGa1−xAs layers, vertically connected QD bilayers are created with a significant increasing in PL intensity. Postprocessing analysis is another issue that frequently arises in QD epitaxial systems. Scanning tunneling spectroscopy (STM) and chemically etched semiconductor nanomembranes have been used to novelly address this problem. Because of their tunable, bright, and narrow PL components, solution-processed colloidal InP QDs have been explored for use in LEDs and displays. Notable improvements are achieved by adding ZnSe/ZnS multicladding to InP cores, which is well suited for wide-color gamut displays. To further this understanding, the surface dangling bonds of InP QDs are investigated using density functional theory (DFT) calculations combined with experimental data.

Several materials are used to create QDs that emit in specific spectral ranges. It is believed that silicene QDs, embedded in multilayer siloxene nanosheets containing $FeCl_36H_2O$ may have potential applications as emitters in blue LEDs in the future. (10) $CsPb(Br/I)_3$ and $CsPbI_3$ QDs subjected to halide exchange showed improved optical and structural performance and can be used in 670 nm LEDs. ZnO nanorods on GaN films can also be decorated with $CsPbBr_3$ QDs to create dual-wavelength green LEDs [29]. One such test is shown in Fig. 2.19 on a grid made from a 4:1 PFDD/SWCNT dispersion deposited on a SiO_2 substrate. 0.2 nN is the lowest force that consistently produced a stable image, regardless of the specific nanotube sample or AFM tip. As shown in Fig. 2.19a, successive images are taken at increasing values of the peak feedback force, using this as the initial force, and the final image is again taken at 0.2 nN. The plot shows that the image remained non-destructive even at the maximum applied AFM force used here (2 nN), since the nanotubes did not appear to move during imaging [10].

Graphene oxide QD/GaN composites have been shown to improve the PL of GaN. Blue-emitting $CsPbBr_3$ QDs can be used as phosphors and white light LEDs, while hexagonal boron nitride sheets are used to improve the thermal stability of CdSe/CdS

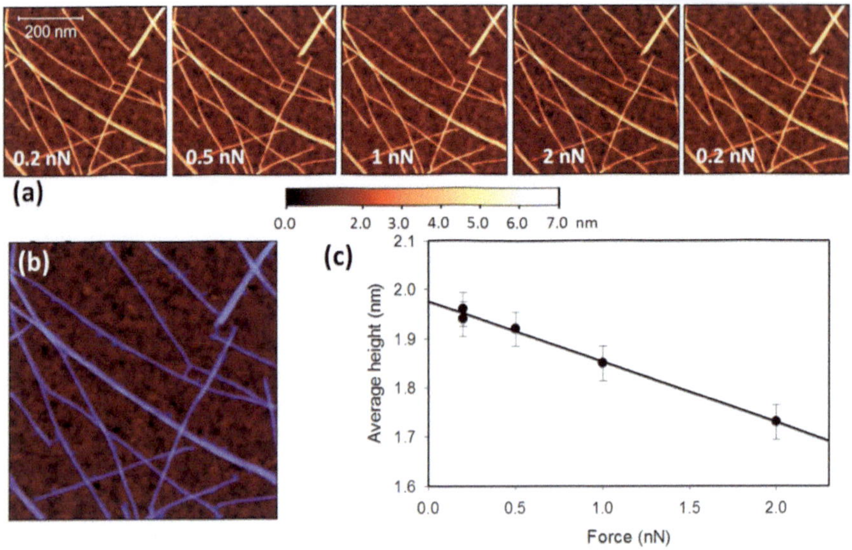

Fig. 2.19 **a** Adjacent images of the same region of the SiO_2-based PFDD/SWCNT network acquired at different peak force feedback levels. **b** Each pixel whose height is considered to determine the average network height is shown in a blue mask. **c** Linear fit of the data together with the average network height as a function of applied force [10] [Open access]

QD-based devices. There is also research into ultrastable QD-based phosphors, as is recently demonstrated using an LED embedded in a monolithic CdSe/CdS/ZnS QD-SiO_2/Al_2O_3 phosphor. Bovine serum albumin (BSA)/QD nanocomposite containing zinc sulfide and zinc quinoline complexes as QDs have also been used to produce white light. In addition, stable graphene QDs produced by microwave irradiation and ultra-bright fluorescence have been proposed as an effective replacement for white light-emitting diodes. Figure 2.20 shows parallel pairs of height profiles that are taken from the measured data, together with a typical 1 μm^2 AFM image of PFDD/SWCNT network on SiO_2. Each pair of profiles has one profile on the substrate near the tube and one profile along the top of the nanotube. The profiles found on the bare SiO_2 substrate (represented by the red dots in the figure) exhibit greater irregularity, with the height difference sometimes exceeding 1 nm. For the four profiles depicted in the figure, the standard deviation of the substrate height profile data ranges from 0.31 to 0.18 nm [10].

However, non-toxic colloidal zinc nitride (Zn_3N_2) QDs offer a viable replacement for LED QD displays. The increase in the optical band gap depending on the size occurs from 1.5 to 3.2 eV. Thin-film LEDs can be based on $CsPbX_3$ QDs obtained using a new solid-state ligand exchange method. It is also recommended to use $CuInS_2$/ZnS QDs with stable thiol- and amine-based ligands to create efficient film-type display devices. CdSe nanoplatelets and polymer encapsulation of PbS QDs are promising as LED composite. In addition, CdSe/CdS and CdSe/ZnS QD films coupled to Au nanorod dimer arrays can be used to modify the PL polarization of

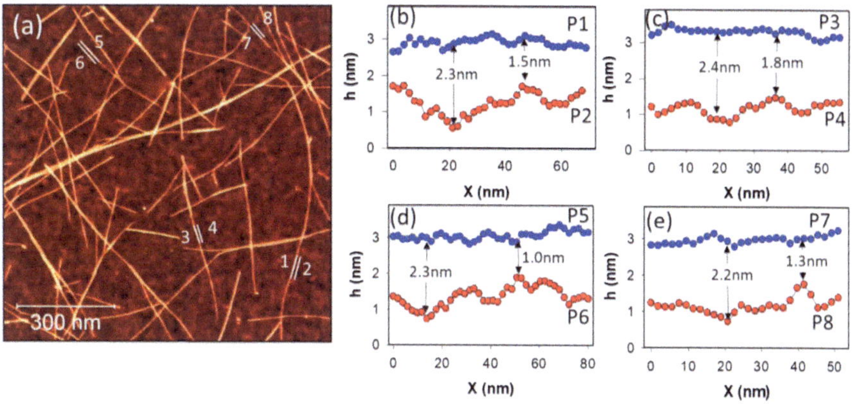

Fig. 2.20 SiO_2 substrate roughness and nanotube profiles. (**a**) A typical 1 × 1 μm square AFM image of the PFDD/SWCNT network with numbered white lines extracted along which profiles P1–P8 are shown (**b**–**e**). Adjacent substrate profiles are shown in red, and data points corresponding to profiles along the top of the nanotubes are shown in blue. Differences in substrate and nanotube heights are shown at specific locations to demonstrate diversity [10] [Open access]

luminous films in the outside and interior walls of TiO_2 nanotubes with PbS quantum dots measuring 4–5 nm, assisted by a bifunctional binder of thiolactic acid. In the photodegradation of organic dyes, the authors found that PbS/TiO_2 nanotubes had increased photocatalytic activity and stability. Moreover, the QDs have increased activity and solar spectrum's useable range. The research is expanded to examine the properties of TiO_2 nanotubes functionalized with 3–4 nm copper sulfide quantum dots. With the aid of cysteine linkers, stepwise crystallization of nanosized CuS on TiO_2 nanotubes allowed for the successful functionalization. The photodecomposition of phenol and malachite green is facilitated by the application of the recently created CuS/TiO_2 catalyst. Three distinct reactors are used for the investigations: quartz, Pyrex (optical filtering).

2.3 Motivational Performance Monitoring

Owing to the discovery that nanostructured materials have unique features compared to their bulk counterparts, nanoscience and nanotechnology have received much attention in recent decades [25]. Due to size effects, the macroscopic quantum tunneling effect, and finite surface effect at nanoscale, nanostructured materials have unique optical, electrical, and magnetic properties. Therefore, the creation of catalysts, fuel cells, gas sensors, photoelectronic devices, energy storage devices, supercapacitors, and lithium-ion batteries become possible due to the wide range of applications that nanostructured materials open up [23, 26]. Currently, the subject of close attention is the production of fine nanopowders of excellent quality in size, shape, and structure. Various methods have been proposed to create materials at the nanoscale, including solvothermal synthesis [28, 29], carbonylation, and the following: anode-arc plasma method [25, 26], co-precipitation [27], microwave irradiation [27], microemulsion sonochemistry [26], ultrasound [24], and sol–gel procedures [25]. Flexibility in changing synthesis parameters and controlling the physicochemical properties of nanopowders without the need for expensive chemicals or complex equipment is a significant advantage of the co-precipitation approach. Due to their exceptional activity, stability, low toxicity, and environmental friendliness, nickel-based nanostructured materials have been widely studied in the field of catalysis [27–2929]. In addition, various materials other than catalysts have been developed using nickel-based nanomaterials, including electrochromic materials [23], p–n heterojunctions [24], gas sensors and magnetic materials [28], optical materials [26] and fuel cell electrodes [27], solar cells [23, 24], batteries [28], and electrochemical capacitors [22]. In the present study, a modified Stöber co-precipitation method is used to create silicate-supported nanonickel catalysts with crystal diameters of about 30 nm [17]. Methane decomposition is carried out using synthesized nickel catalysts supported on silicate, and then the activity and stability of methane conversion are assessed. One of the newest products in the fields of materials science and sustainable energy, respectively, is hydrogen. In modern conditions of significant depletion of fossil fuel reserves and degradation of other energy sources, hydrogen can be considered a viable environmental energy carrier as a result of increasing greenhouse gas emissions [15]. Because it produces no greenhouse gas emissions when burned, hydrogen plays a critical role in solving current energy problems. Using field emission scanning electron microscopy (FESEM), the surface morphology, size, and density of ZnO NRs are investigated and results are shown in Fig. 2.21. ZnO and Ag ZnO NRs grown on Si substrates with constant coating density, smooth top surface, uniform height, and vertical alignment along the C-axis are shown in Fig. 2.21a–d. These views and cross sections confirm and validate the XRD results [11].

Moreover, renewable feedstocks including water, biomass, and biogas can be used to create hydrogen [16, 18]. Since methane constitutes a significant portion of biomass and is available in huge quantities both in industrialized countries and on the deep ocean floor, an all-green methane-splitting technology can be considered a sustainable method for hydrogen production [26, 29]. Therefore, to simplify matters, the automotive industry, academic institutions, and governments have focused more attention on the potential use of hydrogen as an alternative fuel, both in widespread adoption and production.

2.4 Motivational of Applications

Almost all known catalytic processes have used nanostructured materials to study their unique physicochemical properties. This section describes several selected use cases.

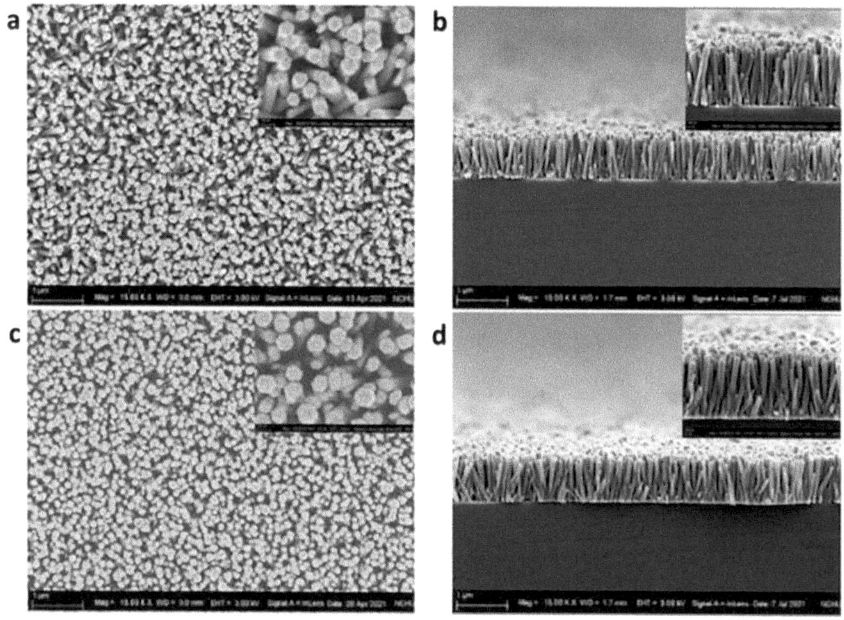

Fig. 2.21 Cross section and top view of Ag@ZnO ZnO nanoparticles (a, b) and nanoparticles (c, d) [11] [Open access]

2.4.1 Producing Hydrogen from Methane

One of the most important challenges we face today is building a sustainable energy system that can meet growing energy demand. Reducing greenhouse gas (GHG) emissions is a cornerstone of this energy system, helping to achieve the goals set out in the Kyoto Protocol and the Paris Agreement (1) by reducing global warming. The diffraction spots in the following figure of the RHEED diagram show that at low I/A ratios, the secondary crystallization of the (111)-oriented z-GaN islands disrupts this 2D evolution. These islands orient in opposite directions to each other in the [111] direction, creating the appearance of twinning. For clarity, the simulated patterns in the third row of Fig. 2.22 show the locations of the reflections in RHEED diagram patterns [34].

The idea of a hydrogen economy is based on the possibility of using hydrogen as a fuel. Hydrogen is considered one of the most promising energy sources, and is considered the cleanest type of fuel because it is primarily burned to produce water. In addition, burning hydrogen releases more energy per unit mass than burning traditional fossil fuels. The use of hydrogen in internal combustion engines and fuel cells has received particular attention, paving the way for development of low-carbon hydrogen economy. The chemical energy of hydrogen can be efficiently converted into electricity and other forms of energy without generating greenhouse gas emissions [7, 8, 10, 11]. Additionally, the development of new methods for converting

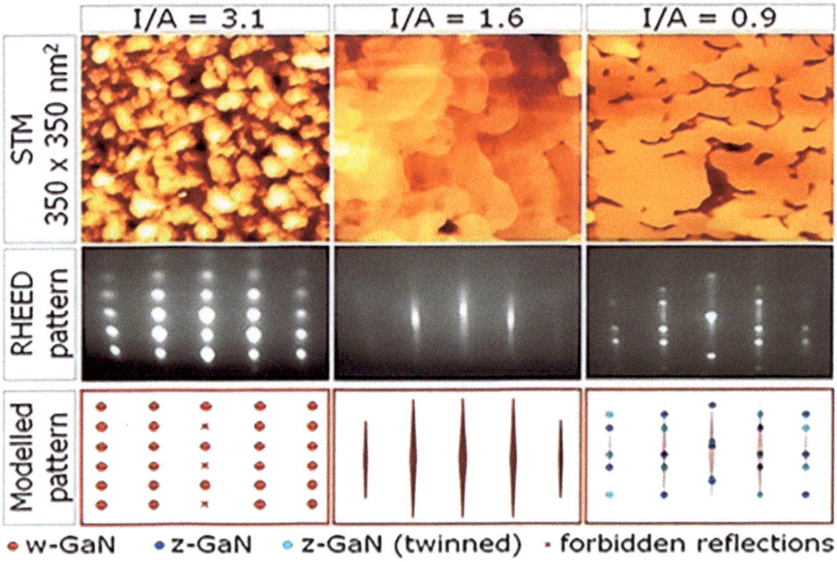

Fig. 2.22 STM images corresponding RHEED patterns, and typical patterns of ~ 10-nm-thick GaN films deposited by IBA-MBE with different I/A ratios. STM, RHEED, ion beam epitaxy with molecular beam activation (IBA-MBE), and ion-to-atom ratio (I/A) [34] [Open access]

industrial CO_2 emissions into important chemicals relies heavily on hydrogen. One of the goals of Carbon2Chem (Carbon2Chem is an innovative German project that aims to convert industrial carbon dioxide (CO_2) emissions into useful chemicals to reduce emissions and achieve sustainable development in industry) initiative is to convert gases from steel mills into basic chemicals such as methanol [12, 13]. Thus, converting industrial pollutants into chemicals helps prevent the release of greenhouse gases into the atmosphere. The preferred growth direction is consistent with these lattice planes. TEM and TEM-SEM are used to characterize the morphology. Nanomaterials are of great importance in the development of hydrogen technologies and their application as a renewable energy source, as they improve their efficiency. For example, zinc oxide, nickel oxide, and other nanomaterials have been used as carriers and catalysts in fuel cells to speed up chemical reactions, distribute nanoparticles uniformly, and improve energy conversion efficiency. Significant progress in using hydrogen in clean energy applications can be achieved by creating these advanced nanomaterials, which will help create a low-carbon economy. According to XRD, ZnO-NR is shown to have well-defined hexagonal rods in SEM (Fig. 2.23). The hexagonal rod arms connected by a distinct core formed a porous network in ZnO tetrapod or T-ZnO. The Pd nanoparticles are consistently dispersed with little agglomeration, as demonstrated by SEM obtained during energy dispersive spectroscopy (EDS) [13].

Hydrogen has no emissions, but its usefulness depends on how cleanly it is produced and how much energy is used [14]. Because water electrolysis has no carbon

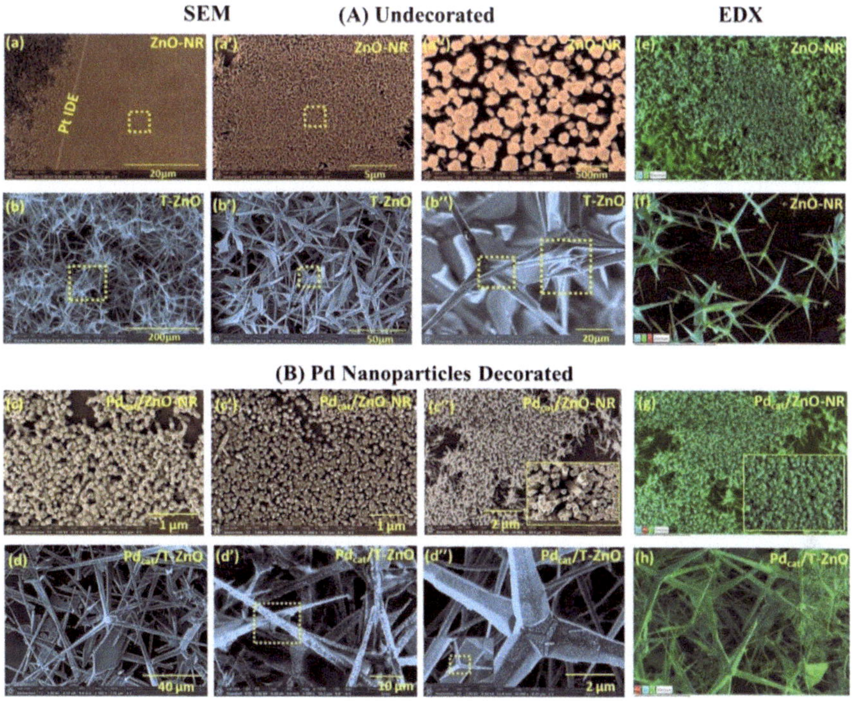

Fig. 2.23 SEM images of T-ZnO and ZnO-NR films at different magnifications show **a** their unstructured surfaces and **b** their palladium-functionalized surfaces. EDX images showing the elemental mapping and distribution of palladium nanoparticles are shown in the right column (green) [13] [Open access]

footprint, it is the most environmentally friendly method of producing hydrogen using electricity generated from renewable sources. Although it is expected to increase in the coming years, global renewable energy consumption is likely to remain modest for the foreseeable future [15]. Barriers include the intermittent nature of renewable energy sources, regional limitations, and the need to address energy storage issues [16, 17]. For example, Germany's onshore wind and solar energy resources can supply electricity to up to 2000 people per year, which is far from sufficient to meet annual electricity demand [18]. Offshore wind energy could provide more electricity, but delivery from the coast to inland regions remains a challenge. Seasons of excess production will alternate with periods of energy shortage due to power fluctuations. The intermittent nature of renewable energy can be addressed by storing excess energy. However, achieving large-scale energy storage is still difficult [19, 20].

After strengthening with porous supports, the studied catalysts have demonstrated long-lasting and greater ability to produce hydrogen [7,7]. The authors selected the most efficient catalytic supports for further research after studying the effects of several catalytic supports, such as MgO, Al_2O_3, SiO_2, TiO_2, ZrO_2, $MgOSiO_2$, $Al_2O_3SiO_2$, H + -ZSM-5, etc. [7]. SiO_2 is the most effective Thermal Conductivity

Detector (TCD) catalyst support for hydrogen production and nanostructure. So, Ashique and Daoud [27] have reported a very stable nanostructured Ni catalyst on porous SiO_2. The catalyst is created using a modified Stober method called co-precipitation. At 600 °C, they have observed the highest hydrogen production at 40.4%, and at 500 °C, the lowest deactivation after 300 min. In addition, Ni/SiO_2 had a higher carbon yield than Fe/SiO_2 [6, 8] confirming this finding.The authors selected the most efficient catalytic supports for further research after studying the effects of several catalytic supports, such as MgO, A12O3, SiO2, TiO2, ZrO2, MgOSiO2, A12O3SiO2, and HZSM-5 [7].

The results of extensive studies have showed the following order of catalytic characteristics in terms of stability and activity: n-Fe/SiO_2 > n-Ni > n-Co > n-Fe [6, 7]; n-Ni/SiO_2 > n-Co > n-Fe. The formation of hydrogen on various catalysts during the temperature-programmed decomposition of methane. In addition to monometallic catalysts, various combinations of multimetallic catalysts have also been investigated for TCD. The addition of Pd to the Ni/Al_2O_3 catalyst improved its reducibility and catalytic efficiency [8]. The authors have explained this greater activity as a result of the relatively faster diffusion of Pd carbon than Ni, which significantly stopped carbon encapsulation and extended catalyst life.

2.4.2 *Cracking of Ammonia*

Hydrogen is a viable replacement for fossil fuels in transitioning to renewable energy to achieve decarbonization goals. However, its physical and chemical characteristics make transportation and storage expensive. Other compounds are considered as H_2 carriers to ensure the economical use of H_2 as an energy vector. Ammonia is the most promising of them. Considering long-distance maritime transport, the value chain of NH_3 as a carrier of H_2 consists of its production, storage at the loading terminal, storage at the unloading terminal, and destruction to release H_2. Cost drivers in the value chain are NH_3 synthesis and cracking. In addition, large-scale NH3 cracking is an immature technique that requires significant effort to speed up the procedure as quickly as possible. In this regard, it examines and critically evaluates the various NH_3 cracking methods that are currently available, taking into account their potential industrial scale. In this regard, hydrogen is a key energy vector that helps provide sustainable and clean energy.

Most of the hydrogen produced and used in industry today comes from steam catalytic reforming of methane, a well-established and affordable commercial method for large-scale applications. However, the steam reforming process directly and indirectly produces significant amounts of CO_2. The two main challenges associated with hydrogen use are transportation and storage. Compared to other fossil fuels, hydrogen has a significantly lower volumetric energy density (1 to 2 kWh/L when compressed between 350 and 700 bar, and its small molecular size means that it tends to diffuse through materials, making them brittle AFM and Vapor Pressure Deficit (VPD) maps of sample ZE41. The arsenic-rich phase (Zr4Zn) and the T phase (Mg_7Zn_3RE) are found to have the largest VPD, measured at about $+180 \pm 10$ mV

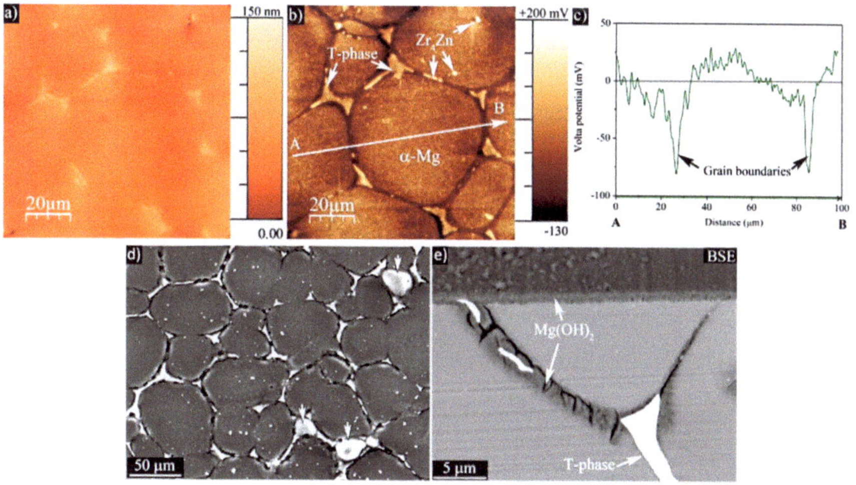

Fig. 2.24 A SEM of as-cast ZE41 alloys after immersion in 3.5% NaCl solution: **d** 60 min (surface) and **e** 60 min (cross section). SKPFM study of as-cast ZE41 alloys: **a** topographic map, **b** surface potential map, and **c** relative stress line profile analysis on α-Mg dendrites [33] [Open access]

and $+100 \pm 5$ mV, respectively, concerning the magnesium matrix (Fig. 2.24a, b) [6]. On another hand, the grain boundaries have reduced VPD of about -80 ± 5 mV for the magnesium matrix (Fig. 2.24c). After corrosion, microscopic studies have revealed significant localized degradation at the grain boundaries, which is aggravated by the T phase (Fig. 2.24d, e) [33].

2.4.3 *Applications of Fuel Cell*

Buses are considered to be the most likely vehicle type to bring fuel cell technology to the urban and regional transportation market early. Most of the topics covered in the automobile section also apply to the use of fuel cells in buses. The catalysts have used a significant impact on the rate of shift reaction [4] and reforming process [5]. Electrons are released during these reactions and then move through an external circuit, creating an electrical current that powers the device. Figure 2.25 shows the complete procedure [32].

A variety of NiPt nanoframeworks are created in another study using a simple OAm/OAc reduction method and acetic acid extraction. By adjusting the ratio of OAm to OAc, a change in morphology from a tetrahedral pattern to a dodecahedral pattern is achieved, resulting in nanoframes with different shapes (Fig. 2.25c) [32]. On the roof of fuel cell buses, hydrogen storage typically uses composite compressed gas cylinders operating at a pressure of 350 °C. Almost all buses produced nowadays use 350 bar compressed hydrogen tanks as there is no need for 700 bar pressure storage tanks as space is not an issue. The roof location is considered extremely safe

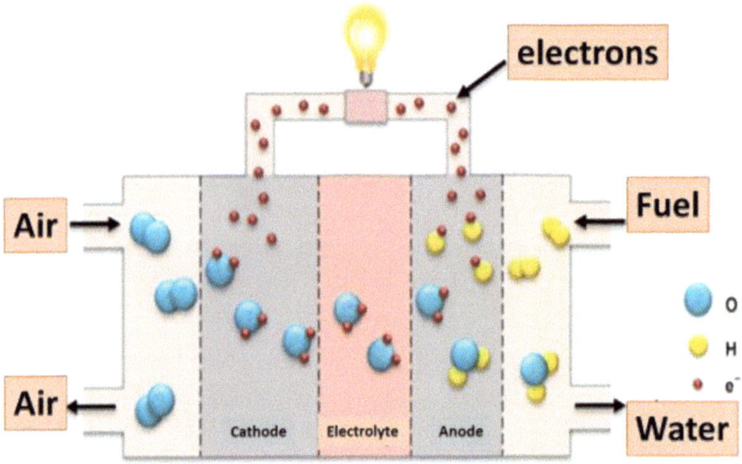

Fig. 2.25 Solid oxide fuel cell operation principle [32]. [Open access]

because hydrogen is significantly lighter than air. The most advanced fuel cell buses that have been built and tested worldwide .

The process for producing this nanocatalyst is discovered by chance during the study of solid polyhedral $PtNi_3$ nanoparticles. When $PtNi_3$ nanoparticles in solvent are examined by TEM 2 weeks after being exposed to an open atmosphere, they have unexpectedly revealed hollow nanoframes. Further studies showed that oxygen dissolved in the solution selectively etch nickel, resulting in the formation of $PtNi_3$. Thermal annealing produced a highly active platinum shell that is strong enough to withstand electrical stress and strain with only little loss of catalytic activity [28–31].

The use of nanoparticles as catalysts, or "nanocatalysts," is a significant step forward in improving the efficiency of chemical reactions. Materials with diameters between one and one hundred nanometers are known as nanocatalysts, and at this scale exhibit special properties such as a large relative surface area. Compared to conventional catalysts, nanocatalysts are more effective in accelerating chemical reactions due to their larger surface area. Due to their small size and unique structure, nanomaterials can increase catalyst dispersion and provide more active reaction sites. This increases productivity in a variety of applications and improves chemical reactivity. The use of nanocatalysts in energy production, water purification, industrial processes, and developments in the medical and environmental sectors demonstrates their adaptability. In addition, the ability to plan and control precise and targeted chemical reaction outcomes is made possible by the size and shape of nanomaterials. Therefore, nanocatalysts have significant advantages for maximizing efficiency and productivity in a range of applications.

Problems

1. Explain metal-based nanocatalysts for nanomaterial development.
2. List some limitations on the development of supported and unsupported nanomaterials.
3. Define the basic types of nanocatalysts.
4. Explain the role of nanoparticles in assessing catalytic performance.
5. How do methane nanoparticles produce hydrogen?

References

1. Huang, S. H., Chen, Y. J., Huang, W. F., & Uan, J. Y. (2022). Electrodeposition of a Li-Al Layered Double Hydroxide (LDH) on a ball-like aluminum lathe waste strips in structured catalytic applications: Preparation and characterization of Ni-based LDH catalysts for hydrogen evolution. *Catalysts, 12*(5), 520.
2. Ntasiou, P., Kerou, A. K., Karamanidou, T., Vlachou, A., Tziros, G. T., Tsouknidas, A., & Karaoglanidis, G. S. (2021). Synthesis and characterization of novel copper nanoparticles for the control of leaf spot and anthracnose diseases of olive. *Nanomaterials, 11*, 1667.
3. Chen, H., Wang, S., Huang, L., Zhang, L., Han, J., Ren, W., & Li, J. (2022). Core-shell hierarchical Fe/Cu bimetallic fenton catalyst with improved adsorption and catalytic performance for congo red degradation. *Catalysts, 12*(11), 1363.
4. Islam, S., Basumatary, B., Rokhum, S. L., Mochahari, P. K., & Basumatary, S. (2022). Advancement in utilization of nanomaterials as efficient and recyclable solid catalyst for biodiesel synthesis. *Cleaner Chemical Engineering, 3*, Article 100043.
5. Ionso, J. P. P., Djimasbe, R., Zairov, R., Yuan, C., Al-Muntaser, A. A., Stepanov, A., & Zinnatullin, A. L. (2023). Use of nickel oxide catalysts (Bunsenites) for in-situ hydrothermal upgrading process of heavy oil. *Nanomaterials, 13*(8), 1351.
6. Fedorov, A. Y., Bukhtiyarov, A. V., Panafidin, M. A., Prosvirin, I. P., Zubavichus, Y. V., & Bukhtiyarov, V. I. (2023). Thermally induced surface structure and morphology evolution in bimetallic Pt-Au/HOPG nanoparticles as probed using XPS and STM. *Nanomaterials, 14*(1), 57.
7. Soler-Morala, J., Jefremovas, E. M., Martínez, L., Mayoral, Á., Sánchez, E. H., De Toro, J. A., & Huttel, Y. (2020). Spontaneous formation of core@ shell Co@ Cr nanoparticles by gas phase synthesis. *Applied Nano, 1*(1), 87–101.
8. Westsson, E., & Koper, G. J. (2014). How to determine the core-shell nature in bimetallic catalyst particles? *Catalysts, 4*(4), 375–401.
9. Mardi, S., Pea, M., Notargiacomo, A., Yaghoobi Nia, N., Carlo, A. D., & Reale, A. (2020). The molecular weight dependence of thermoelectric properties of poly (3-hexylthiophene). *Materials, 13*(6), 1404.
10. Vobornik, D., Chen, M., Zou, S., & Lopinski, G. P. (2023). Measuring the diameter of single-wall carbon nanotubes using AFM. *Nanomaterials, 13*(3), 477.
11. Lin, Y. H., Sivakumar, C., Balraj, B., Murugesan, G., Nagarajan, S. K., & Ho, M. S. (2023). Ag-decorated vertically aligned ZnO nanorods for non-enzymatic glucose sensor applications. *Nanomaterials, 13*(4), 754.
12. Hassan, Q., Ahmadi, S., & Kerman, K. (2020). Recent advances in monitoring cell behavior using cell-based impedance spectroscopy. *Micromachines, 11*(6), 590.
13. Knoepfel, A., Poudel, B., & Gupta, S. (2023). Surface-catalyzed zinc oxide nanorods and interconnected tetrapods as efficient methane gas sensing platforms. *Chemosensors, 11*(9), 506.

14. Łątka, P., Olszański, B., Żurowska, M., Dębosz, M., Rokicińska, A., & Kuśtrowski, P. (2024). Spherical lignin-derived activated carbons for the adsorption of phenol from aqueous media. *Molecules, 29*(5), 960.
15. Iliger, K. S., Sofi, T. A., Bhat, N. A., Ahanger, F. A., Sekhar, J. C., Elhendi, A. Z., Al-Huqail, A. A., & Khan, F. (2021). Copper nanoparticles: Green synthesis and managing fruit rot disease of chilli caused by Colletotrichum capsici. *Saudi J Biol Sci, 28*, 1477–1486.
16. Abdelaziz, A. M., Salem, S. S., Khalil, A. M., El-Wakil, D. A., Fouda, H. M., & Hashem, A. H. (2022). Potential of biosynthesized zinc oxide NPs to control Fusarium wilt disease in eggplant (Solanum melongena) and promote plant growth. *BioMetals, 35*, 601–616.
17. Zaki, S. A., Ouf, S. A., Albarakaty, F. M., Habeb, M. M., Aly, A. A., & Abd-Elsalam, K. A. (2021). Trichoderma harzianum-mediated ZnO NPs: A green tool for controlling soil-borne pathogens in cotton. *J Fungi, 7*, 952.
18. Abdelkhalek, A., Qari, S. H., Abu-Saied, M.A.A.-R., Khalil, A. M., Younes, H. A., Nehela, Y., & Behiry, S. I. (2021). Chitosan nanoparticles inactivate Alfalfa mosaic virus replication and boost innate immunity in nicotiana glutinosa plants. *Plants, 10*, 2701.
19. Abdelrhim, A. S., Mazrou, Y. S. A., Nehela, Y., Atallah, O. O., El-Ashmony, R. M., & Dawood, M. F. A. (2021). Silicon dioxide nanoparticles induce innate immune responses and activate antioxidant machinery in wheat against Rhizoctonia solani. *Plants, 10*, 2758.
20. Hashem, A. H., Abdelaziz, A. M., Askar, A. A., Fouda, H. M., Khalil, A. M. A., Abd-Elsalam, K. A., & Khaleil, M. M. (2021). Bacillus megaterium-mediated synthesis of selenium nanoparticles and their antifungal activity against Rhizoctonia solani in Faba bean plants. *J Fungi, 7*, 195.
21. Kondo, Y., et al. (2008). Preparation, photocatalytic activities, and dye-sensitised solar-cell performance of submicron-scale TiO_2 hollow spheres. *Langmuir, 24*(2), 547–550.
22. Pei, L., Lin, D., Yuan, S., Lu, R., Bai, Z., Sun, Y., Zhu, Y., Jiang, Y., Zhu, J., & Wang, H. (2023). A multifunctional and long-term waterborne anti-corrosion coating with excellent 'hexagonal warrior' properties. *Chemical Engineering Journal, 457*, Article 141158.
23. Zhang, Y., Wu, H., & Guo, S. (2022). Sandwich-structured surface coating of a silver-decorated electrospun thermoplastic polyurethane fibrous film for excellent electromagnetic interference shielding with low reflectivity and favorable durability. *ACS Applied Materials & Interfaces, 14*, 40351–40360.
24. Hernán, D. M. V., Aida, M. E., & Gilberto, B. G. (2020). Detailed study of the electrochemical behavior of low-reflectivity TiAlN coatings. *Surf Innov, 9*, 293–307.
25. Picas, J., Menargues, S., Martin, E., & Baile, M. (2023). Cobalt free metallic binders for HVOF thermal sprayed wear resistant coatings. *Surface & Coatings Technology, 456*, Article 129243.
26. Zudyte, B., & Luksiene, Z. (2021). Visible light-activated ZnO nanoparticles for microbial control of wheat crop. *Journal of Photochemistry and Photobiology, B: Biology, 219*, Article 112206.
27. Dawoud, T. M., Yassin, M. A., El-Samawaty, A. R. M., & Elgorban, A. M. (2020). Silver nanoparticles synthesized by Nigrospora oryzae showed antifungal activity. *Saudi J Biol Sci, 28*, 1847–1852.
28. Sofy, A. R., Sofy, M. R., Hmed, A. A., Dawoud, R. A., Alnaggar, A.E.-A.M., Soliman, A. M., & El-Dougdoug, N. K. (2021). Ameliorating the adverse effects of tomato mosaic tobamovirus infecting tomato plants in egypt by boosting immunity in tomato plants using zinc oxide nanoparticles. *Molecules, 26*, 1337.
29. Kabeerdass, N., Al Otaibi, A., Rajendran, M., Manikandan, A., Kashmery, H. A., Rahman, M. M., Madhu, P., Khan, A., Asiri, A. M., & Mathanmohun, M. (2021). Bacillus-mediated silver nanoparticle synthesis and its antagonistic activity against bacterial and fungal pathogens. *Antibiotics, 10*, 1334.
30. Mostafa, Y. S., Alamri, S. A., Alrumman, S. A., Hashem, M., & Baka, Z. A. (2021). Green synthesis of silver nanoparticles using pomegranate and orange peel extracts and their antifungal activity against Alternaria solani, the causal agent of early blight disease of tomato. *Plants, 10*, 2363.

31. Zhao, Y., Min, X., Ding, Z., Chen, S., Ai, C., Liu, Z., & Fang, M. (2020). Metal-based nanocatalysts via a universal design on cellular structure. *Advancement of Science, 7*(3), 1902051.
32. Helal, H., Ahrouch, M., Rabehi, A., Zappa, D., & Comini, E. (2024). Nanostructured materials for enhanced performance of solid oxide fuel cells: A comprehensive review. *Crystals, 14*(4), 306.
33. Yasakau, K. (2020). Application of AFM-based techniques in studies of corrosion and corrosion inhibition of metallic alloys. *Corrosion and Materials Degradation, 1*(3), 345–372.
34. Rauschenbach, B., Lotnyk, A., Neumann, L., Poppitz, D., & Gerlach, J. W. (2017). Ion beam assisted deposition of thin epitaxial GaN films. *Materials, 10*(7), 690.
35. Gibson, C. T. (2020). The attachment of carbon nanotubes to atomic force microscopy tips using the pick-up method. *Applied Sciences, 10*(16), 5575.

Chapter 3
Map to Nanomaterials Through Green Nanotechnology

The term "green nanomaterials" refers to producing nanomaterials through environmentally friendly processes. Green nanotechnology uses safe, natural materials instead of hazardous chemicals and destructive processes to reduce their negative impact on the environment and improve sustainability. These methods of creating nanomaterials include using bacteria, plant extracts, and biological processes. For example, plant materials can be used to make nanoparticles or improve the eco-efficiency of production. In addition to improving the sustainability of production, these methods offer safer and greener solutions in a variety of industries, including technology, energy, and medicine.

3.1 Introduction

By manipulating materials at the atomic or molecular level, nanotechnology is an interdisciplinary field that enables the creation of new materials and devices with improved characteristics and properties. Green nanotechnology, a sustainable strategy designed to reduce the environmental impact of nanomaterial production while improving their efficiency and applications has gained popularity as nanotechnology advances. Emphasizing resource conservation, energy efficiency, and minimization of hazardous byproducts in the synthesis and using nanomaterials, green nanotechnology combines the concepts of nanotechnology and environmental sustainability [1, 2].

The special qualities of nanomaterials, including nanoparticles, nanotubes, and nanowires, distinguish them from bulk materials. These characteristics include high surface area, strong reactivity, and ability to interact at the molecular level with biological systems. These characteristics make them ideal for a variety of applications such as energy storage, drug delivery systems, and environmental cleanup.However, traditional methods for producing nanomaterials often use hazardous chemicals, a

R. A. Al-Samarai and Y. Al-Douri, *Engineering of Nanomaterials*,
https://doi.org/10.1007/978-981-95-5499-7_3

lot of energy, and non-renewable resources, all of which can be harmful to the environment and public health. Green nanotechnology, which offers a path to sustainable nanomaterial production, enters the picture here.

The goal of green nanotechnology is to provide cost-effective, environmentally friendly processes for creating nanomaterials. A greener option to traditional chemical synthesis is provided by these methods, which involve using biological processes such as plant extracts, microbes, and enzymes. For example, nanoparticle production uses organic matter from microbes or plants to create nanoparticles by reducing metal ions. This method drastically reduces energy consumption and eliminates the need for hazardous chemicals. In addition, green nanotechnology promotes the reuse and recycling of materials ensuring that nanomaterials can be included in circular economy [3, 4].

In addition to providing safer, more environmentally friendly processes, the use of green nanotechnology in nanomaterial production improves the performance of materials in some applications. Green nanotechnology creates new opportunities for using nanomaterials in environmentally friendly and energy-efficient applications such as waste management, water purification, solar energy conversion, and reducing the environmental impact of nanomaterial synthesis. The transition to green nanotechnology will be critical as the need for more environmentally friendly solutions increases, which plays a critical role in ensuring that benefits of nanomaterials can be achieved without compromising environmental safety or health [5, 6].

3.1.1 Nanotechnology

Due to their numerous applications and unique properties, nanomaterials have attracted a great interest in science over the past few decades. Because of the seemingly endless potential of these small materials in healthcare, several sophisticated biomedical instruments and processes have been developed for biomedical applications, such as diagnostics and therapies. Despite their potential, nanomaterials' unpredictability has long been a significant barrier. The lack of knowledge regarding potential adverse effects hinders the efficient application of modern biomaterials in clinical trials. Nonetheless, nanoparticles are progressively making their way into the biomedical implant industry because of their superior mechanical, chemical, optical, and biological qualities. Nanomaterials can pass through the blood–brain barrier and are easily absorbed by cells due to their small size, less than 100 nm. Because nanomaterials are easy to drug load and capable of functionalization, they have become an intriguing area of research in medicine. Their potential in terms of therapeutic potential, targeted delivery, tumor recurrence characteristics, optical, magnetic tumor localization, and diagnostic potential [7].

Among the many applications of nanomaterials in medicine, the studies are underway to determine how they affect cancer cells, understand the molecular mechanisms underlying their interactions, and evaluate their safety and efficacy. To assess their cellular response and toxicity, the effects of gold (AuNP), silver (AgNP), and palladium (PdNP) nanoparticles on PC-3 cells are investigated. Figure 3.1 provides

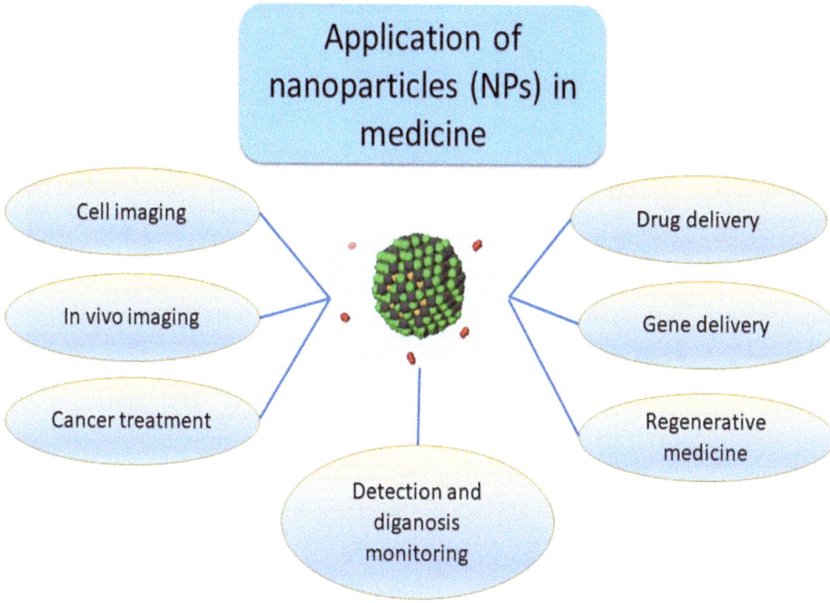

Fig. 3.1 A brief overview of current using nanoparticles (NPs) in medicine, including regenerative medicine, drug delivery, gene delivery, cancer treatment, and cellular and in vivo imaging [1] [Open access]

a schematic summary [1]. Due to their unique characteristics, NPs have promising prospects for cancer treatment. For example, two therapeutic approaches involving optical interference are photodynamic therapy (PDT) and photothermal therapy (PTT) [1].

Only NPs that enter cells via passive intracellular transport are currently approved for therapeutic use. As described previously, the late clinical stage has not yet been completed for NPs that enter cells via active intracellular transport. As a result, current medical applications of NPs focus on regenerative medicine, gene delivery, drug delivery, cell imaging, in vivo imaging, and cancer treatment. These characteristics and information from clinical and preclinical research have led to the commercialization of numerous nanoproducts to enhance human life quality, including bioimplants, medication delivery systems, tissue engineering scaffolds, biomarkers, biosensors, filters, imaging agents, and more [1–6]. However, some studies of long-term interactions between nanomaterials (metals, metal oxides, ceramics, and composites) and body have shown some degree of harmful reactions [7–10]. An inhibitory effect on cell migration in PC-3 cells treated with PVP-coated PdNPs, especially after short treatment periods.

3.1.2 Nanotoxicity

A family of materials ranging in size from 1 to 100 nm are known as nanoparticles. These materials differ from comparable bulk materials or tiny molecules due to their specific physical, chemical, and biological properties [1, 2]. Due to their special material properties, nanoparticles are widely used in various products and applications such as medical devices, cosmetics, industrial catalysis, energy conversion and storage and imaging technologies [3–5]. The main source of BC, a naturally occurring nanopolymer, is bacteria [11–14]. Given the wide variety of nanoparticles and their unique properties, some are naturally occurring such as bacterial cellulose (BC), which is produced by bacteria and has distinct properties that make it different from synthetic and environmental nanoparticles. Bacteria use cellulose synthase complexes to produce grain-like cellulose (Fig. 3.2c). The width of the BC is between 0.01 and 0.10 μm, making it look like a ribbon. This is two to three times smaller than the diameter of plant cellulose, which is typically 10 μm. The length of the fibers varies from a few hundred nanometers to micrometers, and they interact to create a network-like structure (Fig. 3.2e, f) [2].

Unlike plant cellulose, BC is by-product of microbial metabolism that is not a structural component of cell wall, so it lacks contaminants such as hemicellulose and lignin. In addition to these logically engineered nanoparticles, humans are constantly surrounded by significant amounts of naturally occurring and random particles such as airborne nanoparticles from industrial combustion and road traffic and nanoparticles produced by erosion or corrosion in water [6, 7]. Due to nanoparticles are

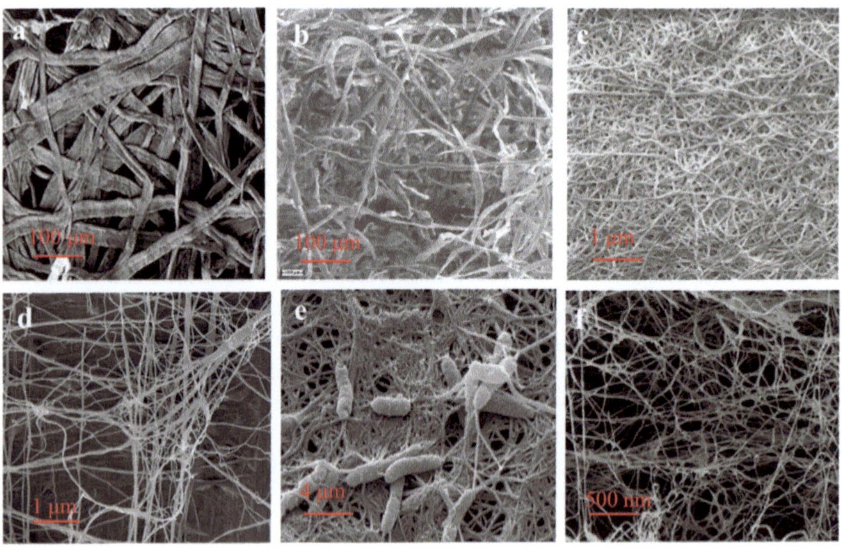

Fig. 3.2 SEM images of cellulose from several. *Sources* cotton, softwood, and (c–f) reproduced from [2] [Open Access]

widespread in the environment and common consumer products, human exposure is inevitable. However, there has been little research and understanding of the possible short- and long-term health hazards that nanoparticles may pose to humans. In this study, we present the main types of nanoparticles and examine how toxicity is influenced by the corresponding physicochemical characteristics of these particles. The amount of fine particles increased from 55 ± 1% after laccase treatment to 62 ± 1% during the treatment of cellulose after enzyme pretreatment. These results indicate that cellulases have the ability to grow fibers by causing damage to both the inner and outer regions, allowing more water molecules to penetrate the fibers [6].

An overview of how nanoparticles interact with the body at the organ, tissue, cellular, and biomolecular levels is given in a discussion of the main processes of nanotoxicity. Numerous experimental and theoretical methods that have been developed and used to assess nanotoxicity enable such mechanistic understanding to be achieved. It is concluded by offering our views on possible approaches to reduce nanotoxicity to fully utilize nanotechnology for human safety. We hope that you will find this review a useful tool to highlight the current state of nanotoxicity research and stimulate new research aimed at expanding our knowledge of nanoparticle toxicity. A better understanding of nanotoxicity may ultimately influence and guide the creation of regulatory frameworks to reduce possible risks to human health. There are two main ways that nanoparticles interact with biological objects: inside and outside. In the external route, the body absorbs nanoparticles from air, water, food, or any other substance through ingestion, inhalation, or skin absorption. Bioimplant nanoparticles are introduced directly into the body in an internal mode.

In addition to investigating the impact of discrimination on human health, there is a need to develop safe nanoproducts, such as nanocellulose, that can be modified with enzymes to improve their properties and impact barriers. Due to high selectivity of the enzyme, the recovered nanocellulose remains largely unchanged after enzymatic hydrolysis (Fig. 3.3h) [2]. The main enzymes used for this are cellulases, which break down β-1,4-glycosidic bonds in the cellulose molecule. More precisely, cellulases are capable of breaking down only the amorphous parts of cellulose, which reduces energy costs for mechanical processing [7, 8].

Primary effects such as oxidative stress, DNA damage, inflammation, and toxicity can result from both intrinsic and extrinsic pathways. These health effects gradually cause secondary effects including tissue and organ deposition, systemic inflammation, dysfunction, and secondary toxicity [11–14]. As was previously noted, a number of factors, such as heredity, pre-existing conditions, exposure level, and nanoparticle properties, are linked to the detrimental impacts of nanomaterials on human health. Studies in animal models and clinical models have demonstrated the ineffectiveness of macrophage purification methods to eliminate nanoparticles. From the body, causing harm to some organs, the lymphatic system, lungs, and nervous system. Nanomaterials can even cross the blood–brain barrier due to their small size, allowing them easy access to most organs.

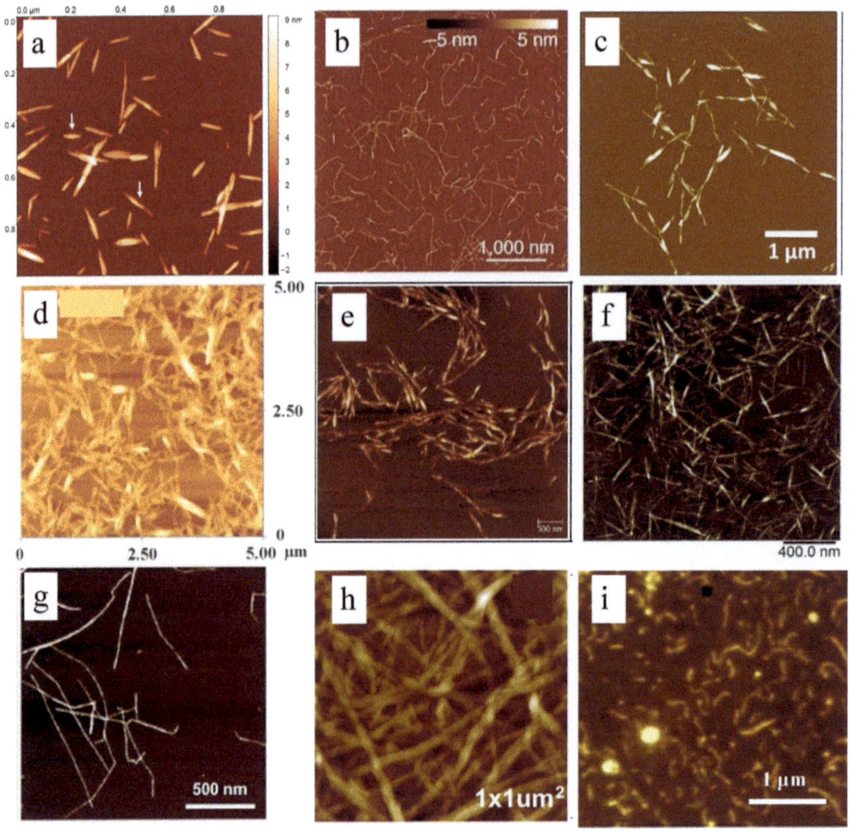

Fig. 3.3 Different methods for producing nanocellulose. The following publications are reprinted with permission: **a** H2SO4, **b** TEMPO, **c** H3PO4, **d** HCl, **e**, **f** Oxalic acid, **g** TEMPO-oxidized CNF, **h** Enzymatic hydrolysis [2] [Open access]

3.1.3 Green Chemistry

A relatively young field called green chemistry, sometimes called sustainable chemistry, promotes the design and development of chemicals to minimize the use and production of harmful chemicals. Atomic scanning microscopy (AFM) is used to image the ultra-thin hybrid cellulose film. Three 1×1 µm (or smaller) surface images are used for each measurement: amplitude, phase, and height images. "The creation of hybrid materials such as nanoparticle-reinforced cellulose films are a promising direction in the context of green chemistry, as their properties can be improved using sophisticated imaging techniques such as atomic microscopy, allowing for precise understanding of their surface and structure." One of the seven surface points is represented by the images shown in Fig. 3.4. Figure 3.4a, b shows the phase and topographic images, respectively. The separated nanoparticles can be seen after the

$AgNO_3$ reaction. The phase image allows for more accurate identification than the topographic image. The surface roughness of the cellulose film makes it impossible to count and measure the nanoparticles in the image using an automated instrument such as the WSxM [3].

Paul Anastas coined the term "green chemistry," which refers to the goal of creating safer and more environmentally friendly chemical processes. It is based on 12 core ideas that are intended to reduce the negative impacts of chemistry on the environment and human health. Green chemistry primarily aims to reduce toxicity by creating safer substances and procedures than its predecessors. These recommendations include reducing waste, using renewable resources, and eliminating hazardous

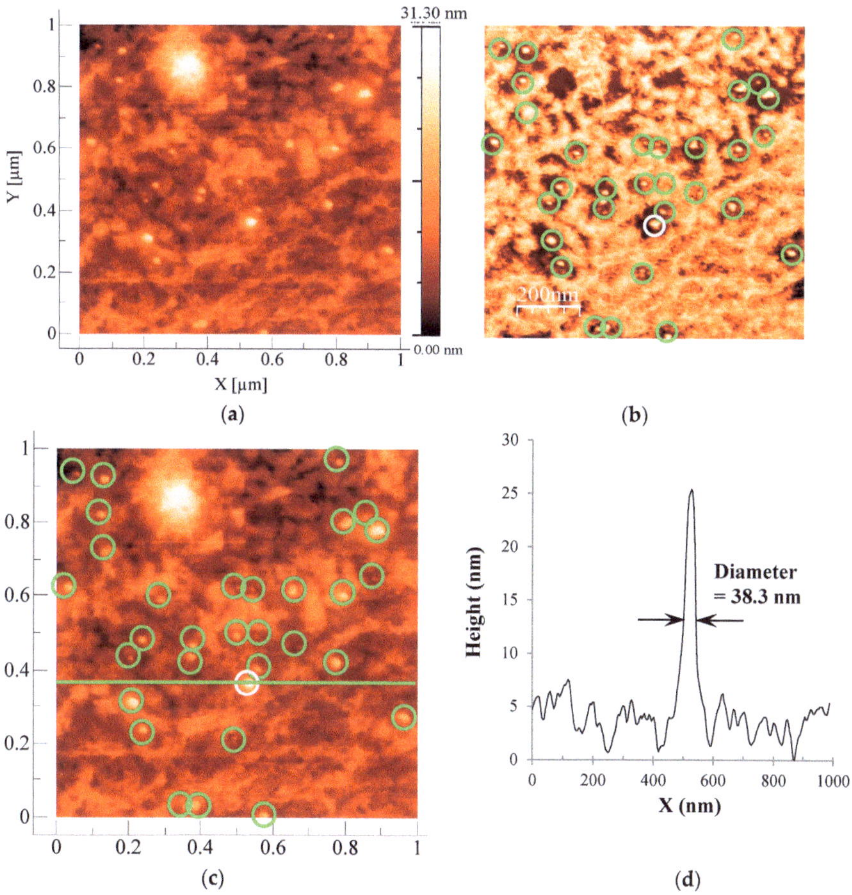

Fig. 3.4 **a, b** AFM records of hybrid cellulose sheets (1000 nm × 1000 nm) in phase and topography modes, respectively. The height profile of the horizontal line crossing the center of the white circle is shown by circles in (**a, c**), which are superimposed AFM images of (**a, d**). Comparable AFM images of the hybrid cellulose film in phase mode are shown by circles in (**a, b**). [3] [Open Access]

or harmful substances. This approach represents a shift toward creating chemistry that is not effective but environmentally friendly and sustainable [10].

The Toxicity, Interaction and Endocrine Disruption (TIPED) testing approach is a significant advancement in green chemistry that is developed to evaluate the potential endocrine disruption of new substances before releasing into the environment. This approach is a collaborative effort between green chemists and environmental scientists to ensure that newly identified compounds as tested for safety, especially for their potential to impact biological systems. This method reduces the likelihood of hazardous chemicals being released into environment by identifying potentially hazardous materials early in process [11].

By encouraging the development of safe alternatives to hazardous chemicals currently in use, green chemistry provides a remedy for chemical pollution. It can lead to creation of substitute, environmentally friendly materials and procedures that reduce the harmful effects of chemicals on ecosystems and human health. Green chemistry makes a significant contribution to the development of more sustainable future through these developments, providing cleaner substitutes for consumer and industrial products, reducing environmental pollution and protecting public health [12].

3.1.4 Green Chemistry: Overview and Foundations

The development of green chemistry has been greatly influenced for decades by environmental pioneers like Rachel Carson, whose groundbreaking 1962 book Silent Spring increased public awareness of the dangers of pesticides and their substantial contribution to environmental contamination. Carson's work is crucial in bringing attention to the ways that chemicals, especially pesticides, damage ecosystems and jeopardize wildlife and human health. This marked the start of a broader environmental movement to address how industrial practices affect the environment [1, 3].

Less than a decade after the publication of Silent Spring, in 1970, the Environmental Protection Agency (EPA) is created in the United States, a major step toward regulatory oversight of chemicals and pollution control. Since then, EPA has positioned itself as successor to Rachel Carson's legacy, recognizing its important role in advancing sustainable chemistry practices and protecting the environment. Its efforts paved the way for the development of green chemistry, promoting safer use of chemicals, reducing waste, and supporting procedures that reduce risks to the environment and public health [4, 5].

The development of modern green chemistry practices can be viewed as an outgrowth of principles that Carson promoted. The fundamental basis for developing effective and environmentally friendly chemicals and materials is now provided by green chemistry. With researchers and businesses embracing ideas that reduce the use of harmful chemicals and seeking more sustainable alternatives, this movement has expanded and influenced a range of industries, reflecting Carson's goal of creating a safer and cleaner world. Because of Carson's long-term work to preserve nature for future generations, green chemistry has become a vital field in addressing global environmental issues (Fig. 3.5) [3].

The Citizens Advisory Committee on Environmental Quality and the Cabinet-level Council on Environmental Quality are created in 1969 as a result of President Richard Nixon's environmental sustainability initiatives. Study the full range of variables affecting environmental quality, which is the committee's mandate. Nixon's initiatives drew criticism for lacking energy. Sometime later, the National Environmental Policy, or National Environmental Policy Act (NEPA), is created, requiring the creation of a Council on Environmental Quality. In the 1980s, the Environmental Protection Agency has created the Office of Pollution Prevention and Toxic Substances, recognizing that pollution prevention is more important than end-of-pipe

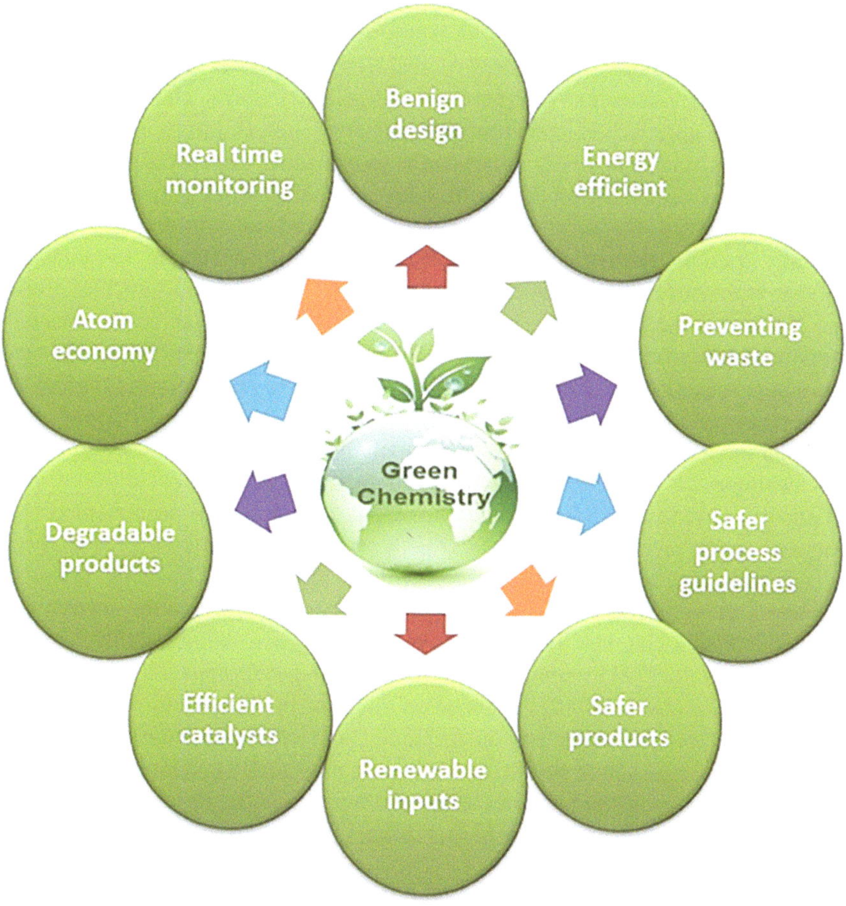

Fig. 3.5 Main components of green chemistry [3] [Open access]

control [12]. During his administration, President Bill Clinton has created the Presidential Green Chemical Challenge Awards to recognize and honor people practicing sustainable chemistry. By end of 1990s, The Twelve published Principles of Green Chemistry. The guidelines became the starting point for procedures and practices designed to reduce adverse environmental impacts.

These principles are closely related to the goals of green chemistry, which emphasizes the prevention of environmental damage by promoting sustainable and safer alternatives in chemical processes. The idea of prevention emphasizes the importance of treating waste and pollution before they occur, rather than simply managing them after created. By using renewable raw materials and focusing on reducing derivatives, green chemistry aims to reduce waste and improve resource use. This approach also calls for the development of safer chemicals and solvents, reducing risks to human health and the environment [7, 8].

In line with these ideas, the nuclear economy, which has traditionally relied on high-risk chemicals and processes, is being replaced by less hazardous methods of chemical synthesis. The concept of real-time analysis is another key aspect, allowing for early detection and prevention of pollution during chemical reactions. This proactive approach is integral to reducing environmental impact [9].

In addition, safer chemistry also extends to providing safer working conditions and preventing accidents, especially in industries that produce polymers and other complex materials. Ongoing efforts to replace hazardous solvents with environmentally friendly alternatives play an important role in achieving sustainability. Solvent-free synthesis methods have gained popularity as more sustainable solution that eliminates the need for harmful solvents while providing an environmentally friendly pathway for chemical reactions. Together, these principles reflect a broader movement toward safer, more sustainable approaches to chemistry that benefit the environment and human well-being [10].

Using safe reagents, photocatalysts, and natural raw materials, a low-energy catalysis technique has been created to further minimize pollution. To cut down on waste production, non-toxic, energy-efficient synthetic methods have been created [12]. These changes have made it possible for the following generation of researchers to comprehend green chemistry's more ecologically friendly methods, practices, and tenets as well as its function in mitigating negative effects on the environment and public health [10, 11]. Twelve core concepts of green chemistry, developed by Anastas and Warner [11], emphasize sustainability and environmental responsibility. Anastas and Zimmerman [12] expanded on this field by introducing the concept of "green chemistry," which focuses on developing and enhancing industrial processes that reduce risks to public health, while maintaining their economic and environmental viability. Waste reduction, energy efficiency, and the development of safer chemical processes are among the concepts that form the basis of a green chemistry strategy for a sustainable future [12].

Sustainable chemistry, sometimes called "green chemistry," is the study and design of chemical processes and products with an emphasis on reducing environmental impacts and improving safety for human health. This field aims to develop chemical products and processes that reduce or eliminate the need to produce hazardous materials. A summary of its basic principles, applications, and benefits is given below:

- Preventive principles of green chemistry: It is better to avoid waste than to treat or clean it up after production. Developing methods that avoid generating waste or pollutants is a fundamental principle.
- Atom economy: All materials used in a process should be used as much as possible in the final product through proper design of synthetic processes. This idea places a strong emphasis on the efficient use of atoms.
- Creating chemical formulations that are less hazardous to use and producing materials that pose less risk to the environment and public health.
- Creating safer chemicals: Chemical products should be as effective as possible with the least amount of toxicity. The goal is to reduce the potential harm from the chemicals used.
- Safer reaction conditions and solvents: Avoid reactants, solvents, or reaction conditions that may pose a potential hazard. Use safer solvents when possible or eliminate them.
- Energy efficiency: Perform processes at room temperature and pressure whenever possible to minimize the amount of energy used. The overall environmental impact is reduced by energy-efficient processes.
- Use renewable raw materials or commodities rather than depleting existing resources. This contributes to the long-term sustainability of chemical processes.
- Minimize derivatives: Avoid unnecessary derivatives or added chemical groups whenever possible. Derivatives often generate more waste and require more reactants.
- Use catalysts to speed up and reduce the number of byproducts produced by processes. Compared to stoichiometric reactants, catalysts are more stable and can be reused.
- Design for decomposition: To minimize environmental impact, chemical products should decompose into safe compounds after use.
- Real-time analysis to prevent contamination: Use real-time monitoring and control strategies in your processes to stop the formation of harmful compounds and reduce contamination.
- Use inherently safer chemicals to prevent accidents: Create procedures and materials with less risk of spills, explosions, and other hazards.
- Reducing environmental impact: Green chemistry techniques help reduce waste and pollution, resulting in cleaner soil, water, and air.
- Improving human health: Green chemistry helps protect human health by reducing the use of and exposure to harmful chemicals.
- Increasing cost efficiency: By reducing waste and using resources efficiently, green chemistry can lead to cost savings.
- Promoting green practices: Sustainable green practices are facilitated in the long term by using renewable resources and developing products that decompose safely.

The 12 pillars of green chemistry are listed in Table 3.1.

Table 3.1 The 12 main principles of green chemistry [15] [Open Access]

No	Principle	Remarks
1	Prevention	It is better to prevent waste than to clean it up after it is formed
2	Atom economy	Synthetic methods should be designed to maximize the incorporation of all materials into the final product
3	Less hazardous chemical syntheses	Design synthetic methods that use and generate substances with little or no toxicity to human health and the environment
4	Designing safer chemicals	Chemical products should be designed to achieve their function with minimal toxicity
5	Safer solvents and auxiliaries	Minimize the use of auxiliary substances (e.g., solvents, separation agents); use safer alternatives whenever possible
6	Design for energy efficiency	Conduct reactions at ambient temperature and pressure to reduce energy consumption and environmental impact
7	Use of renewable feedstocks	Prefer raw materials that are renewable, when technically and economically viable
8	Reduce derivatives	Avoid unnecessary derivatization (e.g., blocking or deprotection steps), which require additional reagents and generate waste
9	Catalysis	Favor catalytic reagents over stoichiometric ones; catalysts are more efficient and generate less waste
10	Design for degradation	Design chemical products to break down into innocuous substances after use, minimizing long-term environmental harm
11	Real-time analysis for pollution prevention	Develop and implement real-time monitoring techniques to detect and control hazardous substances during the process
12	Inherently safer chemistry for accident prevention	Choose substances and process conditions that minimize risks such as explosions, fires, or toxic releases

3.1.5 Applications and Scope of Green Chemistry

It is common knowledge that the chemical industry needs cleaner technologies. This trend toward "green chemistry" [13, 14, 16] or "sustainable technologies" requires a paradigm shift from traditional views of process efficiency that place a heavy emphasis on chemical output to one that places financial value on waste disposal at the source and avoiding the use of toxic and/or hazardous materials. The US Environmental Protection Agency (EPA) is credited with coining the term green chemistry. In response, Japan established the Green and Sustainable Chemistry Network, further advancing the global movement toward environmentally responsible and sustainable chemical practices. Published in 1999, the Royal Society of Chemistry has sponsored the first edition of the journal Green Chemistry. So, it is said that green chemistry is here to stay. Hildebrand and colleagues [12] have performed an atomic-scale microscopic analysis of cell walls of 16 different diatom species to investigate the design concepts involved in the assembly of silica structures and underlying organic matter. Understanding natural processes that create materials at the nanoscale such as the synthesis of silicon structures in diatoms can inspire new sustainable technologies in

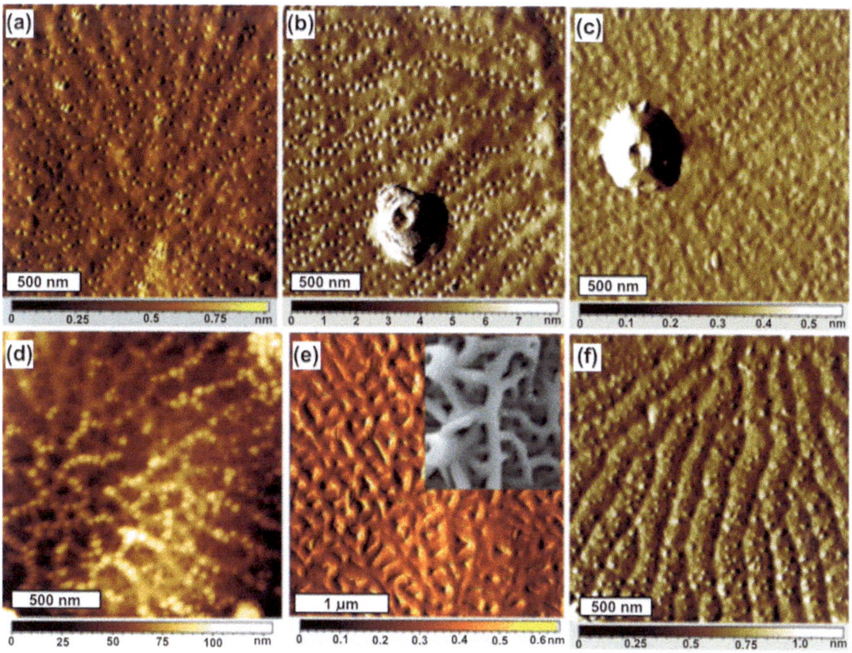

Fig. 3.6 Characteristics of the proximal and distal valves of the three Thalassocera species are similar and distinct, respectively. The proximal valve surfaces represent T. pseudonana (**a**), T. weissflogii (**b**), and T. oceanica (**c**); the distal surfaces represent T. pseudonana (**d**), T. weissflogii with an overlay electron microscope image (**e**), and T. oceanica (**f**). [4] [Open Access]

addition to industrial applications of green chemistry. To prepare samples for atomic-scale microscopy, diatoms are treated with acid to remove organic layers and then samples are air-cleaned. For the same features in different species, as well as for different features within a species and on different faces of the same organism, the silica structures of diatoms varied greatly (Fig. 3.6) [4].

The U.S. Green Chemistry Program, formally adopted by the Environmental Protection Agency in 1993, has led to initiatives such as the Conference on Green Chemistry and Engineering and the Presidential Green Chemistry Challenge Awards. Although research in green chemistry began before 1990s, significant progress has been made in the U.S., Italy, and U.K. Significant progress has been made in the United States, Italy, and the United Kingdom, although green chemistry research began before the 1990s. Green chemistry, which embodies the concept of safety by design, has focused on the efficient use of renewable raw materials, waste removal, and avoidance of hazardous compounds to create environmentally friendly processes and products since the early 1990s. The 12 principles of green chemistry [13] embody this idea and can be summarized as follows:

1. Preventing losses rather than correcting them.
2. Efficiency of atoms.
3. Less toxic/hazardous compounds.
4. Development of safer products.

5. Use of safe solvents and auxiliary substances.

An essential component of using green chemistry is creating novel solvent systems or solvent-free techniques to create chemicals that are acceptable to the environment. Furthermore, lowering waste production and energy use encourages adherence to the 12 green chemical principles. Pharmaceuticals, reaction solvents, biological transformations, biomaterials, safe alternative energy sources (such as photovoltaics, quantum dots, fuel cells, and batteries), innovative catalyst designs, low-risk molecule designs, etc. are among the industries most impacted by green chemistry.

3.2 Green Nanotechnology

Using green chemistry concepts in nanotechnology allows nanoparticles to be produced and processed in a safe, affordable, environmentally friendly, and sustainable manner. This greater understanding of green chemistry in the pharmaceutical sector is reducing or eliminating the use of non-green materials. In this context, top-down and bottom-up manufacturing techniques enable the production of nanomaterials in a sustainable and environmentally friendly manner, making nanostructure fabrication methodologies an essential part of achieving green chemistry goals [14]. The top-down approach is concerned with the creation of nanomaterials from bulk materials, while the bottom-up approach focuses on the creation of nanomaterials from atoms and molecules in the range of 1–10 nm [5]. This contrast between top-down and bottom-up methods highlights how important it is to choose the right synthesis methods to produce nanoparticles in an environmentally responsible and sustainable manner. At normal temperatures, an aqueous extract of Carica papaya leaves is used for the biosynthesis of magnetite (Fe_3O_4) nanoparticles. The X-ray diffraction pattern and SEM image of the obtained magnetite (Fe_3O_4) nanoparticles are shown in Fig. 3.7, which shows a plate-like shape with large grains and encapsulated structures. The crystallite size of magnetite is calculated using the Scherrer formula and is found to be about 33 nm [6].

Treatments are further divided into physical, chemical, and biological categories. Methods used include mechanical milling, laser ablation, sputtering, etching, and electrical explosion. When exposed to 5 μM alamethicin, the lipid membrane has developed large defects or pores that left small lipid islands among the larger bound lipid bilayer regions (Fig. 3.8) [5].

It is unable to record the early stages of peptide binding and membrane detachment due to rapid interaction of peptides with membranes. The first peptide bond reactions between indolicidin and alamethicin occur in less than 10 min according to QCM-D stands for quartz crystal microbalance with dissipation monitoring analysis. In addition, various synthetic methods such as chemical vapor deposition, plant and microbial methods, solvothermal methods, hydrothermal methods, sol–gel methods, and reverse micelle methods are used to prepare nanomaterials [16]. Nanotechnology is further divided into two categories based on how it is used and produced;

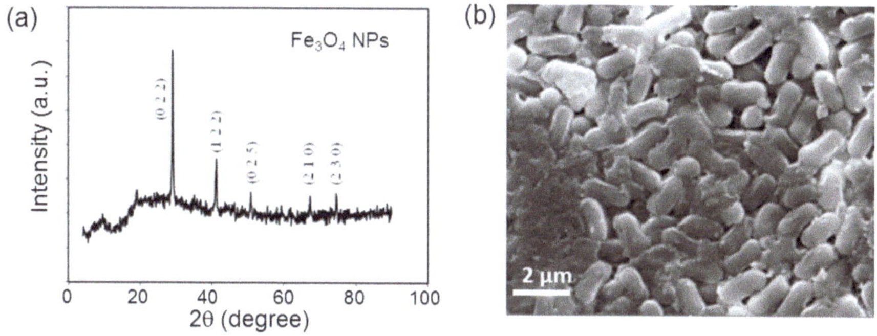

Fig. 3.7 **a** Fe_3O_4 nanoparticles synthesized with papaya leaf extract, **b** SEM image, and XRD pattern. [6] [Open access]

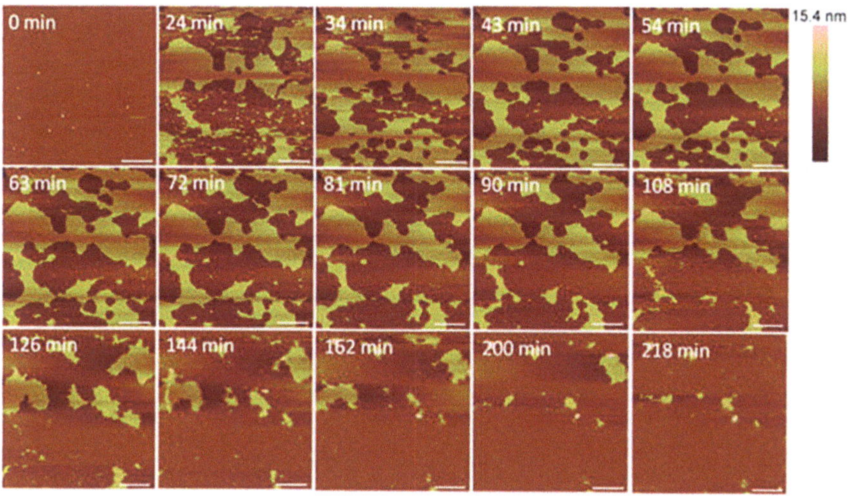

Fig. 3.8 Defects developed in PC bilayer supported after injection of 5 μM alamethicin for 218 min. On the substrate, each AFM image is scanned at same location. One micrometer is represented by the scale bars [5]. [Open access]

wet nanotechnology, which manipulates biological systems such as enzymes and genetic material and dry nanotechnology, which concentrates on the production of silicone, metal, and semiconductors. Important factors to consider when synthesizing nanomaterials include their shape, size, topography, agglomeration, solubility, size distribution, composition, surface chemistry, hydrophilicity, and formula.

In recent years, there has been increased interest in the topics of graphene carbon and energetic nanomaterials, semiconductor devices, green nanotechnology, nanocomposites, films, sensors, nanoencapsulation, and computational nanotechnology [17]. Higher and stronger composite production, larger production scales, improved commercial performance, increased sustainability, and significant implications in the biomedical field are some of the major benefits of using nanomaterials in applications [18]. The fact that nanoparticles have quantum effects that make the particles effective in drug delivery, catalysis, and sensing devices is one of their main advantages. The energy industry is creating high-efficiency nanobatteries and storage systems that deliver performance and durability, while requiring less charging time. Nanomaterials have proven to be very revolutionary as used in many different fields. It is expected that further research in the field of nanotechnology will solve important global problems and lead to innovative solutions.

Using the above methods, organic or inorganic materials can be used to create nanomaterials. Numerous types of nanomaterials are used for good. However, due to their hazardous nature and possible side effects, especially when used in chemical processes, their usage is limited. Green nanomaterials are one of the new synthesis methodologies used in recent decades to reduce negative impacts and expand the range of uses. To reduce the dangers associated with nanoparticles, scientists are focusing on using natural resources to combine them. They can be used in a wide variety of applications without endangering living beings [19]. Green technology reduces the risks arising from chemical reactions and is environmentally friendly. Sustainable synthesis of particles with less harmful effects by using green chemicals in the synthesis process is the main goal of green nanotechnology. The environmental impact analysis includes studies on the life cycle of green nanoparticles, including raw material processing, synthesis, application, and disposal. Green nanotechnology implements effective methods for recycling products and significantly reduces the amount of waste. Numerous nanoproducts are used in building materials, renewable energy, and water treatment among other environmental applications. Green products guarantee the environment and consumer well-being as well as innovation, efficiency, sustainability, green solution to solve a variety of problems arising in various sectors.

3.2.1 Creation of Green Nanomaterials

Nanotechnology-based methods are becoming increasingly popular in many industries. The phrase "green nanomaterials," which refers to the nanotechnological features of green chemistry, originated from the use of nanomaterials in environmentally friendly processes and products, from production to disposal. The most common methods for producing environmentally friendly green nanomaterials are bottom-up methods, using plant extracts or microbes. This technology produces green nanomaterials with special properties including antibacterial, catalytic, optical, magnetic, and optical properties. Nanomaterials have a wide range of applications due to their special properties that make them less harmful to the environment and easier to use. Green nanomaterials, safe for humans are used in medicine, where nanotechnology has several applications, to speed up the corresponding technological chains [8–12]. They can be used in industrial and environmental applications. Green nanomaterials also have certain disadvantages, including cost, resistance to implementation, and technological limitations. This suggests that if additional funds are allocated for research and development, and appropriate legislative measures are taken to encourage the use of green nanomaterials in processes, green nanomaterials could find wide application. This chapter discusses green nanomaterials in great detail, including their production, characteristics, and industrial uses. In addition, the silica-supported silver nanoparticles are subjected to atomic force microscopy (AFM) observations. Nanotechnology-based approaches are becoming increasingly popular in various fields. The use of nanomaterials in environmentally friendly production and waste disposal processes have led to the emergence of the term "green nanomaterials" to describe the properties of nanomaterials in green chemistry. Figure 3.9a shows the contact mode images. By masking the contact mode images and determining the maximum height of the selected grains, the heights of 222 silver nanoparticles were measured. Figure 3b shows a histogram of the silver nanoparticle height distribution. A mean height of 6.3 nm and a standard deviation of 1.4 nm were obtained by Gaussian fitting this distribution. Quantum information theory, or QITM for short, was used to quantify individual silver nanoparticles using the knock mode. Figure 3.9c–e show the adhesion, elasticity, and height values, respectively, that can be obtained. This single silver nanoparticle is approximately 5 nm high, and its elastic constant is slightly larger than that of the silica background. In addition, the adhesion of the silver nanoparticles is smaller than that of silica background. It is verified that the nanoparticles are made of a different material than background by looking at Fig. 3.9d, e [7].

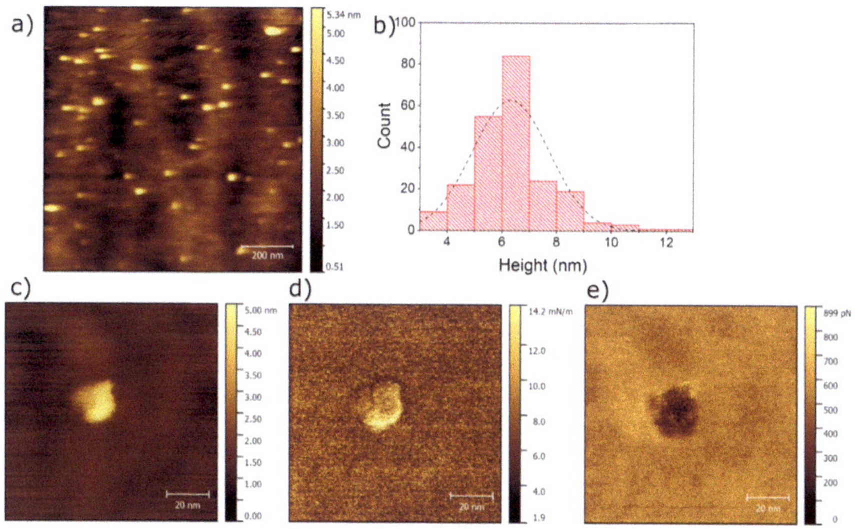

Fig. 3.9 Atomic energy microscopy (AFM) images of silver clusters deposited on a silicon dioxide wafer substrate. Histogram of silver nanoparticles from contact mode images; topographic height determined in contact mode; elastic modulus measured in QI mode; height recorded in QITM mode; adhesion measured in QI mode [7] [Open Access]

The following methods are used in the development of green nanomaterials to apply green chemistry concepts:

- Focusing on desired properties, using safe raw materials, and assessing the biological components of the material (such as size, structure, and surface function) are essential for designing safer nanomaterials.
- Environmental design includes studying the toxicity of degraded subunits, understanding the degradation properties of nanomaterials, and utilizing biosynthetic methods to minimize environmental impact.
- Waste prevention should be a key consideration in the design process. This can be achieved by reducing the formation of unwanted derivatives or byproducts, developing alternative purification methods (such as nanofiltration), and eliminating the need for solvents during purification.
- Design for process safety: replacing harmful solvents and reagents with safe alternatives and renewable raw materials.
- Minimizing energy losses during reactions by optimizing process parameters under ambient conditions.
- Implementing real-time monitoring capabilities to further enhance process control and minimize unnecessary energy expenditure [11].

These design approaches can be used in nanofabrication to create environmentally friendly and sustainable nanomaterials.

3.3 Preparations of Green Nanomaterials

Materials with at least one dimension in the range 1 to 100 nm are known as nanomaterials. Their compact size and high surface area-to-volume ratio give them special qualities. They can be created in a variety of ways such as nanoparticles, nanotubes, and nanowires, which can be made from a wide range of materials, including metals, metal oxides, polymers, and ceramics. These materials show signs of quantum confinement, which limits the behavior of electrons and particles and produces different characteristics. Nanomaterials are used in a wide range of industries, including energy, electronics, cleaning, cosmetics, diagnostics, imaging, targeted drug/gene delivery in medicine, advanced nanocarriers, and cancer immunotherapy. The main uses of nanoplatforms are shown in Fig. 3.10. Although there may be advantages to the use of nanomaterials, there may be disadvantages and further research is needed to fully understand these risks.

Cells grown on tissue culture plastic have showed a significantly higher viability compared to 2.5, 5, and 10% SrHA paste in indirect tests. Additionally, cells grown on tissue culture plastic are viable compared to pastes containing 0 and 100% SrHA. Although all media are centrifuged before use, varying amounts of SrHA molecules are detected in pre-prepared media across the various materials tested. Fluorescence images depicting live and dead cells have shown that the majority of cells are live and very few visible dead cells are present (Fig. 3.10) [8].

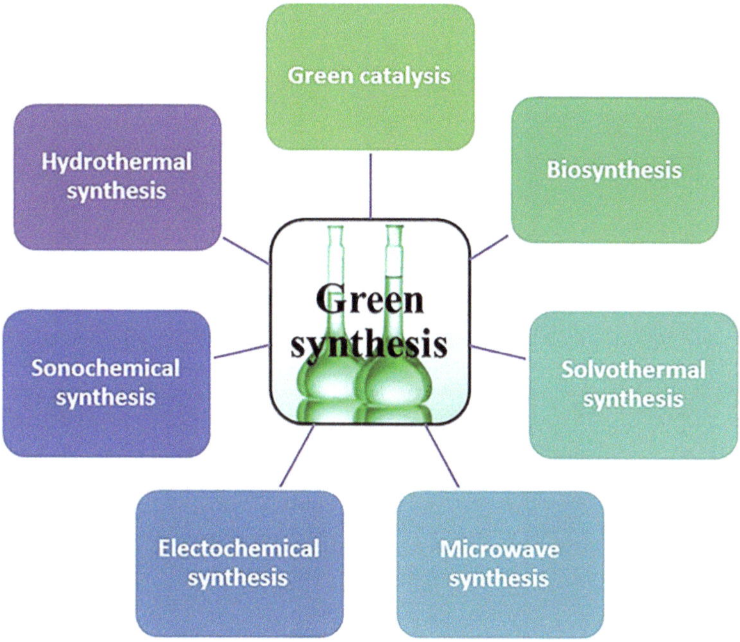

Fig. 3.10 Different methods for producing environmentally friendly nanomaterials [8] [Open access]

From a broader perspective, nanoparticle synthesis involves many routes, one of which is green pathway processes. Below is a brief overview of the pathways involved in their synthesis. There are two main methods for producing nanoparticles: top-down and bottom-up synthesis. Using a chemical or physical process, large materials are reduced to smaller particles, typically less than 100 nm in size, in a top-down manner. Top-down methods include chemical etching, laser ablation, and mechanical milling. While the top-down strategy has its advantages such as designing huge sizes, it is limited in its ability to control the size, shape, and purity of large quantities of nanoparticles. Commonly used green methods for creating nanomaterials are described below: (1) Replacing hazardous solvents with environmentally friendly solvents such as ionic liquids. (2) Replacement of traditional methods of chemical synthesis with environmentally friendly methods such as sonochemical synthesis, microwave synthesis, hydrothermal synthesis, solvothermal synthesis, electrochemical synthesis, and biosynthesis [13, 14, 16–20]. Nanomaterials are defined as materials with at least one dimension between 1 and 100 nm. They have special properties that make them useful in a variety of industries, including energy, electronics, cleaning, cosmetics, and medicine due to their small size and high surface area-to-volume ratio. Figure 3.11 shows a variety of "green" methods for creating nanomaterials [8].

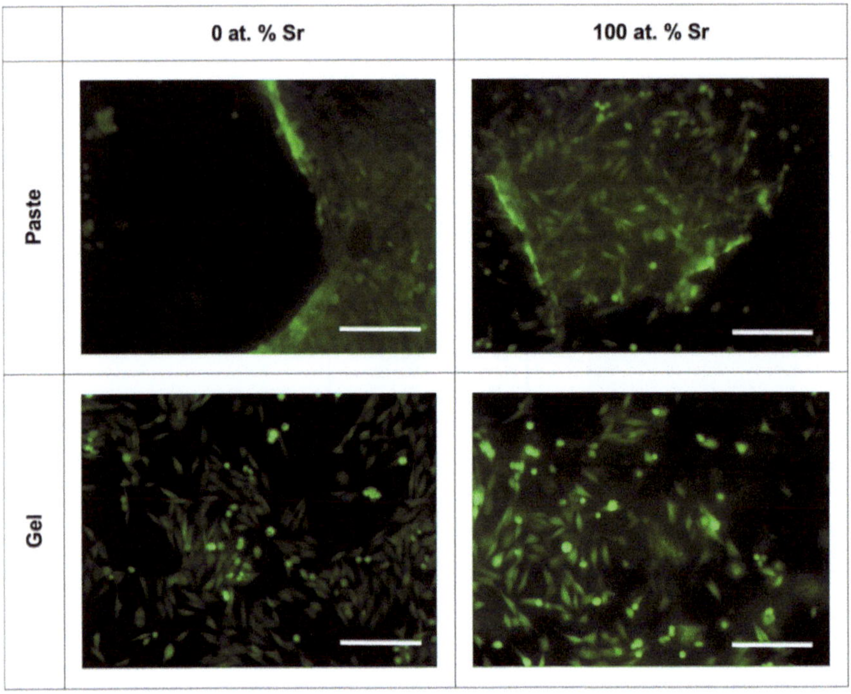

Fig. 3.11 Fluorescence imaging of MG63 cells cultured indirectly with SrHA paste and gel (0 and 100 at Sr ratio) for live/dead analysis. 200 μm is the scale bar [8]. [Open access]

Using both techniques, particles containing 100% SrHA had a significantly higher aspect ratio than any other degree of Sr substitution (Fig. 3.12). The samples are generally uniform, with smaller, rounded particles in 0% Sr samples and longer, rectangular particles in the 100% Sr samples. For the wet deposition method, the aspect ratio shifted from 1.7 for 0% Sr to 4.5 for 100% Sr. In particular, the average particle size is 50 × 30 nm at 0% Sr and 90 × 20 nm at 100% Sr [8].

In contrast, the bottom-up approach uses chemical reactions to assemble atoms, molecules, or ions into nanoparticles. Using this method, the size, shape, and purity of nanoparticles created can be precisely controlled. Sol–gel synthesis, hydrothermal synthesis, and chemical vapor deposition are some examples of bottom-up methods. The required qualities of the finished nanoparticle product determine that synthesis method is best since each method has its pros and cons.

3.4 Fabrication of Green Nanomaterials Using Biomolecules

In contrast to the nanoparticles of the bulk materials from which they are derived, the nanoscale allows for the creation of high surface area-to-volume ratios and quantum effects that provide distinctive features. Outstanding catalytic activity, optical and

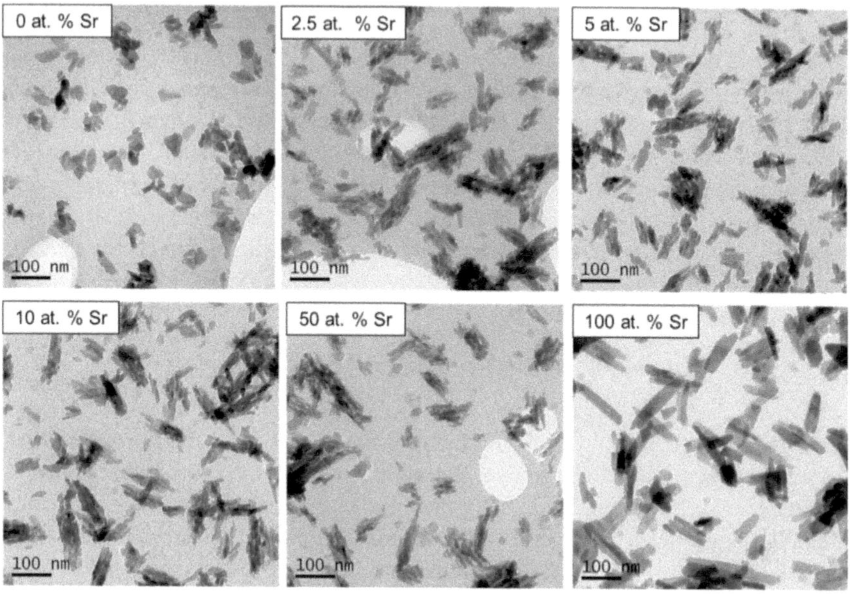

Fig. 3.12 Transmission electron microscopy of unsintered SrHA powders made by wet deposition (0, 2.5, 5, 10, 50, and 100% Sr). [Open Access] [8]

electrical activity, and antibacterial activity are some of the amazing qualities of these nanomaterials. Remarkable catalytic activity, optical properties, electrical properties, and antibacterial activity are some of the exceptional properties of these nanomaterials. Nanomaterials can be created using various synthetic methods. In the case of metal nanoparticles, the synthesis may start with ions that will be reduced and capped to create a nanoscale material or the starting material may be large, which can be reduced using top-down approaches. To create nanoparticles, chemical, physical, physicochemical, or environmental process can be used. The latter, associated with bionanotechnology, are currently paying more and more attention to biogenic synthesis. Biosynthesis is more cost-effective, environmentally friendly, and does not involve the use of hazardous chemicals like chemical or physical synthesis. As a result, there is a strong possibility that the production of green nanomaterials using waste, plants, bacteria, fungi, and algae will be able to provide the necessary biocompatibility and a wide range of applications. Thus, biogenic nanomaterials provide a wide range of possible applications, such as the detection of environmental pollutants (nanosensors) and treatment for ecosystem restoration.

The potential use of algae as a raw material for the synthesis of iron, silver, and gold nanoparticles [16]. They have discussed the ability of these nanoparticles (NPs) to deal with environmental pollutants, their advantageous characteristics and biocompatibility as well as the major barriers to their widespread use and prospects.

Wet co-precipitation is used [17] to create disk-shaped zinc oxide nanoparticles functionalized with β-cyclodextrin (β-CD) and niacin (vitamin B3). For 190 min at pH 6, the ZnO/β-CD/nicotinic acid nanocomposite exhibited a photocatalytic effect against methylene blue, resulting in a blue-colored solution. It has exhibited antibacterial and antifungal properties against Streptococcus aureus and Aspergillus niger. The anticancer properties of ZnO/β-CD/niacin nanocomposite on MCF-7 cell lines have also attracted interest. Thus, these materials have a range of biological applications, including serving directly as active agents (e.g., antimicrobial, catalytic, anticancer, nano adsorbent, and antioxidant structures) or offering potential for nanodelivery, analyte detection, and diagnostic bioimaging.

3.5 Biosynthesis Methods for Green Nanomaterials

Due to their exceptional physical and chemical characteristics such as strong mechanical bonds, large surface areas, optical properties, low melting points, surface plasmon resonance, effective antibacterial properties, and magnetic properties, the production of metal nanoparticles has recently attracted much attention. As a result, it has many applications in biotechnology, agriculture, antimicrobials, anticancer agents, catalysis, and photography. Green synthesis is recognized as the most important strategy for the sake of the environment, and green nanotechnology is becoming increasingly relevant as a cost-effective, environmentally friendly, efficient, and non-toxic replacement. Plants, fungi, and algae are just some of the natural retentions and reducing agents used to create these environmentally friendly nanoparticles. Plant extracts are a low-cost, natural blocking reducing agent for the environmentally friendly production of nanoparticles, which also ensures stability. Furthermore, research has been done on how plants including Festuca rubra, Brassic juncea, and Medicago sativa produce gold and silver nanoparticles inside their cells [14, 16, 17]. Other instances of metal nanoparticles produced using green biosynthesis.

3.6 Green Nanomaterials' Applications

With innovative uses in a wide range of industries including medicine, engineering, food, textiles, and environment, nanotechnology has become a leading field. In recent years, there has been a noticeable increase in demand for nanotechnology [14]. In 2020, the global nanomaterials industry generated revenue of approximately US$38.5 billion, and the revenue growth rate has increased by approximately 12% [2]. The geometry, structure, and morphology of materials at nanoscale determine their characteristics, so nanoscale materials are necessary to achieve high performance in modern technologies. The materials created by this process with a size of less than 100 nm are called nanostructures, nanoaggregates, and nanocomposites as the name suggests [13]. The idea of nanotechnology is put forward in 1959 by Nobel Prize-winning scientist Richard Feynman.

There are two main methods for creating nanostructures: top-down and bottom-up [14]. The top-down approach concerns the creation of nanomaterials from bulk materials, while the bottom-up approach focuses on the creation of nanomaterials from atoms and molecules in the range, 1–10 nm. Treatments are further divided into physical, chemical, and biological categories. Techniques used include mechanical milling, laser ablation, sputtering, etching and electrical explosion. Among the techniques frequently used to manufacture nanomaterials are: chemical vapor deposition, plants,microorganisms, hydrothermal methods, sol-gel methods, and reverse micelle methods [16]. Nanotechnology is divided into two categories: dry nanotechnology, which concentrates on the synthesis of silicone, metals, and semiconductors and wet nanotechnology, which deals with biological systems, including genetic material and enzymes. Important factors to consider when synthesizing nanomaterials include their shape, size, topography, agglomeration, solubility, size distribution, composition, surface chemistry, hydrophilicity, and formula. In recent years, there has been increased interest in the topics of graphene carbon and energetic nanomaterials, semiconductor devices, green nanotechnologies, nanocomposites, films and sensors, nanoencapsulation, and computational nanotechnologies [17]. Broader and more powerful composite production, larger production scales, improved commercial performance, increased sustainability, and significant implications in the biomedical field are some of major benefits of using nanomaterials in applications [18]. The fact that nanoparticles have quantum effects that make the particles effective in drug delivery, catalysis and sensing devices is one of their main advantages.

The energy industry produces nanotechnology-based batteries and storage devices that are durable, highly efficient, and require less charging time. Nanomaterials have proven to be very revolutionary and are used in many different fields. It is expected that further research in the field of nanotechnology will solve important global problems and lead to innovative solutions. Nanomaterials are currently used in many different industries. Commercially available products include water purification systems, energy storage devices, detergents, cosmetics, pharmaceuticals, and electronics. This chapter focuses on the biomedical applications of green nanomaterials since biocompatible materials are much sought after in the medical arena. Nanomaterials are employed in a wide range of fields and have shown themselves to be highly revolutionary. It is expected that further research in the field of nanotechnology will solve important global problems and lead to innovative solutions. Using the above methods, organic or inorganic materials can be used to create nanomaterials. Numerous types of nanomaterials are used for benefited applications. However, due to their hazardous nature and possible side effects, especially when used in chemical processes, their use is limited. Green nanomaterials are one of the new synthesis methodologies used in recent decades to reduce negative impacts and expand the range of uses. The researchers are focusing on using natural resources to synthesize nanomaterials to reduce their associated hazards. This methodology allows them to be used in various fields without endangering living beings [14]. Green technology reduces the risks arising from chemical reactions and is environmentally friendly.

Sustainable synthesis of particles with less harmful effects by using green chemicals in the synthesis process is the main goal of green nanotechnology. The environmental impact analysis includes studies on the life cycle of green nanoparticles, including raw material processing, synthesis, application, and disposal. Green nanotechnology implements effective methods for recycling products and significantly reduces the amount of waste. Numerous nanoproducts are used in building materials, renewable energy, and water treatment, among other environmental applications. Green products guarantee a healthy living environment, as well as innovation, efficiency, sustainability, and environmentally friendly and sustainable solutions to solve a wide range of problems encountered in various industries. Consequently, there is a strong focus on environmentally friendly methods for creating nanoparticles and how they are used in various industries.

Similar properties are found in metallic Probabilistic Finite Automata (PFA) samples, on which a silver layer is deposited for 30–240 s. By comparing the surface structures of these samples at a wavelength of 300 nm (Fig. 3.13), it is clear that the metal deposition time affects the size of the resulting structures. It was discovered that as the deposition time increases, the surface properties improve. This effect is verified by comparing the surface roughness of the PFA samples with that of the silver-leached surface. The surface roughness of the silver-leached PFA samples increased with increasing metal deposition time (Table 3.2).

3.7 Green Nanomaterials: Applications in Biomedicine

Due to their enhanced biocompatibility, green nanoparticles are widely used in biomedical industry. A variety of synthesis methods allow the creation of a wide range of MnO_2 nanostructures, highlighting the important role that fabrication methods play in tailoring their morphology to specific biomedical applications. Various morphologies of MnO_2 nanostructures including nanosheets, nanotubes, nanorods, nanowires, nanoflakes, nanoneedles, nanoflowers, nanoedges and nanospheres are shown in Fig. 3.14 [6]. The diverse morphological forms of MnO_2 structure are determined by the synthesis methods used [6].

Green nanomaterials are used to track stem cells in nanomaterial-based diagnostics, biosensing, and bioimaging (MRI and CT). The therapeutic potential of green materials has been investigated in the areas of antibacterial and anticancer effects, drug delivery, and hyperthermia. Over the past few decades, the scientific community has shown great interest in the synthesis of nanomaterials with unique properties and their use in the medical field. This gives rise to an advanced field of therapeutic science known as nanomedicine. Nanomaterials are small mono- and polydisperse particles of various shapes and sizes from 1 to 100 nm. Nanoparticles have several unique properties, including greater tensile strength, flexibility, high surface-to-volume ratio and optical and electrical capabilities. These new materials, including drug development, bioimaging materials, therapeutic agents in the form of nanoemulsions or nanoformulations, reagents for diagnostic kits and components

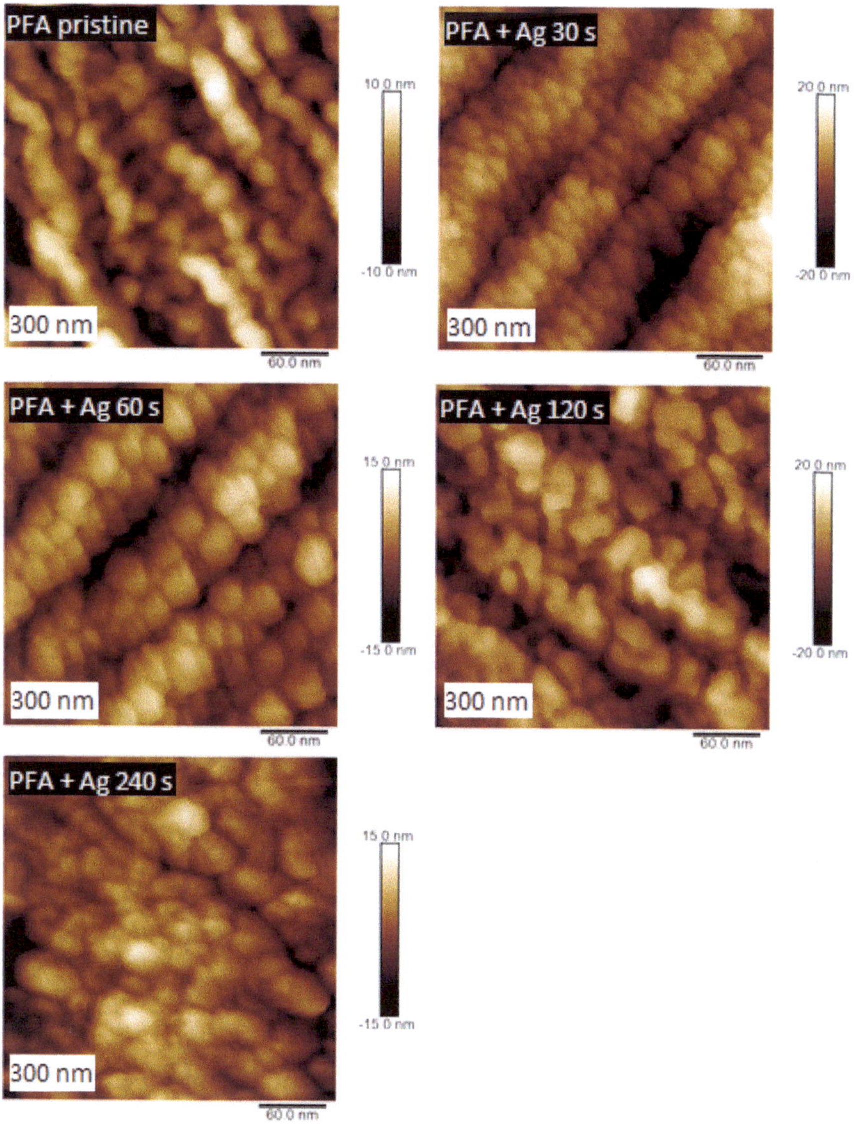

Fig. 3.13 AFM images at 300 × 300 nm square showing surface morphology of both pure PFA and silver-coated PFA for 30, 60, 120, and 240 s [9]. [Open access]

of clinical implants are important for the diagnosis and treatment of diseases. Additional intriguing applications include bone regeneration such as the creation and modification of nanoscaffolds to improve their stability and biocompatibility. This chapter describes an understanding of using various nanomaterials for biological applications, including dendrimers, carbon nanotubes, metals and their oxides. The

Table 3.2 Various green biosynthesized nanomaterials [15]. [Open access]

Nanomaterial	Biosynthesizing agent	Property	Remark
Gold nanospheres	*Eucalyptus globulus* bark extract	Morphology controlled by pH; LSPR at 525 nm	Good control over morphology
Gold nanoparticles	*Pseudomonas denitrificans* (bacterium)	Stable; presence of sulphydryl, amido, and carbonyl groups	Stable GNPs; good control over size and shape
Gold nanoparticles	*Prasiola crispa* (freshwater green alga)	SPR at 535 nm; FCC crystal structure	Nearly spherical; crystallite size ≈ 9.8 nm
Gold nanoparticles	*Sargassum swartzii* (marine alga)	High surface charge (27.6 mV); high stability	Induces cytotoxicity; mitochondrial damage
Gold nanoparticles	*Marsilea quadrifolia* (herbaceous plant)	SPR at 530 nm	Dose-dependent cytotoxicity on HeLa cells
Anisotropic gold nanoparticles	*Cordia myxa* (fruit extract)	SPR at 569 nm	Enhances glucose uptake in 3T3-L1 adipocytes; antidiabetic potential
Silver nanoparticles	*Crotalaria verrucosa* (leaf extract)	High stability	Effective in sensing, catalysis, and cancer diagnosis
Silver nanoparticles	*Bacillus licheniformis* (soil bacterium)	SPR at 422 nm	Cost-effective method to reduce dengue vector larvae; strong antibacterial and antibiofilm activity
Silver nanoparticles	*Camellia sinensis* (plant extract)	SPR at 394–400 nm	Sensing of Cu^{2+} and Pb^{2+} ions in aqueous solution
Silver nanoparticles	*Caulerpa racemosa* (marine green alga)	SPR at 441 nm	Excellent catalytic activity; effective methylene blue degradation
Silver nanoparticles	*Bacillus methylotrophicus* (methanol-utilizing bacterium)	SPR at 416 nm	Notable antimicrobial activity
Platinum nanoparticles	*Azadirachta indica* (plant extract)	SPR at 241 nm	Useful for catalytic applications

(continued)

Table 3.2 (continued)

Nanomaterial	Biosynthesizing agent	Property	Remark
Platinum nanoparticles	Quail egg yolk	SPR peak at 320–340 nm	Potential in catalysis, pharmaceuticals, and energy applications
Palladium nanoparticles	*Chlorella vulgaris* (single-cell green alga)	SPR at 370–440 nm	Rapid synthesis; good particle stabilization
Iron nanoparticles	*Sesamum indicum* (seed extract)	Well-dispersed particles; BCC crystal structure	Shows significant antityphoid activity
Iron nanoparticles	*Citrus maxima* (peel extract)	Iron in mixed oxidation states (Fe^{2+}/Fe^{3+})	Effective in Cr removal
Cu, Ag, and Fe nanoparticles	*Stenotrophomonas maltophilia* (multidrug-resistant bacterium)	SPR: Ag and Cu at 400 nm, Fe at 275 nm	Antibacterial; inhibits biofilm formation

chapter also discusses routes for the synthesis of nanomaterials, which are valuable for various biomedical applications due to their special characteristics, including biocompatibility and immunocompatibility.

3.7.1 Green Nanomaterials: Applications in Diagnosis

Since green synthesized nanomaterials have better biocompatibility even at high concentrations and after long incubation times, they are widely used in biomedical disciplines including diagnostics and therapeutics. This chapter elaborates the use of green nanomaterials for nanomaterial-assisted diagnostics in the fields of biosensing and bioimaging. Among the green synthesized nanomaterials reviewed in this chapter, metals and metal oxide nanoparticles such as gold (Au), silver (Ag), and iron oxide (Fe_3O_4) nanoparticles, as well as magnetic nanoparticles have shown great promise for bioimaging using Computed Tomography (CT) and magnetic resonance imaging (MRI). In addition to bioimaging, the development of biosensors using environmentally friendly synthesized nanomaterials for diagnostic purposes is also being considered. The conclusion is that some nanomaterials, including gold, silver, and certain metal oxides such as ZnO and CeO_2 nanoparticles can be used in electrochemical methods for the determination of analytes such as lactose, hydrogen peroxide, and glucose [18], depending on the synthesis methods used [19].

Fig. 3.14 Scanning electron microscope images of several manganese oxide nanomaterials, including nanosheets, nanotubes, nanorods, nanowires, nanoflakes, nanoneedles, nanoflowers, nanorims, and nanospheres. These micrographs show the morphology of several polymanganese oxide samples described in the text. [6] [Open Access]

Biosensors made using nanomaterials are used not for bioimaging but also for diagnostic purposes. Reduced graphene oxide and gold nanocomposites were fabricated using rose water to develop biosensors capable of electrochemically detecting glucose levels (via electrochemical reduction of glucose oxidase) [19].

In a similar vein, another study used reduced graphene oxide and gold nanoparticles to produce a self-assembled membrane using a one-pot green synthesis process. Hydrogen peroxide was detected amperometrically as a cancer biomarker. In a similar vein, a third-generation cellobiose dehydrogenase-based lactose biosensor, featuring high detection stability and sensitivity, was developed using gold and silver nanoparticles in a quercetin-based synthesis process [14].

3.7.2 Green Nanomaterials: Applications in Therapeutics

An emerging area of research in nanomedicine is the use of green chemistry for the synthesis of metal nanoparticles (MNPs) and their biological use. The green route, which uses plants, bacteria, algae, and fungi as biological sources, has several advantages over the current synthetic route. It is a simple, one-step process that is fast, effective, inexpensive, and environmentally friendly. It also does not require any special tools or catalysts. Compared to chemically synthesized nanoparticles, most biologically synthesized metal nanoparticles (b-MNPs) are more therapeutically effective and biocompatible. Due to their unique physicochemical properties, synthesis, biocompatibility, biosynthesized silver nanoparticles, or b-AgNPs are one of the most promising options among metal nanoparticles in biomedical research. In addition, since ancient times, silver has been widely used as a medicinal substance to treat various diseases. Moreover, the overall therapeutic efficacy of biosynthesized silver nanoparticles (b-AgNPs) is enhanced by active compounds derived from biological sources.

The biological use of b-AgNPs for the treatment of various diseases, including cancer, infectious infections, diabetes, inflammation, wound healing, and neurodegeneration, has been investigated by several researchers over the years. This chapter depicts the recent advances of b-AgNPs in biomedical research, including their use in drug transport, bioimaging, anticancer, antiangiogenic, antibacterial, and biosensing applications, as well as their use as therapeutic agents that can replace more traditional methods. In addition, we will review the biodistribution, pharmacokinetics, and clearance of b-AgNPs in animal models in light of their encouraging clinical results. Finally, we will discuss the challenges and potential applications of b-AgNPs for future therapeutic applications in healthcare settings. These biocompatible gold particles have demonstrated good radiopaque properties, making them suitable for use in computed tomography [142]. Furthermore, gold nanotriangles derived from yam bean extract are used to evaluate in vitro photodynamic and photothermal therapy of HeLa cells, a human cervical cancer cell line widely used in scientific research. A hybrid of gold, copper, and selenium is produced in a heterogeneous dimeric nanoparticle by a green reduction process using ascorbic acid.

Plasmonic interaction between copper and gold caused a remarkable response in Hep-2 cells treated with photothermal therapy [13]. Using mouse fibroblast cell line 3T3, silver nanoparticles with excellent biocompatibility are synthesized using egg white medium. This silver-protein biocomposite exhibited increased radioactivity under X-ray irradiation, making it suitable for radiotherapy in cancer treatment [14]. In addition, iron oxide nanoparticles prepared with chitosan derivative have shown a strong biocompatibility and efficient absorption with natural lactose [16].

These cells include human lung cancer cells A549 and mouse embryonic fibroblast cells L929. For cancer therapy, these magnetic nanoparticles have demonstrated magnetic hyperthermia and stimuli-responsive cellular localization [16]. Magnetotactic bacteria (geobacter sulfur-reducing) biologically produce ferrite nanoparticles, which have promising uses in cancer therapy and magnetic hyperthermia-induced

MRI [17]. Biosynthesized silver nanoparticles have been the focus of numerous research studies due to their potential to combat cancer, viruses, fungi, and bacteria (both Gram-positive and Gram-negative) [14, 16–18]. Drug delivery using green nanomaterials is another well-established therapeutic strategy that utilizes metal, metal oxide, carbon, liposomes, micelles, quantum dots, dendrimers, and polymeric nanoparticles to actively or passively deliver drugs to affected areas. Due to their functionality and biocompatibility, the produced green gold nanoparticles have found fascinating applications as drug and gene delivery vehicles [11, 13]. Due to their large surface area and high drug-carrying capacity, graphene and its derivatives serve as drug and gene carriers [13]. The polycrystalline polymer-based nanomicellar system is another excellent example of a green synthetic polymer-based drug delivery system [14]. Fluorescent and biocompatible gold nanoparticles have been produced and used as biomarkers in conjunction with liposomes to track endothelial cells in vitro and in vivo [16]. Similarly, cell-targeting peptides and polyethylene glycol (PEG)-modified gold nanoparticles efficiently target specific cell types and are easily internalized [17].

Some more examples of applications of green nanomaterials in biomedicine are shown in Table 3.3. The desire to develop green nanotechnology, that is, to combine nanotechnology with the principles of green chemistry is driven by concerns about the safety of nanoparticles affecting human health and the environment. Many of the major problems associated with nanotoxicology could be addressed sustainably through green nanotechnology. Significant safety improvements can be achieved through changes in the design of materials synthesis and proper handling. Safer products have been developed, while reducing production costs, raw material consumption, harsh chemicals and energy consumption through the adoption of environmentally friendly chemical technologies. With recycling/reuse provisions, efficient use of resources has been achieved through the application of nuclear economics. Process efficiency is ensured by minimizing unnecessary intermediates and preventing waste generation. Product quality is raised to safer levels when biosynthetic technologies are used instead of hazardous synthetic or chemically modified nanoparticles. Green nanotechnology is being used to produce more biocompatible materials that can be safely injected into the body for medical and diagnostic uses. Furthermore, green nanotechnology offers a bright future for pharmaceutical, cosmetic, electrical, and other commercial companies without harming the environment [21, 22].

Green nanomaterials, which use environmentally friendly and sustainable technologies to produce nanoparticles, represent a breakthrough in nanotechnology. Green nanotechnology increases sustainability and reduces its environmental impact by using biological processes such as plant extracts and bacteria and reducing reliance on hazardous chemicals. In addition to encouraging safer manufacturing processes, this move toward greener approaches combines creativity and environmental responsibility, providing long-term solutions for sectors such as technology, energy, and healthcare. Ultimately, green nanomaterials represent the incorporation of environmentally friendly methods into scientific and commercial breakthroughs.

Table 3.3 Biomedical applications of green synthesized nanomaterials [15]. [Open Access]

Nanomaterial	Synthesis approach	Application
Carbon dots	*Ocimum sanctum*-mediated hydrothermal method	Fluorescent sensing of Pb^{2+} ions in breast cancer cells (live imaging); detection of Pb^{2+} in water
Reduced graphene oxide	*Ocimum sanctum* hydroalcoholic phytochemical synthesis	Biocompatible in animal models; potential drug/gene carrier
Silver–chondroitin sulfate nanocomposite	Thermal treatment or UV irradiation	Antimicrobial scaffolds
Silver/ carboxymethylcellulose–poly(acrylamide) hydrogel nanocomposite	*Plant extract*-based in situ reduction	pH-responsive drug delivery systems
Silver–TiO_2 nanocomposite	*Euphorbia heterophylla* leaf extract-mediated reduction	Catalytic applications
Iron oxide–gold core–shell nanoparticles	Sonochemical synthesis	Targeted drug delivery and contrast agents for MRI/CT
Fe_2O_3/$CuFe_2O_4$/chitosan nanocomposites	Sol–gel auto-combustion route	MRI imaging
Au@C@Pt core–double shell nanocomposite	Microwave-assisted synthesis	Biosensing of H_2O_2
ZnSe/Au nanocomposites	Low-temperature aqueous synthesis with direct coupling	Cellular bioimaging
ZnO–carbon dot nanocomposites	Hydrothermal method	Photocatalysis and photoreduction of CO_2
Chitosan–gold nanocomposite	Solvothermal synthesis	Antifungal activity
Carbon dots	Egg yolk-mediated solvothermal method	Bioimaging
CQD–polymer composite	Microwave-assisted pyrolysis method	Bioimaging and biolabeling
Fe_3O_4–Ag core–shell nanoparticles	*Vitis vinifera* (grape stem) extract-mediated reduction	Reusable antibacterial agent
Iron oxide–pectin hybrid nanocomposite	Co-precipitation method	Targeted drug delivery, molecular separation, and ion exchange
Silver–graphene oxide nanocomposite	Glucose-mediated reduction	Tryptophan biosensor

(continued)

Table 3.3 (continued)

Nanomaterial	Synthesis approach	Application
Fluorescent carbon dots	Garlic extract-mediated hydrothermal synthesis	Multicolour cellular imaging; radical scavenging
Silver-activated charcoal nanocomposite	*Malpighia emarginata* (West Indian cherry) extract-mediated synthesis	Microbial filtration efficiency

Problems

1. Explain the green nanotechnology roadmap for nanomaterials.
2. What's the point of incorporating some green chemistry?
3. How will green nanomaterial characterization and biomolecule synthesis techniques work?
4. Why are green nanomaterials important, and how significant are their diagnostic applications?
5. Can you explain the process of microwave-assisted pyrolysis?

References

1. Pavelić, K., Pavelić, S. K., Bulog, A., Agaj, A., Rojnić, B., Čolić, M., & Trivanović, D. (2023). Nanoparticles in medicine: Current status in cancer treatment. *International Journal of Molecular Sciences, 24*(16), 12827.
2. Li, J., Zhang, F., Zhong, Y., Zhao, Y., Gao, P., Tian, F., & Cullen, P. J. (2022). Emerging food packaging applications of cellulose nanocomposites: A review. *Polymers, 14*(19), 4025.
3. Carapeto, A. P., Ferraria, A. M., & Botelho do Rego, A. M. (2019). Silver nanoparticles on cellulose surfaces: Quantitative measurements. *Nanomaterials, 9*(5), 780.
4. Mišić Radić, T., Vukosav, P., Čačković, A., & Dulebo, A. (2023). Insights into the morphology and surface properties of microalgae at the nanoscale by atomic force microscopy (AFM): A review. *Water, 15*(11), 1983.
5. Kareem, M. A., Bello, I. T., Shittu, H. A., Awodele, M. K., Adedokun, O., & Sanusi, Y. K. (2020). Green synthesis of silver nanoparticles (AgNPs) for optical and photocatalytic applications: A review. *IOP Conference Series Mater Science Engineering, 805*, 12020.
6. Abuzeid, H. M., Julien, C. M., Zhu, L., & Hashem, A. M. (2023). Green synthesis of nanoparticles and their energy storage, environmental, and biomedical applications. *Crystals, 13*(11), 1576.
7. Mery, M., Orellana, N., Acevedo, C. A., Oyarzún, S., Araneda, F., Herrera, G., & Romero, C. P. (2018). Fabrication of silver nanoparticles using a gas phase nanocluster device and preliminary biological uses. *Materials, 11*(12), 2574.
8. Harrison, C. J., Hatton, P. V., Gentile, P., & Miller, C. A. (2021). Nanoscale strontium-substituted hydroxyapatite pastes and gels for bone tissue regeneration. *Nanomaterials, 11*(6), 1611.
9. Slepička, P., Rimpelová, S., Slepičková Kasálková, N., Fajstavr, D., Sajdl, P., Kolská, Z., & Švorčík, V. (2021). Antibacterial properties of plasma-activated perfluorinated substrates with silver nanoclusters deposition. *Nanomaterials, 11*(1), 182.

10. Hernández-Morales, L., Espinoza-Gómez, H., Flores-López, L. Z., Sotelo-Barrera, E. L., Núñez-Rivera, A., Cadena-Nava, R. D., Alonso-Núñez, G., & Espinoza, K. A. (2019). Study of the green synthesis of silver nanoparticles using a natural extract of dark or white Salvia hispanica L. seeds and their antibacterial application. *Applied Surface Science, 489*, 952–961.
11. Soto, K. M., Quezada-Cervantes, C. T., Hernández-Iturriaga, M., Luna-Bárcenas, G., Vazquez-Duhalt, R., & Mendoza, S. (2019). Fruit peels waste for the green synthesis of silver nanoparticles with antimicrobial activity against foodborne pathogens. *LWT, 103*, 293–300.
12. El-Sherbiny, I. M., & Salih, E. (2018). Green Synthesis of Metallic Nanoparticles Using Biopolymers and Plant Extracts. In S. Kanchi & S. Ahmed (Eds.), *Green Metal Nanoparticles* (pp. 293–319). Scrivener Publishing LLC.
13. Adyani, S. H., & Soleimani, E. (2019). Green synthesis of Ag/Fe_3O_4/RGO nanocomposites by Punica Granatum peel extract: Catalytic activity for reduction of organic pollutants. *International Journal of Hydrogen Energy, 44*, 2711–2730.
14. Ahmed, R. H., & Mustafa, D. E. (2020). Green synthesis of silver nanoparticles mediated by traditionally used medicinal plants in Sudan. *International Nanotechnology Letter, 10*, 1–14.
15. Foday, E. H., Jr., Bo, B., & Xu, X. (2021). Removal of toxic heavy metals from contaminated aqueous solutions using seaweeds: A review. *Sustainability, 13*(21), 12311.
16. Ahmed, T., Noman, M., Shahid, M., Niazi, M. B. K., Hussain, S., Manzoor, N., Wang, X., & Li, B. (2020). Green synthesis of silver nanoparticles transformed synthetic textile dye into less toxic intermediate molecules through LC-MS analysis and treated the actual wastewater. *Environmental Research, 191*, Article 110142.
17. Banasiuk, R., Krychowiak, M., Swigon, D., Tomaszewicz, W., Michalak, A., Chylewska, A., Ziabka, M., Lapinski, M., Koscielska, B., Narajczyk, M., et al. (2020). Carnivorous plants used for green synthesis of silver nanoparticles with broad-spectrum antimicrobial activity. *Arabian Journal of Chemistry, 13*, 1415–1428.
18. Bindhu, M. R., Umadevi, M., Esmail, G. A., Al-Dhabi, N. A., & Arasu, M. V. (2020). Green synthesis and characterization of silver nanoparticles from Moringa oleifera flower and assessment of antimicrobial and sensing properties. *Journal of Photochemistry and Photobiology, B: Biology, 205*, Article 111836.
19. Celebioglu, A., Topuz, F., Yildiz, Z. I., & Uyar, T. (2019). One-step green synthesis of antibacterial silver nanoparticles embedded in electrospun cyclodextrin nanofibers. *Carbohydrate Polymers, 207*, 471–479.
20. Gomes, P., Tama, D., Carvalho, H., & Souto, A. P. (2020). Resistance variation of conductive ink applied by the screen printing technique on different substrates. *Coloration Technology, 136*(2), 130–136. https://doi.org/10.1111/cote.12451
21. Kiani, M., Rabiee, N., Bagherzadeh, M., Ghadiri, A. M., Fatahi, Y., Dinarvand, R., & Webster, T. J. (2020). High-gravity-assisted green synthesis of palladium nanoparticles: The flowering of nanomedicine. *Nanomedicine: Nanotechnology, Biology and Medicine, 30*, Article 102297.
22. Rashmi, B. N., Harlapur, S. F., Avinash, B., Ravikumar, C. R., Nagaswarupa, H. P., Anil Kumar, M. R., Gurushantha, K., & Santosh, M. S. (2020). Facile green synthesis of silver oxide nanoparticles and their electrochemical, photocatalytic and biological studies. *Inorganic Chemistry Communications, 111*, Article 107580.

Chapter 4
Glucose Sensor Using Metal Nanoparticles

Recently, the researchers have shown a great interest in the noble metal nanoparticles because they promise to create label-free biological and chemical sensors. The unique characteristics of nanostructured noble metal arrays such as high surface-to-volume ratio, localized surface plasmon resonance, exceptional conductivity, and anomalous transmittance as well as light reflection are what give these sensors their high capabilities. The most commonly used method for measuring glucose is amperometric approach. Conversely, there have reported some localized surface plasmon resonance (LSPR) sensors that demonstrate good sensitivity to refractive index changes, which occur at metal/dielectric interface. In addition, several researchers have considered the fiber-based glucose sensor and phenomenon of attenuated total reflection. The chapter discusses in detail both enzymatic and non-enzymatic sensors for silver, gold, and copper nanoparticles. The interests of researchers are taken into account when discussing the production of glucose sensors.

4.1 Introduction

Glucose measurements included only few detection methods. Spectrophotometry, thermometry, and optical and electrochemical sensors are a few examples of these measurement methods. Of these methods, the most impressive is electrochemical sensors. This method allows measurements to be made with a great accuracy and precision at low glucose concentrations. It has the advantage of being cheaper than other approaches and easier to use. Electrodes modified with glucose oxidase enzymes have been used in most glucose-sensing studies. Diabetes is becoming a global problem, especially in countries with low and middle living standards. The World Health Organization (WHO) stated in its 2016 global diabetes report that there has been a steady increase in prevalence of diabetes over the past 30 years [1–4].

In 2014, there are 422 million cases, which is much higher than estimated people with diabetes need to monitor their blood sugar levels to treat the disease created

R. A. Al-Samarai and Y. Al-Douri, *Engineering of Nanomaterials*,
https://doi.org/10.1007/978-981-95-5499-7_4

by the first enzyme-based glucose sensor. In addition, new electrochemical glucose sensors are constantly being developed, due in part to the ability to measure blood glucose concentrations using the catalytic properties of glucose oxidase (GOx) or glucose dehydrogenase (GDH) [5]. Projected detection performance glucose could not be achieved even using enzymatic sensors. Enzyme-based glucose sensors have made significant advances, but there are still significant challenges in their performance and stability in certain scenarios. As a result, research has begun to focus on developing new technologies that use non-enzymatic electrodes such as using nanoparticles (NPs) that are deposited on electrodes. Due to their uniform size and structure, these nanoparticles, such as surface-functionalized gold nanoparticles (SP-AuNPs) and surface-functionalized silver nanoparticles (SP-AgNPs), have demonstrated exceptional sensitivity capabilities, increasing their potential for accurate and long-term glucose detection. Thus, key advances in glucose sensing technology have been made possible by the transition from enzyme-based electrodes to nanoparticle-based electrodes. The results of field emission scanning electron microscopy-energy dispersive X-ray spectroscopy (FESEM-EDX) study of size and shape of SP-AuNPs and SP-AgNPs are shown in Fig. 4.1a, b. Although some irregular particles are found in the size range of 5–30 nm, SP-AuNPs are found in monodisperse spherical forms. Similarly, monodisperse spherical forms of SP-AgNPs with sizes ranging from 10 to 40 nm are found. These results are consistent with earlier studies [6] on spherical nanoparticles (NPs) with sizes ranging from 35 to 50 nm, which are obtained from Catharanthus roseus leaf extract [1].

The main reason for the failure of enzymatic sensors to produce desired results is their numerous disadvantages, which include complex immobilization processes, expensive operating requirements, poor stability, and critical operating conditions. In addition, the repeatability, selectivity, and sensitivity of enzyme electrodes are influenced by several environmental parameters, including temperature, humidity, pH, and instability of enzyme molecules. These disadvantages negatively impact the applicability of enzyme-based sensors in applications such as glucose sensing, food monitoring, and bioprocess control. Difficulties with enzyme-based glucose sensors have made it possible to explore electrodes lacking enzymes. Because intermediates are absorbed throughout the reaction process, electrochemical analysis of nanoparticles used in enzyme-free sensors results in performance degradation, which is characterized by surface wear or surface contamination. In addition, sensor studies become more expensive due to high cost of some metal particles. Therefore, it is a critical to create glucose sensors that are affordable, sensitive, and robust to electrochemical assays performed in the absence of enzymes. Metals corresponding to these properties are represented by Ni-based nanoparticles. For this reason, natural electrocatalytic activities and oxidation reactions based on the $Ni(OH)_2$/NiOOH redox reaction in an alkaline environment have been studied.

This electrocatalytic activity of Ni-based catalysts is inherently due to physical or non-covalent adsorption. A fragile structure is created from a nickel film and placed on an electrical mechanism of gold and glucose nanoparticles. The materials for research, monitoring, and experimental procedures are difficult [7]. Thus, the development of new sensor materials is required. To increase electrocatalytic

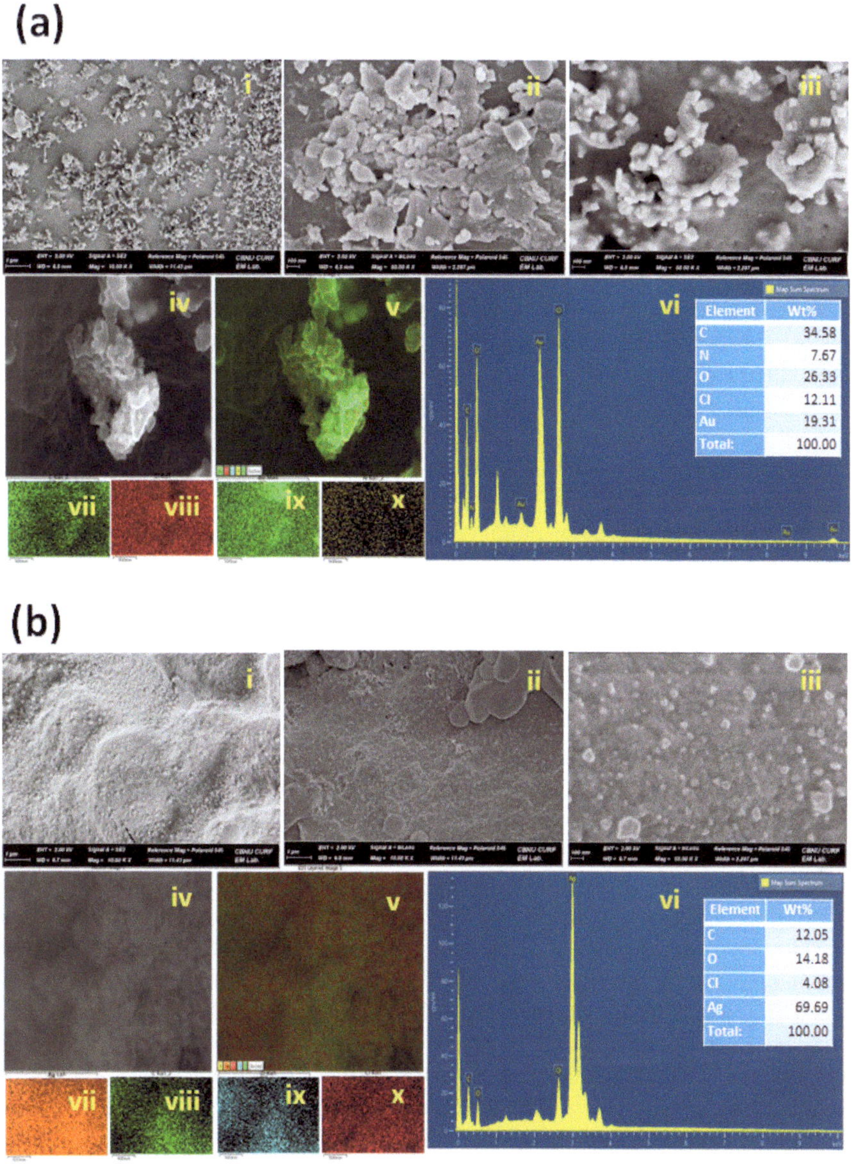

Fig. 4.1 SEM of AuNPs and AgNPs prepared by the SP method in the field. **a** (i–iii) Size and shape of synthesized AuNPs; scale bar = 100 nm. (iv–ix) Elemental mapping for distribution and purity measurements in the selected area of the electron microscopy of nanogold sample. (x) Energy dispersive X-ray spectra of the synthesized AuNPs (the elemental ratios in weight percent are shown in inset table). **b** (i–iii) Size and shape of synthesized AgNPs; scale bar = 100 nm. (iv–ix) Elemental mapping for distribution and purity measurements in selected area of electron microscopy of nanosilver sample. (x) Energy dispersive X-ray spectra of synthesized AgNPs (the elemental ratios in weight percent are shown in the inset table) [1] [Open access]

activity, it is also necessary to find effective ways to create strong contacts between the electrode surface and material being measured. A biosensor is a small analytical instrument that uses a recognition element coupled to a transducer (reader) to identify and measure a specific chemical (analyte). The development of new biosensor assemblies, immobilization methods, and new material applications that can translate these approaches into practical applications have been the focus of countless studies [7–9]. For accurate and sensitive detection of a target molecule, a reliable biosensor must have a low-cost, fast, stable and inert platform [10, 11].

Noble metals such as platinum [10], palladium [11], gold [12], and their metal alloyshave been widely investigated for using in the development of non-enzymatic glucose sensors. However, the economic use of precious metals is severely limited due to their rarity and high cost. On another hand, in an alkaline environment, transition metals such as copper [13], cobalt [12], nickel [8], and their metal oxides or hydroxides [7, 9] exhibit exceptional electrocatalytic abilities toward glucose. Transition metals are superior to noble metals in that their synthesis is easier to control, they are more active and less expensive. In light of these elements, the development of non-enzymatic glucose sensor has received great attention using transition metals.

4.2 A Glucose Sensor Based on Silver Nanoparticles

Biosensors offer a wide range of potential applications in environmental monitoring, food analysis, bioprocess monitoring, clinical diagnostics and prognosis and bioassays [11]. Because glucose detection has a wide range of clinical and experimental applications, many laboratories around the world are interested in developing fast, accurate, and affordable biosensors. Glucose biosensors are, in fact, one of the few sensing devices that are successfully sold today [12]. The history of commercial blood glucose monitoring strips has been extensively studied for a long time [14]. The development of more accurate and efficient sensors for clinical and experimental applications has been made possible by studies such as one on the modification of gold electrodes with a GO/Co/chitosan composite, which has been shown to increase sensing efficiency. The results show that at constant glucose concentration, dip transmittance for the proposed structural model gradually decreases as the amount of AgNPs increases. FESEM characterization is performed on the surface of Au film electrode modified with GO/Co/chitosan composite with the elemental distribution (Fig. 4.2). The GO/Co/chitosan composite is successfully deposited on the surface of Au film electrode as shown in SEM images in Fig. 4.2a. The composition morphologies of composite are different from those of the bare electrode (Fig. 4.2b). Figure 4.2c–f displays the elemental composition of this intermediate, indicating that the GO/Co/chitosan composite is structured with effective compositions that are consistent with our previous analysis of the structure of this intermediate [15].

Medical testing requires sensors that can measure glucose concentrations. Here, we use finite-difference time domain (FDTD) and experimental techniques to explore the feasibility of glucose detection by introducing silver nanoparticles (AgNPs)

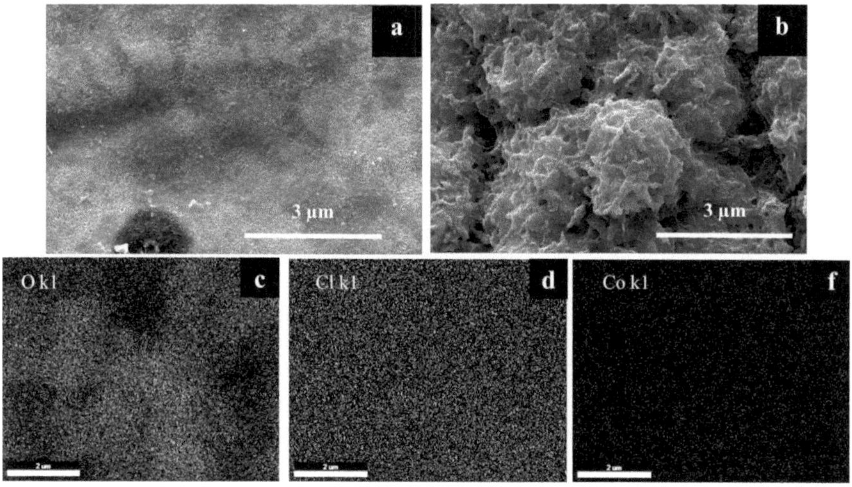

Fig. 4.2 Surface characterization of GO/Co/chitosan-modified electrode in fabricated biosensor using FESEM images and elemental mapping: **a** bare and **b** GO-, Co-, and chitosan-treated electrodes. EDS analysis and elemental mapping images of **c** O, **d** Cl, and **f** Co [15] [Open access]

uniformly distributed in space. Given a great interest in creating accurate, rapid, and affordable sensors for glucose detection, efforts have been directed toward improving performance through the use of nanoparticles. The using of silver nanoparticles (AgNPs) has shown great promise in improving glucose concentration detection using advanced methods such as the finite-difference time-domain (FDTD) method. In addition, they have measured any potential interference by measuring 20 μM glucose that contained predetermined amounts of uric acid (UA) and ascorbic acid (AA). At concentrations of 20 and 80 μM for UA and AA, respectively, there is no noticeable interference for sensors. When the glucose concentration of sensor is changed step by step, the current–time (I-t) curve shows a fast and sensitive response to the change in glucose concentration at the sensor.

The decreasing current and fluctuations in glucose concentration have increased to an almost stable value. After optimization of the results, the estimated linear range of glucose concentration is determined to be from 0.5 to 50 μM with a correlation coefficient of R2 = 0.9990 and a detection limit of 0.1 μM at S/N = 3. The sensitivity of sensor is determined to be 135.9 μA mM-1. The detection process for glucose values greater than 50 μM required appropriate dilution of serum sample. The decreasing current and fluctuations in glucose concentration have increased to an almost stable value. After optimization of the results, the estimated linear range of glucose concentration is determined to be from 0.5 to 50 μM with a correlation coefficient, R2 = 0.9990 and a detection limit of 0.1 μM at an S/N ratio of 3. The sensitivity of the sensor is determined to be 135. 9 μA mM-1.

Appropriate dilution of serum sample is necessary for the detection process for glucose values greater than 50 μM. Glucose biosensors use glucose as a source of detection signals, either directly from glucose or by stimulating the conversion of glucose into other detectable electroactive compounds. Today, glucose levels are determined by both enzymatic and enzyme-independent methods. However, the most commonly used method is immobilized glucose oxidase (GOx)-based detection. New immobilization methods, electrode surface modifications, and immobilizer ants are welcome since enzyme immobilization undoubtedly affects the biosensor response [1–5, 16–18].

In this context, the focus has improved the performance of sensors using advanced techniques such as finite-difference time-of-flight (FDTD) and nanoparticles such as AgNPs. The studies have shown the ability of nanoparticles to improve the accuracy of glucose concentration sensing, as they have shown a rapid and sensitive response to changes in glucose concentration. Although many conventional methods such as enzymatic sensing methods are available, efforts to use nanomaterials have witnessed a remarkable development in increasing the sensitivity and accuracy of sensing. At the same time, the focus on surface modification methods of electrodes and immobilization tools has become an essential tool for improving the response of biosensors. As these researches advance, we find that the combination of nanomaterials and novel immobilization methods represents a major step toward developing more efficient and stable sensors. The goal of current research is to incorporate nanomaterials into biosensor designs to improve their sensitivity, stability, and detection accuracy. Although there are no distinct peaks, the spectral signal of algae has shown a distinct profile that distinguished it from the spectra of the synthetic particles. FT-IR spectra of SP-AuNP, SP-AgNP, and SP extract are shown in Fig. 4.3i. OH (hydroxyl) stretching vibration is represented by a large peak in FT-IR spectrum of SP extract at about 3392 cm^{-1}, while the C=C (carboxylic acids) vibration is represented by a sharp narrow peak at 1650 cm^{-1}. The carbonyl derivatives and C–H (amide) stretching are represented by peaks at 1611 cm^{-1} and 2938 cm^{-1}, respectively. The peaks located at 1436 and 1239 cm^{-1} indicated that the methyl (CH_3) group is moderately stretched. Meanwhile, the stretching vibration of the C–O group in alcohol and ether molecules is associated with bands at 1049 and 1184 cm^{-1}, respectively. Carbonyl functional groups (C=O) in ester bonds of fatty acids and triacylglycerols are represented by the absorption peak at 1239 cm^{-1} of SP extract [1].

The selection of surface modifications and immobilizers for an enzyme biosensor is critical not only to improve the sensitivity and stability of device since an effective modification or immobilizer can stop enzyme leakage, preserve the bioactivity of the biosensor, and extend its shelf life. Many substances such as polymers, phospholipid bilayers, clay, porous alumina, and zeolite have been used as immobilizers in enzyme-based biosensors. Using the unique characteristics of homogeneous silver nanoprisms, the group [7] has reported another surface plasmon resonance (SPR)-based glucose sensor (Fig. 4.2). The team has developed a workable plan to use colorimetric technology to analyze serum glucose at the micromolar level in real time. The sensitivity of colorimetric sensor is confirmed by testing for influence of several interfering substances such as fructose, lactose, and maltose. The authors have investigated the sensitivity of developed colorimetric glucose sensor and several analytical reactions.

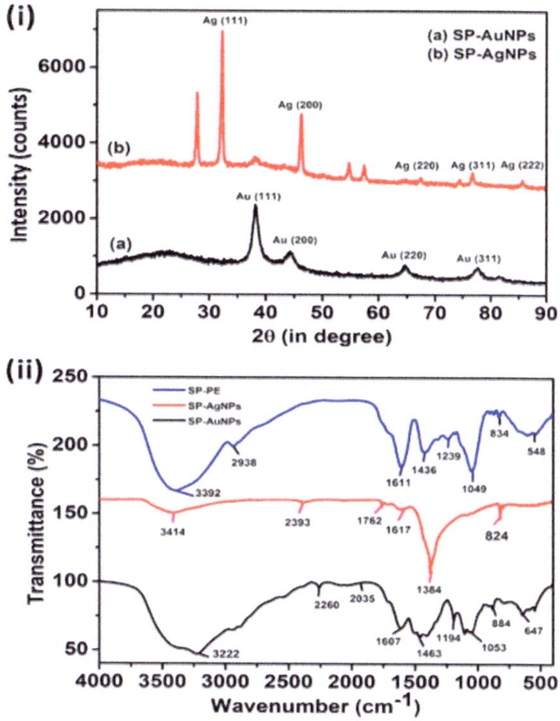

Fig. 4.3 (i) SP-AuNPs and SP-AgNPs' powder XRD patterns. (ii) Functional groups in the plant extract and nanoparticle samples are indicated by the FT-IR spectra of SP-AuNPs, SP-AgNPs, and SP extract. [1] [Open access]

To do this, a homogeneous system is created, a certain amount of glucose solutions is added and system is incubated for 40 min. Because glucose serves as a clinical diagnostic tool for diabetes that is critical for regulating blood sugar levels, glucose determination is an exciting area of biochemical research. In addition, glucose determination plays a role in pharmaceutical monitoring, bioprocess control, and food and environmental management [7, 8]. The high sensitivity, excellent selectivity, and good stability of glucose determination have attracted great interest in this context. Since the first enzyme-based electrode [9], a large amount of research has been conducted on electrochemical glucose sensors to achieve the goals of rapid response, ease of use, and affordability. However, due to complex immobilization process and ease of deactivation of glucose oxidase (GOX), which interferes with the binding of glucose to active site of GOX, the enzyme-based sensor always experiences low sensitivity or loss of activation. Consequently, further efforts have been made to develop highly sensitive and stable non-enzymatic glucose sensors. It is well known that since they have more active sites and a larger surface area than other types of metal nanoparticles, nanosized metal nanoparticles (NPs) generally exhibit enhanced electrocatalytic activity. However, high surface energy promotes easy agglomeration of nanoparticles, which is detrimental to the effective targeting of active sites.

One of the commonly used methods for stabilizing metal nanoparticles (NPs) in electrochemical sensor systems is their immobilization on solid supports such as carbon nanotubes [10], graphene [11], and mesoporous carbon [12]. These combinations of metal NPs and solid media can provide ideal stability, large loading quantities, and good electrical conductivity. However, the size of metal NPs is usually unpredictable. Metal–organic frameworks (MOFs) are developed class of hybrid porous materials composed of organic and metal ligands. MOFs have shown significant promise in various applications such as separation [14], gas storage [13, 19], drug delivery [8, 14], and heterogeneous catalysis [7, 9] because of their porosity, large surface area, and chemical resistance. Reports of using MOFs in the field of electrochemical biosensing are becoming increasingly common. In these reports, MOFs have different responsibilities: (1) MOFs can be immediately used to create sensors as an electrocatalyst [10, 11]. MOFs have been used as a sacrificial template to create metal oxide nanoparticles (NPs) for detection of glucose and hydrogen peroxide [12]. Cu-MOF is a matrix to support Au-SH-SiO_2 for the electrocatalytic oxidation of l-cysteine [1] and hydrazine [1, 17]. Several strategies have been developed to improve the stability and functionality of non-enzymatic sensors to overcome the challenges associated with the enzymatic sensing of glucose. The use of metal nanoparticles (NPs) has been highlighted among these strategies due to their ability to enhance photoelectrochemical activity due to their huge surface area and accessible active sites. However, a major hurdle is the rapid aggregation of these NPs. To stabilize the metal nanoparticles and stop their aggregation, metal–organic compounds (MOFs) have emerged as a viable option. This will increase the stability and photoelectrochemical activity of particles. By combining structural advantages such as high porosity and large surface area, MOFs allow for creation of sensors that are more robust and efficient. They have proven their effectiveness in various applications such as hydrogen peroxide and glucose sensing, showing their significant potential in enhancing the performance of biosensors in numerous applications. In general, MOFs are an excellent place to permanently house an electrocatalyst due to the cages that exist in crystalline porous structures. The size of metal NPs is generally unpredictable, although the aforementioned carbon–metal NP hybrids are widely used. On other hand, MOFs with porous structures can stabilize NPs from migration and aggregation while sterically confining metal NPs to a limited size. Therefore, the special structure of metal nanoparticles enclosed in MOFs is expected to significantly enhance the electrocatalytic activity and stability. In the Spotlight After co-cultivation for 24 h, the resulting nanoparticles are evaluated for their ability to inhibit the growth of E. coli MG1655 strain biofilms. Atomic energy microscopy is used to visualize the formed biofilms (Fig. 4.4). Before testing, the algae extract, silver, and iron nanoparticles are inoculated onto a solid agar medium to assess the microbiological purity of the materials [1].

According to the study, the maximum level now causes the ciliary dipole band to move toward blue, changing the difference of silver NANC (AgNP) to a distinct color improvement. For personal computer (PC) users, the relationship between SPR transitions and their concentration are decreased from 0.2 to 100 μM. The analysis of several chemical species, including AA, DA, and UA, in serum and 17 amino acids, where distinctive elements with the same SPR transition band are found, based on the results of target detection test. In addition, binding to amino groups that affect binding

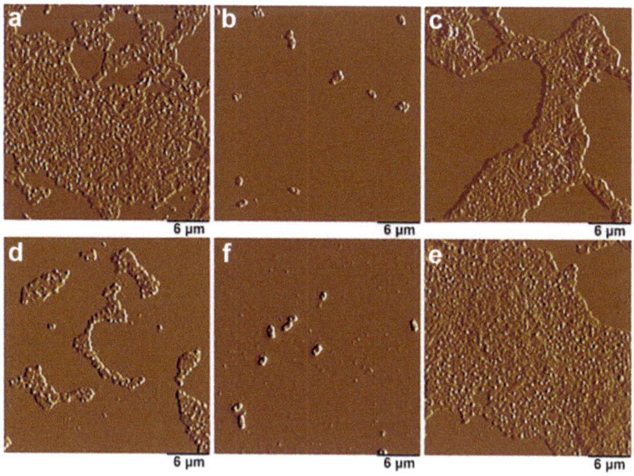

Fig. 4.4 AFM images of biofilms obtained by E. coli MG1655 after 24 h of co-culture with iron and silver nanoparticles. The experimental results are shown below: **a** control, **b** 10% algae extract, **c** citrate-stabilized silver nanoparticles, **d** extract-stabilized silver nanoparticles, **e** unstabilized iron nanoparticles, and **f** extract-stabilized iron nanoparticles [1] [Open access]

through the synthesis of amino groups and the effective activity changed by AgNP are identified. According to these recent results, adjustments such as electrodes seem to sensitize. The near-field interference pattern obtained by incident pulsed UV laser on a diffraction grating is used to construct permanent surface relief gratings on azo polyimide films of different compositions. A 1000 l/mm phase mask (Edmund optics) is used to project the third harmonic of a Nd: YAG laser (Brilliant, Quantel; 1064 nm, pulse duration 10 ns, diameter 6 mm) incident on three-fold Galilean expander. A schematic representation of the experimental setup used is previously shown in Fig. 4.5. [20].

According to the study, high glucose levels cause a blue shift of the dipole band, which significantly changes the color of silver nanoparticles and improves the optical efficiency of the sensor. A strong and accurate correlation between SPR peak shift and glucose concentration is observed from 0.2 to 100 μM. In addition, the interaction of chemical species in serum, including AA, DA, and UA, is investigated; the results are consistent with SPR shift spectrum. This means that the interaction between amino groups and glucose can improve the effective activity of the modified electrodes. In this case, the near-field interference pattern of ultraviolet laser is used to create continuous guidelines on films, which improves the functionality and sensitivity of device.

When various glucose concentrations are administered one after other, a straight line connecting the correlation coefficient R2 = 0.999 is created. The calibration range for glucose determination is determined by the author's cover values from 0.14 to 4.11 mmol/L. To further test the specificity of the sensor, various interfering moieties that typically coexist with glucose in serum are considered. Physiological

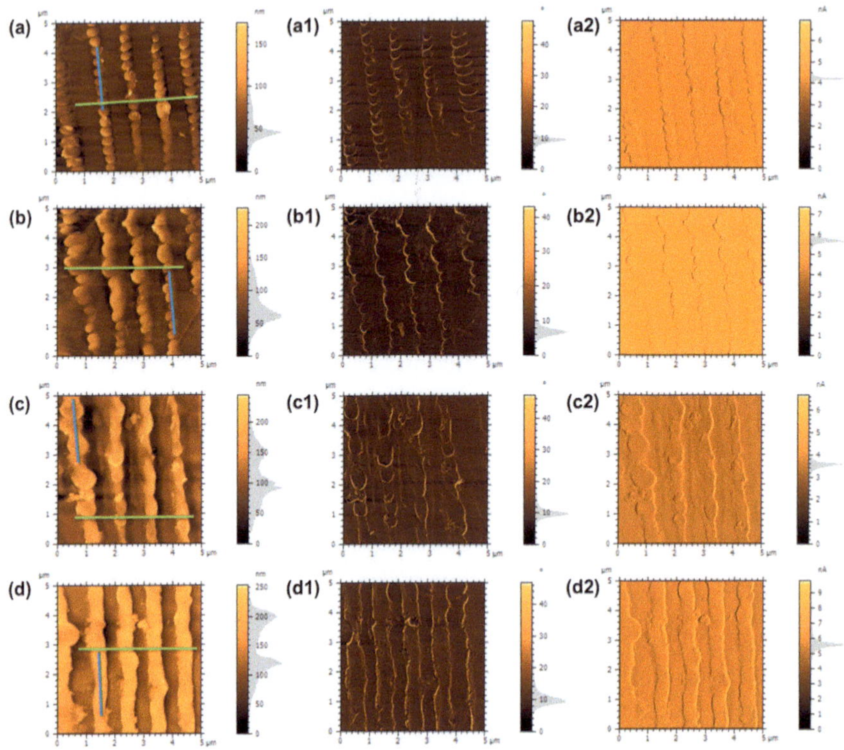

Fig. 4.5 To perform simultaneous AFM topography, phase contrast, and amplitude imaging, Az0.25PSS-L (a, a1, a2), Az0.50PSS-L (b, b1, b2), Az0.75PSS-L (c, c1, c2), and Az1.00PSS-L (d, d1, d2) irradiated with a phase mask laser are used. Green lines indicate the locations of cross-sectional profiles taken perpendicular to the SRG, while blue lines indicate those taken along the SRG. The unit of amplitude images determined by the procedure is nA [20]. [Open access]

plasma glucose levels range from 3 to 8 mmol/L, whereas the amounts of UA, AA, and DA are 0.1 mmol/L, which means that the ratio of glucose to interfering substances is 30:1 [12, 14].

Nanocellulose can be used as a matrix to support silver ions because, in general, its molecular chains, if left unmodified or otherwise modified, contain many functional groups such as hydroxyl groups, which have many negative charges in water and used for adsorption cations [9]. Moreover, in production of AgNPs, nanocellulose has been found to act as a stabilizing, dispersing, and/or reducing agent [14]. Metal nanoparticles are well stabilized by nanocellulose, which can stimulate particle nucleation and prevent their aggregation [19], conjunctive normal form (CNF) can significantly improve the stability and dispersibility of AgNPs in physiological environments. They have confirmed that CNF-AgNPs have a certain level of antimicrobial ability. Functional groups such as hydroxyl groups found in nanocellulose create an ideal environment for supporting silver ions, promoting stability and dispersion of metal nanoparticles while preventing their agglomeration. Additionally, when creating silver nanoparticles (AgNPs), nanocellulose exhibits the ability to function as a stabilizing, dispersing, and even reducing agent. Electrocatalyst deposition using sophisticated techniques such as three-electrode voltammetry improves this stability in electrochemical situations, providing better performance for the catalytic material, including Co-P and Pt. To illustrate, an indium oxide-coated glass electrode is used, the voltage is adjusted to create effective catalyst layers, expanding the use of materials in many settings. A three-electrode voltammeter (AMETEK Princeton Applied Research, PARSTAT MC) is used for electrodeposition of Co-P and Pt electrocatalysts as shown in Fig. 4.6a. Indium oxide-coated glass in an aqueous solution of 0.025 M $CoSO_4$ $7H_2O$, 0.5 M NaH_2PO_2 H_2O, and 0.025 mM CH_3COONa for Co-P electrocatalysts or 0.5 mM H_2PtCl_6 $6H_2O$ for Pt electrocatalysts is used as the working electrode. Ag/AgCl in 3 M KCl(l) and Pt wire are used as a reference and counter-electrode, respectively. An electrical potential of −0.9 to −1.2 V (relative to the Ag/AgCl electrode) for Co-P electrocatalysts or −0.35 V (relative to the Ag/AgCl electrode) for Pt electrocatalysts is applied for ten minutes [21].

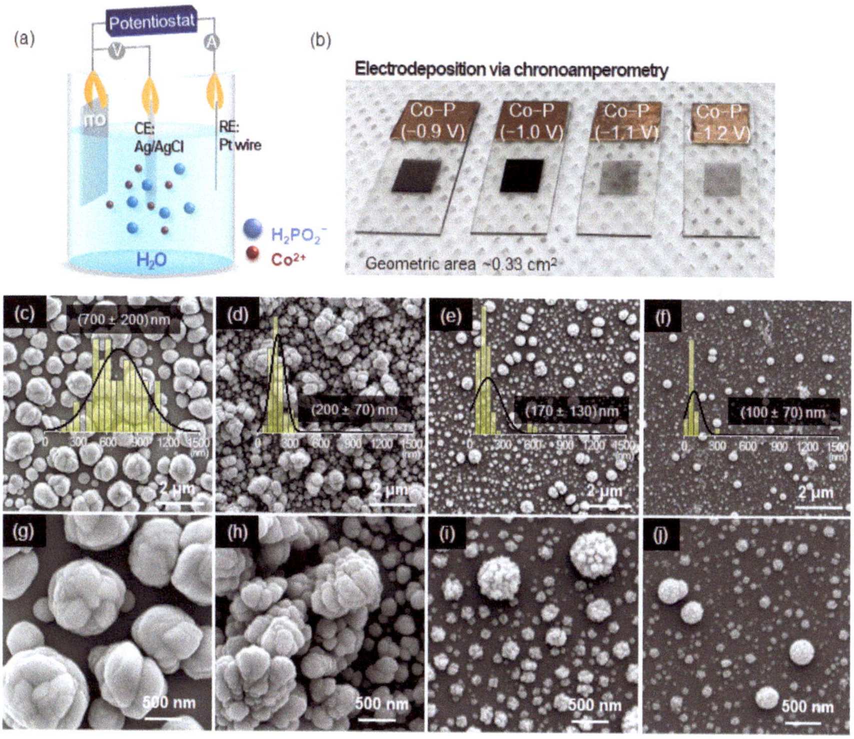

Fig. 4.6 **a** Schematic diagram showing how electrodeposition creates cobalt and phosphate nanoclusters. **b** Figure and SEM images (c–c) of the resulting cobalt and phosphate nanostructures on ITO grown at different applied potentials: − 0.9 V for (**c**, **g**), −1.0 V for (**d**, **h**), −1.1 V for (**e**, **i**), and −1.2 V for (**f**, **j**) (Appendix: Histogram of nanocluster size distribution) [21] [Open access]

Figure 4.7a shows XRD patterns of Co-P nanostructures electrodeposited on a bare ITO substrate at applied voltages of −0.9 and −1.0 V. Three major diffraction peaks are presented at 41.8°, 44.8°, and 47.6°. They can be attributed to the hexagonal compact Co (JCPDS No. 04-003-3863) (100), (002), and (101) planes. The absence of diffraction peaks associated with various Co-P forms is probably due to the nature of electrodeposition method which sometimes produces a lower hybridization ratio [22]. Understanding the performance of deposited materials and their potential applications requires analyzing their chemical composition and crystal structure. As shown in Fig. 4.7a, where the diffraction patterns show large peaks attributed to the compact hexagonal structure of cobalt, X-ray diffraction (XRD) analysis in this context provides accurate information about the crystal structure of deposited materials. In addition, the presence of charged Co_2+ and other oxidized species is confirmed by X-ray photoelectron spectroscopy (XPS), which provides a more complete understanding of the chemical composition. These properties help the

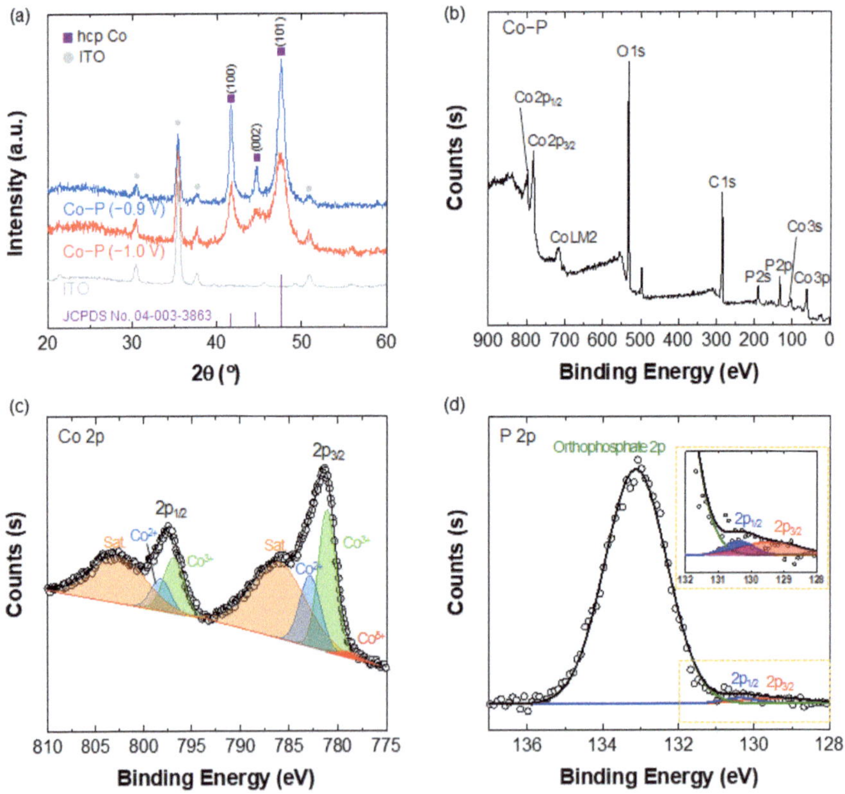

Fig. 4.7 **a** Electrodeposited XRD patterns of Co-P on ITO substrate at applied voltages of −0.9 and −1.0 V. **b** High-resolution spectra of **c** P 2p and **d** Co 2p from Co-P nanoclusters, as well as a survey scan [21] [Open access]

deposited materials perform better, making them suitable for various electrochemical applications. On other hand, XPS spectra of Co-P nanostructures in Fig. 4.7b–d indicate the presence of the material. The high-resolution Co 2p spectra in Fig. 4.7c that shows a small peak at 778.9 eV, which is associated with the positively charged Coδ + species in Co-P, as well as the moon elements (Sat) and the oxidation state peaks in Co (Co^{2+} and Co^{3+}) [21].

Once the crystal structure is determined and Co-P compounds are characterized, the electrochemical organic surface area (ECSA) needed to be evaluated for practical performance of these components in electrochemical applications. The selected area electron diffraction (SAED) pattern of content management system (CMS) nanoparticles illustrates a clear ring pattern as shown in Fig. 4.8c. This sequential arrangement of rings indicates the presence of crystalline Cu nanoparticles. The lattice edges at intervals of 0.25 nm, 0.2271 nm, and 0.1261 nm are in remarkable agreement with XRD pattern and correspond to the Cu (1 1 1), (2 0 0), and (2 2 0) planes [17].

After growing AgNPs using the silver mirror reaction on cylindrical surface of sensor, it is seen that the sensitivity of IR absorption signal increases much more. The

Fig. 4.8 SEM images of **a** MS, **b** 10-CMS NP, **c**, **d** comparison of elemental maps, and **e** TEM image of 10-CMS NP [23] [Open access]

infrared absorption signal is significantly increased when the sensitivity of sensor is enhanced using silver nanoparticles (AgNPs) through interaction with silver mirror, which increases the efficiency of sensor in detecting glucose. Compared with traditional sensors, the higher absorption has demonstrated a reliable response to glucose, demonstrating these advantages when using the modified CO_2 laser ATR sensors. In addition, the deposition quality of thin film can be affected by the characteristics of substrate, including its composition and surface roughness level. Therefore, scanning electron microscopy (SEM) is used to study the morphology of steel substrate. The results show that the root mean square (RMS) roughness is 18 nm with a small number of surface defects, suggesting a limited impact on deposition process and improving the performance of sensor. When AgNP-modified ATR sensors are used in combination with CO_2 laser, the gain ranged from two to six times at different absorption wavelengths. In addition, the modified electrode produced a signal with high absorption in the presence of glucose compared to conventional ATR sensors. Large and isolated glucose uptake peaks in the fingerprint region help distinguish glucose in vitro from interfering species in human blood [2]. The growth, structure, and properties of thin films depend not only on the deposition method but also on other variables such as substrate composition, temperature, and deposition rate. Errors in the substrate surface morphology can have a significant impact on how the film grows. As a result, it has examined the morphology of steel substrate and calculated that root mean square (RMS) roughness is 18 nm as shown in Fig. 4.9a. SEM examination of the top surface of steel substrate shows a small number of defect spots, but they are not numerous (Fig. 4.9b). Therefore, it can be expected that the surface roughness will have little impact on the deposition procedure [2].

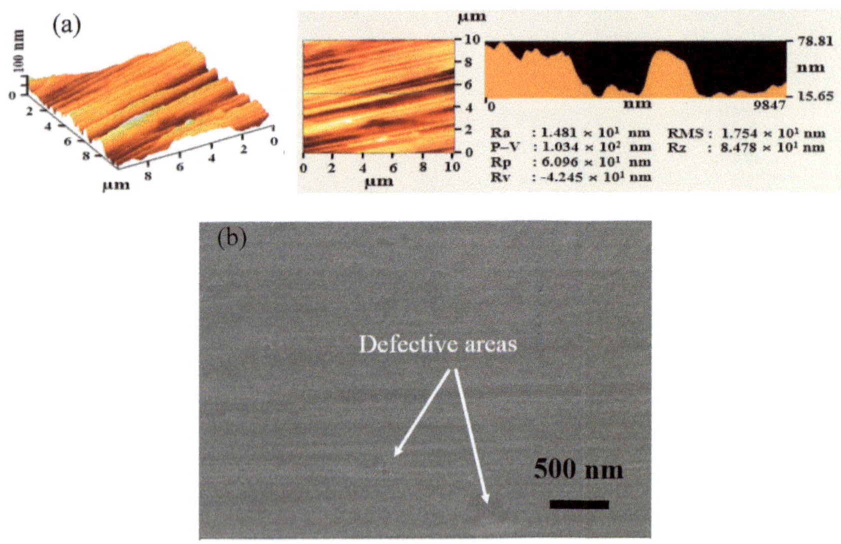

Fig. 4.9 Images of a steel substrate are obtained **a** with AFM and **b** with SEM ([2] [Open access]

In addition to enhancing the sensitivity of the sensor by using silver nanoparticles, studies have also been conducted on the nanostructure of materials used in the creation of complex sensors. For example, SEM is used to study the morphology of GaAs nanowires, which revealed a certain regularity with a diameter of up to 200 nm and length of approximately 8.5 μm. Transmission electron microscopy (TEM) and single-area electron diffraction (SAED) images have confirmed the conformity of crystals with the structure of zinc-doped GaAs and their purity. In addition, two-dimensional GaAs films (2DNLS) obtained by atomic electron microscopy (AEM) have shown a uniform thickness of 18 nm and clean surfaces, indicating the complete elimination of AlAs sacrificial layers. Both GaAs nanowires and regular crystalline characteristics of 2D GaAs layers improve their p-type semiconductor properties, indicating their potential for use in complex sensors. The morphology of GaAs nanowires determined by SEM is shown in Fig. 4.10a. With a diameter, 200 nm and length, 8.5 μm, the nanowires are obviously uniform. The single-zone selected area electron diffraction (SAED) image of GaAs nanowires and transmission electron microscope (TEM) are shown in Fig. 4.10b. With a lattice pitch of 0.325 nm along the (111) planes, the GaAs nanowires clearly have good crystallinity, which is consistent with Zn-doped GaAs. The atomic number electron microscopy (AEM) image of GaAs 2DNLS is shown in Fig. 4.10c. The clean surface of this sheet and its thickness of about 18 nm indicate that AlAs sacrificial layers have been completely eliminated.

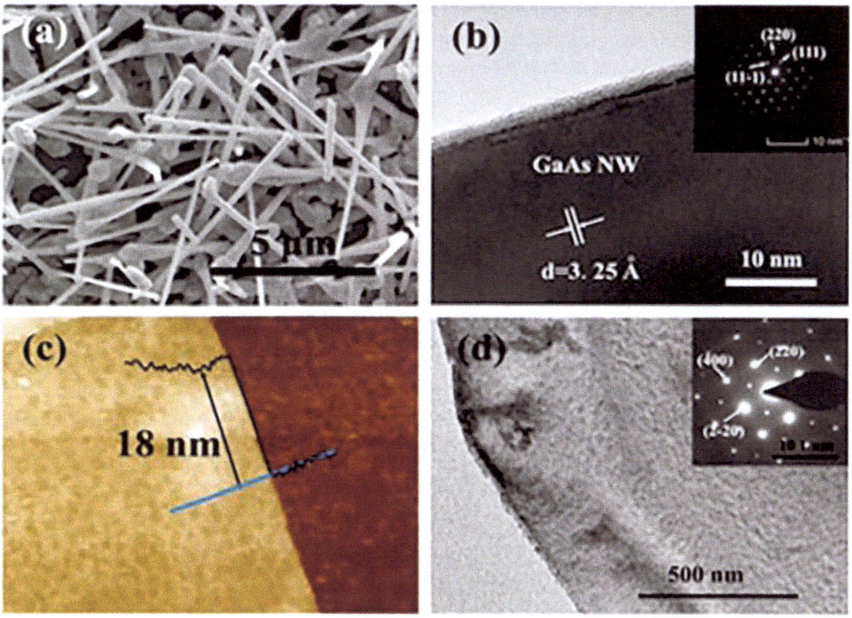

Fig. 4.10 SEM images of the GaAs NW tip are shown in **a**–**c** along with TEM and AFM images of GaAs 2D NLS and GaAs 2D NLS, respectively. The SAED of GaAs 2DNLS is inset into the TEM image [6] [Open access]

As reported by SAED and TEM, GaAs 2DNLS with Zn-doped crystal in Fig. 4.10d shows a constant thickness and good crystallinity. Both GaAs NW and GaAs 2DNLS exhibit p-type semiconductor properties [6].

Over the last 10 years, nanotechnology has introduced innovative nanomaterials such as nanotubes (NTs), quantum dots, nanosilica, nanofibers, nanoparticles (NPs), and nanorods, opening up interesting opportunities in the field of biosensing [14]. Due to many significant advantages over traditional materials, nanoparticles can be integrated.

4.3 A Glucose Sensor Using Gold Nanoparticles

In humans, diabetes is a common health problem. The development of a highly sensitive and selective biosensor for measuring glucose levels in the human body is a critical. The proposed sensor is based on localized surface plasmon resonance (LSPR) technology. The proposed sensor is designed using a tapered optical fiber design. AuNPs or gold nanoparticles are used to improve the sensitivity of the sensor. For this purpose, AuNPs with size, ~10 nm, is synthesized. TEM, scanning electron microscope (SEM), and UV–vis spectroscopy are used to characterize AuNPs. In addition, the enzyme glucose oxidase is used to improve the selectivity of sensor. The selectivity of the sensor is assessed using a variety of analytes, including L-cysteine, urea, cholesterol, galactose, and ascorbic acid, which are commonly found in serum. The potential of AuNP/NG nanohybrid as a useful electrochemical catalyst for glucose sensing is evaluated [14]. The text describes an analytical experiment using amperometric measurement (amperometric response) of the whole material at a concentration of 4.0 mM at a voltage of 0.0 V. The response time is measured and is about 10 times. The level of detection (LOD) is expected to be 12 μm at a noise-to-noise ratio of 3 (S/N = 3). The results are compared with AFM data obtained in conventional 2 μm × 2 μm fields, where the height change (red to white) is shown over the magnification range from 0 to 125. Bearing analysis is used to calculate the surface volume of gold tin oxide nanoparticles and coverage could be calculated (supplementary information). Using procedures 1, 2, and 3, Fig. 4.4a–c illustrates 99%, 99%, and 97%, respectively, coverage of electrodeposited gold nanoparticles. Figure 4.11a–c shows that gold nanoparticle films are obtained on the gold tin oxide and tin surface. The average sizes of gold nanoparticles are obtained using the continuous component method [3].

The AuNP/NG-decorated ITO electrode exhibited anti-interference behavior toward AA (0.05 mM), UA (0.05 mM), and DA (0.005 mM), exhibiting exceptional sensitivity values. The scientists have concluded that non-enzymatic glucose detection by the fabricated electrode is performed at potential, 0.0 V with negligible interference. The amperometric response is examined to evaluate the long-term stability of sensor. After 700 s of operation, current losses can reach 4.5% of the initial data (Fig. 4.11c). Cyclic scanning voltammetry (CV) testing is performed using 4 mM glucose in 0.1 M NaOH pulling the AuNP/NG patterned electrodes onto

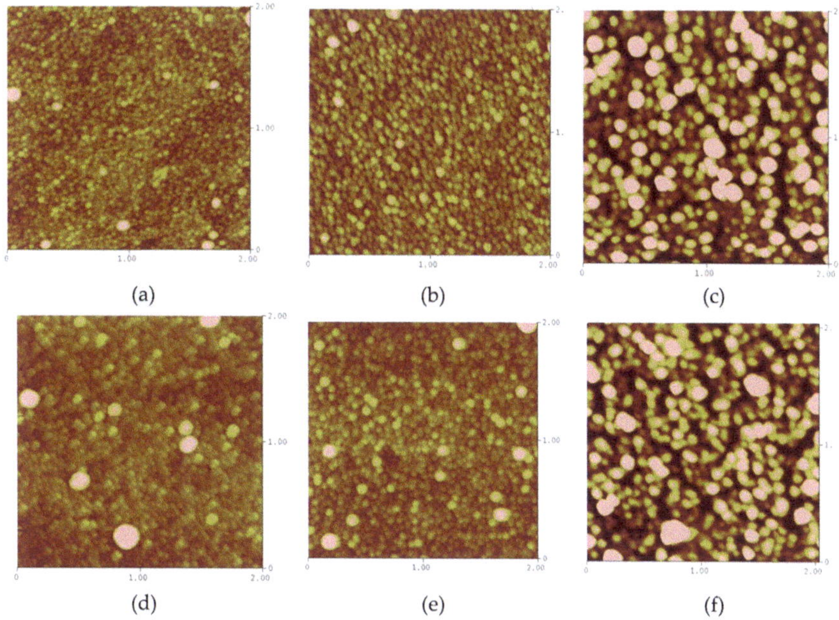

Fig. 4.11 AFM scans of gold nanoparticles electrospun with **d** BSA, **e** GOx, and **f** casein-coated gold nanoparticles using **a** Method 1, **b** Method 2, and **c** Method 3. AFM scan area is 2 μm by 2 μm with a color scale from 0 to 125 nm [3] [Open access]

ITO. After saccharification, 90% of the peak current is measured, indicating good stability of the electrode compared to the initial glucose value, as shown in Fig. 4.11c. When the AuNP/NG-decorated ITO electrode is tested for authenticity, the recovery rate is found to be between 93.1 and 94.6%. The researchers' new glucose sensor can be used to estimate glucose levels in real samples into advanced micro-, nano-, and tiny gadgets. Large surface-to-volume ratio, tiny dimensions, variable behavior, high chemical reactivity, and tunable properties are just some of these characteristics [14]. Nanomaterials represent an intriguing prospect in biotechnology and bioanalytical chemistry due to their unique properties. AFM image of γ-APTES + NPs particles near the tip coated by the spin-impact method and corresponding surface cross-sectional profile are shown in Fig. 4.12a. AFM scan shows that the thickness of γ-APTES + NPs particles coated by the spin-impact method ranges from about 450 to 470 nm [4].

The maximum surface profile is used as a film profile because it is difficult to obtain a smooth film surface profile with large detection thickness using AFM. The AFM image of γ-APTES + NP-coated PSW particles with C-AFM probe as cross-sectional surface profile is shown in Fig. 4.12b. The surface profile of PSW without γ-APTES + NPs coating is shown for comparison. It is found that the film coated with C-AFM tip could have a smooth profile with a thickness ranging from 21 to 23 nm. Metallic poly(NP)s are widely used in many biomedical fields. They are invented

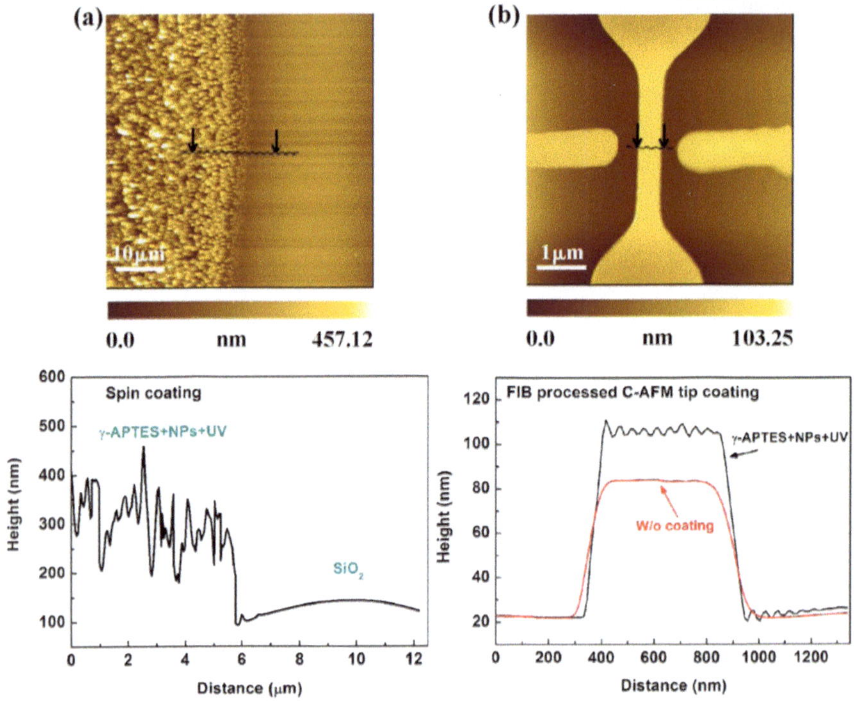

Fig. 4.12 For PSW coated with γ-APTES + NP film, AFM images, and surface characterization are obtained using **a** micropipette/spin-coating and **b** C-AFM tip/coating scanning using FIB [4] [Open access]

long ago in sensor devices a little later due to the increase in the surface area of medical lines and the transfer of electrons from the enzyme to the electrode, which leads to detection results. In addition, the assembly of GOx-bound NPs on electrode surface is facilitated by magnetic properties of MNPs [4]. They can create conductive wires on the surface of electrodes [13]. As the particles develop, the surface area of MPTS-nAuE electrode increases, as glucose content increases. It is believed that the catalytic function of AuNPs could be explained by emerging concept of water oxide/adatom mediator [19]. AC measurement is used to analyze true discrimination and interference using a potentiometric method to determine how gold nanoparticles (AuNPs) affect the protein adsorption capacity of GC electrodes. The binding of pods to AuNP surface is read by applying a higher negative voltage during electrode deposition (as shown in Fig. 4.13). The results show that the voltage drop during deposition is significant and harsh for AuNPs since germination and growth are two fundamental principles that must be met in establishing deposition independence.

The shape of the polyethylene terephthalate (PET) film due to the printing pressure and mechanism is explained in the next issue. The area around the nanopore is subjected to tensile stress, which is known to be due to the transfer printing pressure. A schematic diagram of compression process attributed to the microdisk in a thin

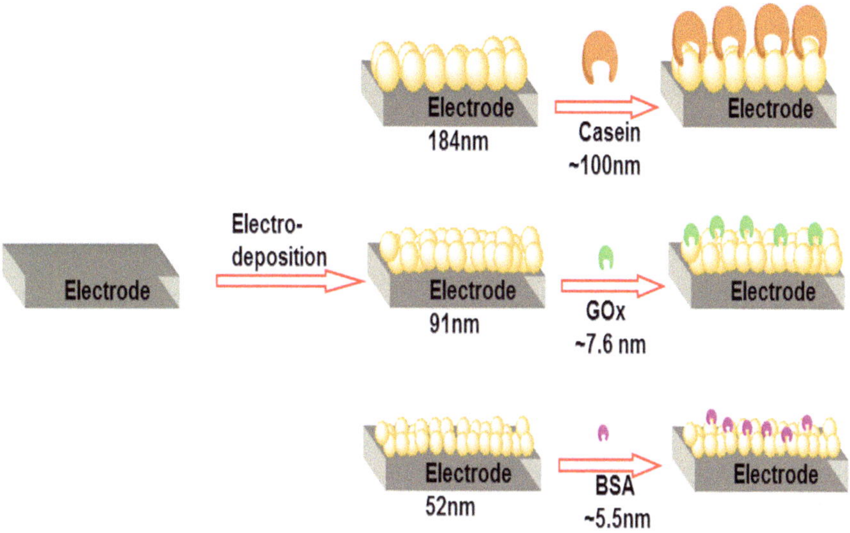

Fig. 4.13 Schematic representation of protein adsorption on AuNP surfaces of different sizes [24] [Open access]

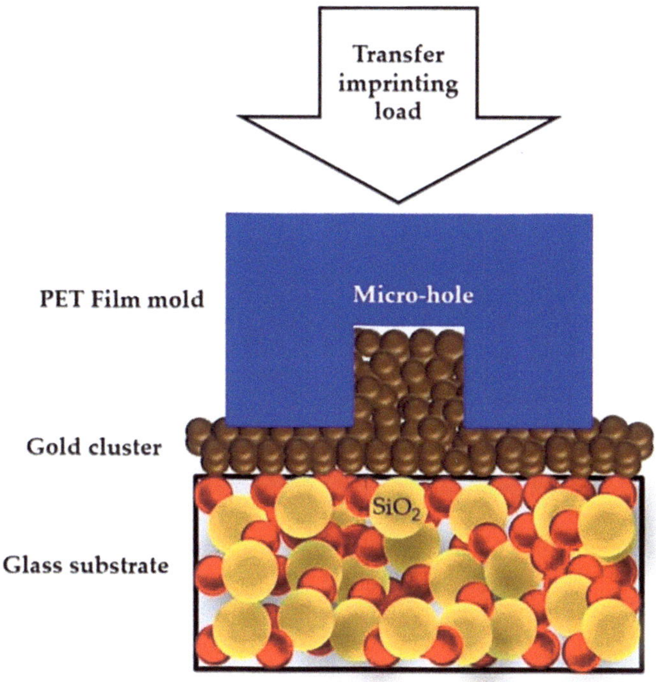

Fig. 4.14 Schematic illustration of the compression mechanism of transfer printing process [25] [Open access]

gold film on substrate as shown in Fig. 4.14. The holes in PET film shape used in this study are about 1000 nm deep.

The thickness of gold coating on quartz glass substrate is 40 nm. 40-nm-thick layer of deposited gold sheet is directly pressed onto 1000 nm deep PET holes. Some of the gold clusters spilled into PET hole chambers after the holes reached a certain penetration depth through the gold layer. As a result, the height of gold layer in cavities has increased steadily during the transfer imprinting procedure. The formation of gold microdisk formations on a glass substrate with a thickness, 50 nm, is verified.

In 2007, several carbon nanotube-nanoparticle composites are used in the following high-throughput electrochemistry. After the launch of carbon nanotube devices, their sensitivity is investigated as a function of the amount of gold nanoparticles (AuNPs) on different tubes carbon nanotubes (CNTs). The difference in line width recognition is found between spin-coated and tip-coated synthetic films. Figure 4.15a, b shows the AE microscope images of γ-APTES + NP-coated PSW particles with different sizes after spin-coating and tip-coating after 120 s UV irradiation. Figure 4.15a shows the large solid particles collected in the spin-plunger product, while the differences between ΔJ and surface-to-volume ratio in Fig. 4.15 show the different feature sizes. As for tip-coated film, many small clusters are uniformly distributed on the surface of PSW as shown in Fig. 4.15b. The range of root mean square deviation of surface roughness of UV-modified PSW nanocomposite with different γ-APTES + NP values. The channel width ranges from 1.67 to 1.75 nm [4].

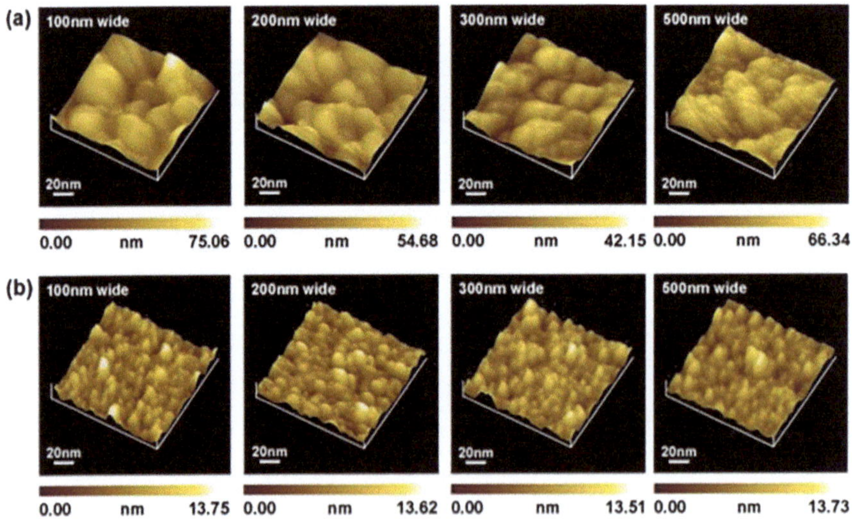

Fig. 4.15 Using micropipette/spin-coating and scanning/probe-coating C-AFM, AFM images of surface morphology of γ-APTES + NPs deposited on PSW surface with 100 nm, 200 nm, 300 nm, and 500 nm are obtained under UV illumination for 120 s [4] [Open access]

The reason for the increased sensitivity is an increased rate of electron transport from GOx to the electrode used for sensing. Figure 4.15 depicts the production process of AuNP-CNT composites. Step 1 involves the creation of an AuNP-CNT composite for which the AuNP surface is modified using cysteamine based on strong thiol–gold interactions. In step 2, the surface of the composite became positively charged due to the amino group in cysteamine, which attracted the GOx core at neutral pH. Step 3 consisted of spreading the modified electrode on a polished graphite disk and drying it at 4°C. Finally, the fourth part with full coverage of the electrode is shown in [8]. Increased electrocatalytic activity is demonstrated by measurements of peak anodic current for glucose oxidation. The electrochemical reaction of a created electrode is further studied using a solution containing 5 mmol/L glucose/0.1 mol/L NaOH at a scan rate of 0.002 to 0.100 V/s (Fig. 4.15). The anodic potential is switched to a more positive value when the anodic and cathodic peak glucose currents are increased progressively with an increasing scanning speed. The peak current is dependent on the scanning speed. The anodic peak is coupled to obtain linearity [4].

Use the best-fitting standard error band (inset in Fig. 4.15), correlation coefficient R2 = 0.996 and linear regression to determine the current value and square root of scan rate. In another important work, a technique is developed in which scientists have used covalent bonds between GOx and AuNPs to create a modified gold electrode as a glucose sensor [9]. At an actuation potential, 0.3 V compared to saturated calomel electrode (SCE), the fabricated nanoparticle-based enzyme sensor exhibited outstanding electrocatalytic performance, and fast response sensitive to changes in glucose concentration, achieving 95% of steady-state current in just 8 s (Fig. 4.16). Current time course data on sequential changes in glucose concentrations indicate the same.

The stability of the fabricated glucose sensor is verified by monitoring the drop in voltammetric current occurring during a constant potential operating cycle. When the sensor is exposed to 30 voltage cycles ranging from 0 to 0.50 V at scan rate, 10 mV/s, no decreasing in catalytic current (Icat) is observed. This indicates the stability and durability of nanoparticle-based enzyme sensor as it maintains consistent electrocatalytic performance even after repeated voltage cycling. The lack of degradation in the catalytic current suggests that the sensor's electrocatalytic efficiency remains high, further reinforcing its reliability for long-term use in glucose-sensing applications. The team stored the enzymatic sensor in 0.1 M phosphate buffer solution at neutral pH of 4 °C and measured glucose concentrations every 2 to 3 days to assess the lifespan of sensor. The current sensor's response to glucose remained virtually unchanged for the first 30 days, and after 2 months, the initial response remains at 70%. Therefore, maintaining GOx activity can be quite effective using AuNP-modified electrode. The excellent biocompatibility of AuNPs and strong covalent cross-linking may be the reasons for storage stability. The simulation results of channel electric field strength at a drain-to-source voltage, 5 V, are shown in Fig. 4.16 for (a) 100 nm, (b) 200 nm, (c) 300 nm, and (d) 500 nm PSWs. As shown in Fig. 4.16e, the channel electric field strengths for 100 nm, 200 nm, 300 nm, and 500 nm PSWs are approximately 0.47 mV/m, 0.35 mV/m, 0.28 mV/m, and 0.23 mV/m, respectively. The channel electric

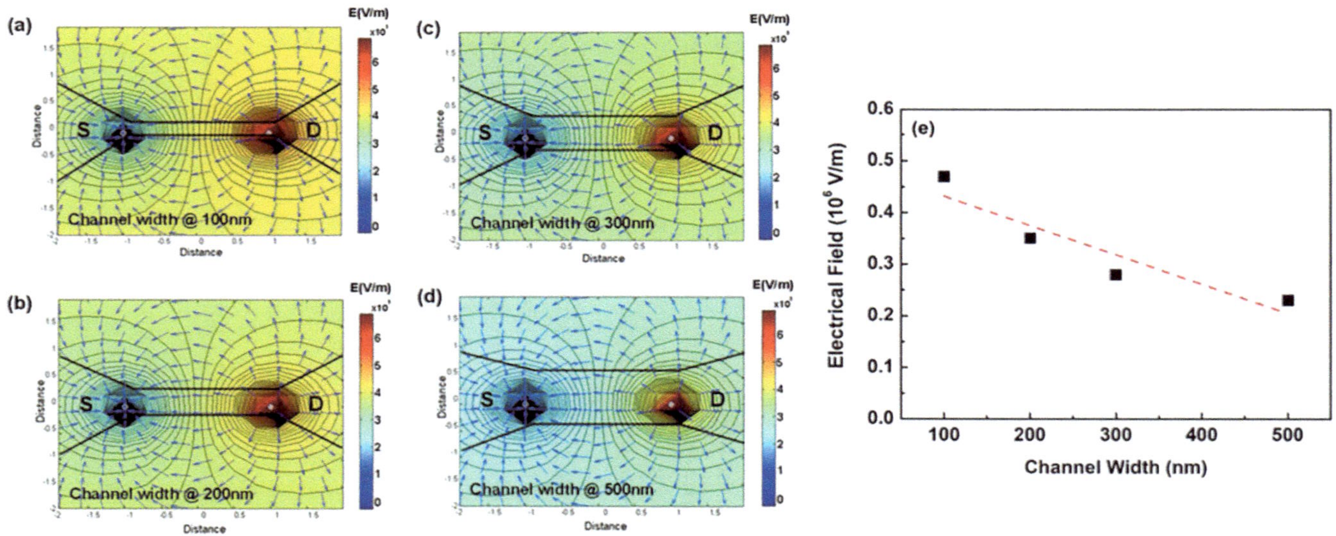

Fig. 4.16 For PSWs, the channel electric field simulations are performed with channel widths of 100 nm, 200 nm, 300 nm, and 500 nm. The electric field strength of corresponding simulations is linearly fitted in (e) [4] [Open access]

field strengths of the PSWs are in the following order: E > E(200 nm) > E(300 nm) > E(500 nm) for each wavelength [4].

4.4 Glucose-Sensitive Copper Nanoparticles

Diabetes is one of the most unpredictable diseases in modern medicine. For continuous diabetes detection, a wide range of low-cost, highly sensitive, and fast-response analyte sensors are available on the market [8]. Depending on the detection method and enzymes used, there are two types of glucose sensors: enzymatic and non-enzymatic. Due to its high selectivity, low detection limit, and high sensitivity, glucose oxidase has been used in the development of glucose biosensors over the past 10 years. This explains the development of nanomaterial-based glucose biosensor, which forms a glucose bioelectrode by coating zinc oxide and metal oxide. However, their reliability is compromised due to a lack of stability caused by the enzyme's inherent sensitivity to pH and temperature. In this regard, there is an increase in the creation of enzyme-free glucose biosensors. The main source of raw materials used in the construction of bioelectrodes for non-enzymatic communication with others is the use of various metal oxides as a sensing layer in their senses. These electrodes are composed of a variety of nanomaterials, including aluminum-based nanodots used as a sensing layer on a fluorine-doped tin oxide (FTO) substrate [7, 9, 10]. The development of metal oxide sensors with nanostructures that act as a bridge between the electrode and enzyme has attracted the interest of many experts. Copper has a small band gap of 1.2 eV [10]. The oxide has received much interest for using in various possible applications such as catalysts, chemical and physical sensors, biosensors, temperature superconductors, and optical and physical sensors [11]. According to some studies, the formation of CuO nanoparticles is considered an attractive material for glucose sensing because it directly electrooxidizes glucose through repeated electrooxidation on the surface oxide layer. Studies of CuO nanoparticles and their mechanical, chemical, and physical characteristics are covered in works [11, 12, 14]. Fine-tuning the characteristics of nanoparticles is necessary to study their morphological and structural characteristics. CuO nanoparticles can be prepared in a variety of ways, and their morphology and shape have a significant impact on the electrochemical performance of this type of glucose determination [19]. The hydrothermal approach for synthesizing CuO nanoparticles gave better results with a wide detection range, low detection limit, and high sensitivity. SEM images shown in Fig. 4.17 present copper prepared by direct magnetic sputtering on a glassy carbon (GC) substrate at pressure, 4.0×10^{-2} mbar, in an argon environment. SEM examination from above shows that the resulting deposits have a nodular shape. The characteristic columnar shape is visible in the cross section. This shape can be assigned to the first region of theoretical model (T/Tm = 0.22 and 30 mTorr) based on Thornton's classification. This region is defined by open boundaries that define the columnar grains [26, 27]. According to graph in Fig. 4.17d, e, the length of copper columns formed under the applied preparation conditions is 2.52 μm ± 0.23 and average diameter is 121.0 nm

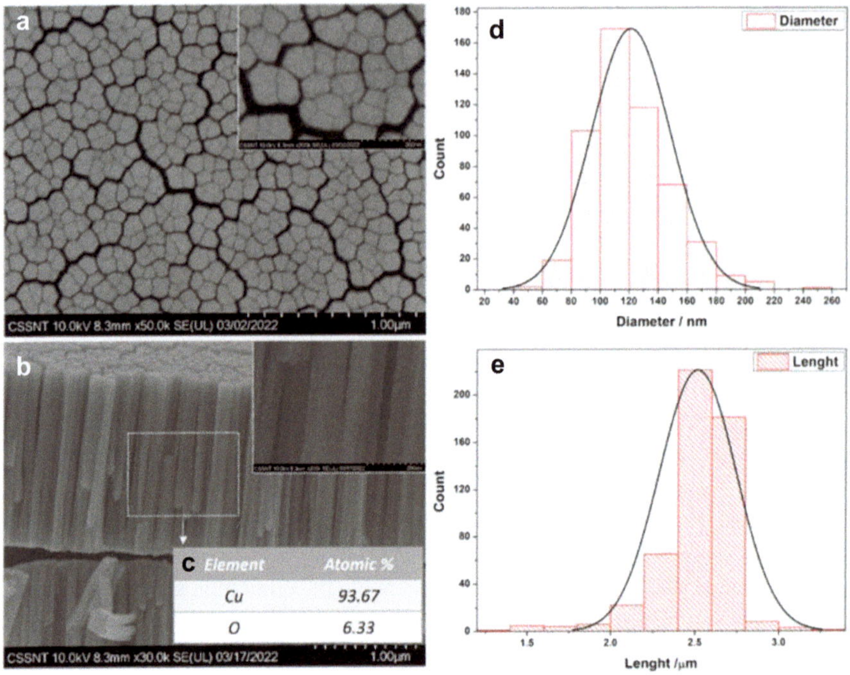

Fig. 4.17 Elemental analysis, top view, cross section, and corresponding diameter and length distribution of CuNC in SEM images (**a**), (**b**), and (**c**) [5] [Open access]

± 27.2. A top-down study of copper pillar dimensions is conducted, Fig. 4.17, which shows a correlation between the pillar dimensions measured in cross section and the diameter dimensions examined in SEM images from above [5].

This material is found to be suitable for use in non-enzymatic glucose biosensors. It has created copper nanoparticles using three different chemical synthesis methods to take advantage of the non-enzymatic glucose biosensor and compare catalytic and enzymatic activities; one approach is based on glycine [7], another is based on hydrazine, and third is based on colloidal solution combustion technology [10, 12] and combustion of solution [14]. So, the optical, morphological, structural, physical, and chemical properties of fabricated nanocomposite electrode are investigated using various characterization analyses. Figure 4.17 at a scan rate of 25 mVs-1 shows an electrochemical behavior of both modified and unmodified CuNPs using potential sweep method. The anodic peaks of the cyclic voltammograms (A1, B1, A2, B2) show how Cu1 + and Cu_2 + are formed, while the cathodic peaks (C1, D1, C2, D2) show how Cu^{2+} is converted to Cu1 + and Cu1 + to Cu0.

Copper oxide nanoparticles are produced in alkaline environments and have diverse shapes, from floral to peony like structures which are synthesized and fabricated using an alkaline environment [14, 19]. More recently, a facile approach is used to synthesize copper nanoparticles with microstructures based on nanoflowers and

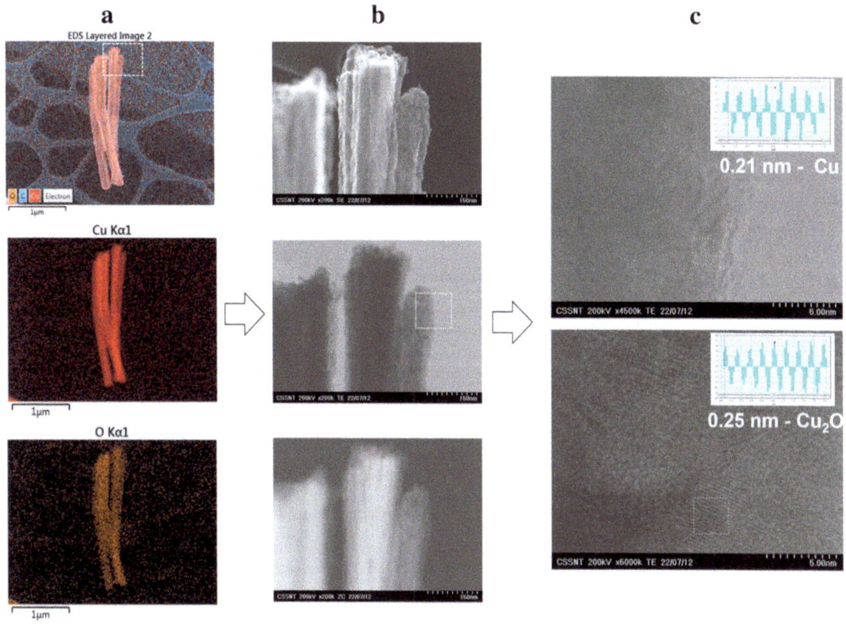

Fig. 4.18 Elemental analysis (EDX); **b** SEM, TEM, ZC studies; **c** interplanar spacing of 0.21 and 0.25 nm attributed to Cu and Cu2O, respectively, HR-STEM micrographs of Cu–NCs [5] [Open access]

Cu_2O precursor. To study the synthesized materials in more detail, high-resolution scanning transmission electron microscopy (HR-STEM) images are obtained as shown in Fig. 4.18. The inter-plane spacings of 0.21 and 0.25 nm are assigned to (111) planes of Cu and Cu_2O, respectively, and columnar structures are crystalline [5]. It is important to note that low oxygen concentration, 6.33%, indicates that NCs have only a modest level of oxidation. In an argon environment with less than 2 vpm oxygen, columnar structures are obtained [27–29].

The SEM image of strip electrode is shown in Fig. 4.19a; the line length across the strip width is approximately 1.021 μm. The presence of gold layer is verified by CV in sulfuric acid, which showed typical anodic (or formation of oxides on the surface) and cathodic (or removal of oxides from the surface) scans [24]. The oxidation peak at 0.9 V can be detected against platinum RE electrode on the strip in Fig. 4.19b, which shows a typical voltage trace recorded on a bare Au electrode. Furthermore, the behavior of the microelectrodes in the presence of probe electrode (1 mM FcCOOH) is investigated by CV. The obtained CVs, which arise due to radial mass transfer to the electrode and exhibit quasi-static behavior (Fig. 4.19c), indicate ultra-thin microelectrodes [24].

These predictions are consistent with other previously published articles [17]. The process involves modifying a glassy carbon electrode (GCE) with carbon nanotubes (CNTs) followed by electrodeposition of copper nanoparticles (CuNPs) dispersed

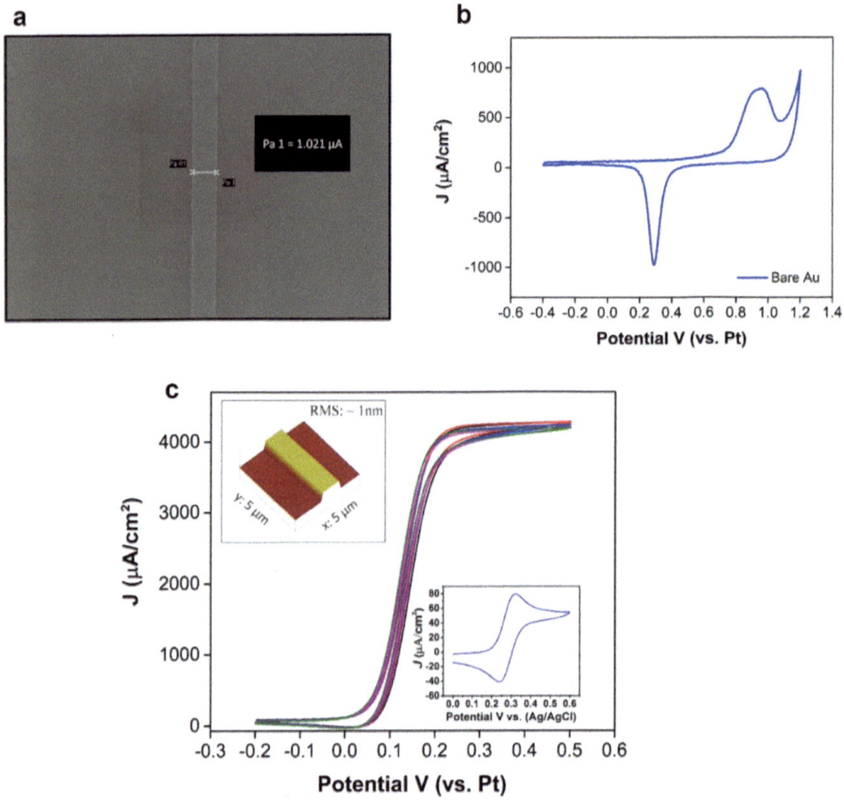

Fig. 4.19 **a**, **b** SEM image of 1.021 μm cyclic voltammograms obtained in 0.1 M H_2SO_4 at scan rate, 50 μV s-1, and **c** cyclic voltammograms obtained in 1 mM FcCOOH for n = 5 electrodes on a single slide (RSD < 0.1%), with an inset graph showing the cyclic voltammograms obtained on a large-scale gold electrode. The inset shows 3D AFM image of 5 μm × 5 μm region at scan rate, 0.7 Hz with background subtraction applied to image [24] [Open access]

in chitosan (CHI). This setup aims to improve the electrochemical stability of the electrode.

DMH, a chelating agent (possibly dimethylhydrazine or a similar compound), exhibits a strong tendency to bind to copper ions as shown in the cyclic voltammetry plots. This chelation affects the electrochemical behavior of copper ions by influencing their redox processes in the system.

To prevent copper ions from leaking out during repeated electrochemical cycles, DMG (dimethylglyoxime) is used. DMG binds to copper ions, stabilizing them in the oxidized form. This prevents copper ions from leaking out of the electrode and allows the system to maintain a robust electrochemical performance for 500 cycles over a voltage range of −1.0 to 0.8 V.

GCE, due to its high conductivity, stability, and wide voltage window, is ideal for electrochemical experiments. Carbon nanotubes increase the surface area and conductivity of GCE, providing a better platform for the deposition of copper nanoparticles. The copper nanoparticles are dispersed in CHI (most likely chitosan, a biopolymer), which acts as a dispersing agent and a stable, biocompatible material for the electrode surface. Together, this combination of materials ensures the stability and long-term performance of the electrode in electrochemical applications, with DMH influencing the behavior of copper ions and DMG preventing ion leakage during long-term cycling. In this chapter, the fabricated electrode is referred to as Cu/Ch/CNT. The electrochemical performance of electrode is tested using a current–time curve and electrochemical impedance spectroscopy (EIS). Various types of electrodes have been developed to evaluate electrochemical impedance characteristics.

In the future, the scientists will focus their efforts on detecting tiny concentrations of glucose and implantable miniature, subcutaneous, and wearable information sensors that can be placed inside the body. As a result, the researchers are faced with numerous obstacles that may arise shortly. In parallel with the efforts to prepare these sensors for continuous glucose monitoring, there will be a search for new materials that have excellent selectivity, low detection limit, wide detection range, and fast response time. Under certain changes, the metal nanoparticles can exhibit glucose concentrations ranging from micromolar to nanomolar. The researchers have examined the stability of sensor and influence of interfering species. Several fabricated sensors have demonstrated excellent stability and anti-jamming properties. It is possible to continue the development and testing of this kind of sensor for any danger to the human body. Overall, metal nanoparticles, which can improve existing technologies and reduce the cost of sensors, are the way of future for glucose detection [6, 15, 21, 24, 30].

As can be seen from the above description, metal nanoparticles have a wide range of applications in glucose determination. Many nanomaterials have been studied as glucose sensors as technology has advanced, without the intervention of other species. Compared to bulk metal, metal nanoparticles have a significantly larger surface area, allowing them to offer a larger number of active sites for GOx immobilization and high electron transfer rates. Amperometric sensors are critical for glucose detection. As more as people have diabetes, the researchers are finding this area very attractive [5].

As expected, the glucose-ampere reaction at 0.1 M NaOH, 100 nM is very sensitive with a reaction time of 1 s when glucose is added. The large surface area of CuNP/PEDOT/GO electrode and excellent conductivity may be responsible for the fast diffusion of glucose and ultrafast reaction. The created sensor electrode has demonstrated a linear response to glucose concentration in the range, 0.1–1.3 mM, with correlation coefficient, R2 = 0.9986. The sensitivity of sensor is 911.96 μA mM^{-1}, and detection limit is 47 nM. After 30 days, the sensor retained almost 92% of its original response, demonstrating a high degree of sensor stability. For electrochemical measurements, a nanometer-thick, ~30 nm gold layer is deposited using electron beam evaporation before the fabrication of Cu-NCs using DC sputtering. This is done to improve the adhesion of Cu-NCs to a glassy carbon (GC) substrate in alkaline environments. The EDX diagram shows the presence of a gold layer, as illustrated in Fig. 4.20 [5].

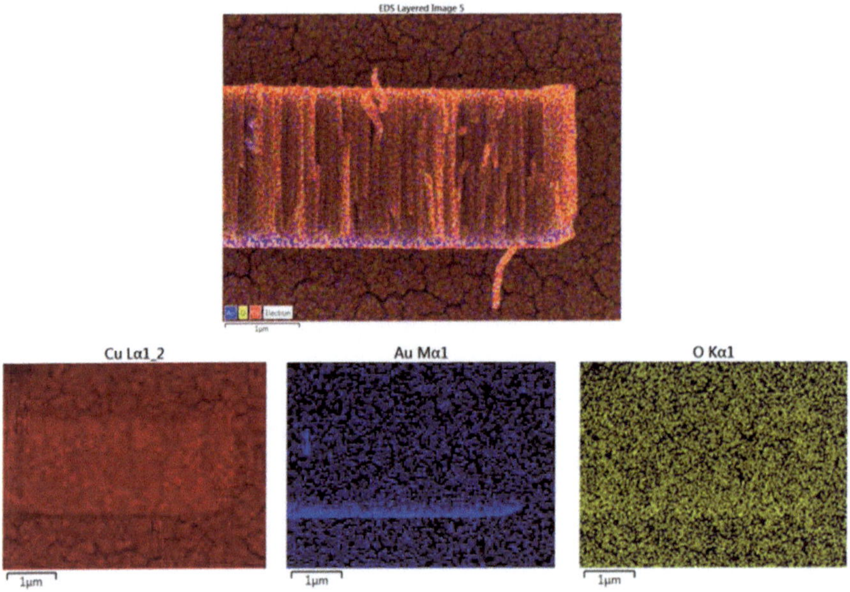

Fig. 4.20 EDX cross-sectional study of Au–CuNC [5] [Open access]

According to the recent studies, noble metal nanoparticles hold great promise for the development of advanced label-free biological and chemical sensors. These sensors exploit the special properties of nanostructured noble metals, including their excellent conductivity, high surface-to-volume ratio, localized surface plasmon resonance, and outstanding light transmittance and reflectivity. These properties collectively enhance the performance of the sensors. The amperometric approach remains the most widely used method for glucose measurement. Alternative measurement capabilities, on the other hand, are provided by localized surface plasmon resonance (LSPR) sensors, which have shown remarkable sensitivity to refractive index changes at metal/insulator interface. In addition, the sensor development capabilities have been expanded by techniques such as attenuated total reflection and fiber-based glucose sensors. This chapter provides a comprehensive summary that includes enzymatic and non-enzymatic sensors based on copper, gold, and silver nanoparticles. This covers the current and future development of noble metal nanoparticle-based sensors in various application areas, reflecting the current interests and developments in this field.

Problems

1. What is the purpose of the material's metal nanoparticle-based glucose sensor?

2. Enumerate a few significant applications: For the silver nanoparticle-based glucose sensor.
3. How would the nanofiber and characterization of a glucose sensor employing gold nanoparticles work?
4. What prohibits ligament tissue engineering from utilizing nanotechnology?
5. In what ways are the various mechanical strengths of the glucose sensor utilized in nanoparticles applied?

References

1. Kuppusamy, P., Kim, S., Kim, S. J., Park, M., & Song, K. D. (2023). Sedeveria pink ruby extract-mediated synthesis of gold and silver nanoparticles and their bioactivity against livestock pathogens and in different cell lines. *Antibiotics, 12*(3), 507.
2. Khachatryan, H., Lee, S. N., Kim, K. B., & Kim, M. (2018). Deposition of Al thin film on steel substrate: The role of thickness on crystallization and grain growth. *Metals, 9*(1), 12.
3. Chiang, H. C., Wang, Y., Zhang, Q., & Levon, K. (2019). Optimization of the electrodeposition of gold nanoparticles for the application of highly sensitive. *label-free biosensor. Biosensors, 9*(2), 50.
4. Lin, J. J., Hsu, P. Y., Wu, Y. L., & Jhuang, J. J. (2011). Characteristics of polysilicon wire glucose sensors with a surface modified by silica nanoparticles/γ-APTES nanocomposite. *Sensors, 11*(3), 2796–2808.
5. State, S., Enache, L. B., Potorac, P., Prodana, M., & Enachescu, M. (2022). Synthesis of copper nanostructures for non-enzymatic glucose sensors via direct-current magnetron sputtering. *Nanomaterials, 12*(23), 4144.
6. Guo, S., et al. (2021). Fast response GaAs photodetector based on constructing electron transmission channel. *Crystals, 11*(10), 1160.
7. Doshi, J., & Reneker, D. H. (1995). Electrospinning process and applications of electrospun fibers. *Journal of Electrostatics, 35*(2–3), 151–160.
8. Ma, Z., et al. (2005). Potential of nanofiber matrix as tissue-engineering scaffolds. *Tissue Engineering, 11*(1–2), 101–109.
9. Huang, L., et al. (2001). Engineered collagen–PEO nanofibers and fabrics. *Journal of Biomaterials Science, Polymer Edition, 12*(9), 979–993.
10. Geng, X., Kwon, O.-H., & Jang, J. (2005). Electrospinning of chitosan dissolved in concentrated acetic acid solution. *Biomaterials, 26*(27), 5427–5432.
11. Um, I. C., et al. (2004). Electrospinning and electro-blowing of hyaluronic acid. *Biomacromolecules, 5*(4), 1428–1436.
12. Jin, H.-J., et al. (2002). Electrospinning bombyx mori silk with Poly(Ethylene Oxide). *Biomacromolecules, 3*(6), 1233–1239.
13. Augustin, A., Udupa, K. R., & Bhat, K. U. (2019). Characterization of DC magnetron sputtered copper thin film on aluminium touch surface. *Transactions of the Indian Institute of Metals, 72*, 1683–1685.
14. Lee, T. D., & Ebong, A. U. (2017). A review of thin film solar cell technologies and challenges. *Renewable and Sustainable Energy Reviews, 70*, 1286–1297.
15. Kim, D. S., et al. (2022). Development of GO/Co/chitosan-based nano-biosensor for real-time detection of D-glucose. *Biosensors, 12*(7), 464.
16. Hutmacher, D. W. (2000). Scaffolds in tissue engineering bone and cartilage. *Biomaterials, 21*(24), 2529–2543.
17. Lu, L., et al. (2000). In vitro and in vivo degradation of porous Poly(DL-lactic-co-glycolic Acid) foams. *Biomaterials, 21*(18), 1837–1845.

18. Oh, S. H., et al. (2003). Fabrication and characterization of hydrophilic poly(Lactic-Co-Glycolic Acid)/Poly(Vinyl Alcohol) blend cell scaffolds by melt-molding particulate-leaching method. *Biomaterials, 24*(22), 4011–4021.
19. Siampour, H., Abbasian, S., & Moshaii, A. (2020). Copper columnar nanostructures fabricated by glancing angle deposition as a robust and scalable method for high sensitive non-enzymatic glucose detection. *Applied Surface Science, 518*, Article 146182.
20. Stoica, I., Barzic, A. I., Ursu, C., Stoian, G., Hitruc, E. G., & Sava, I. (2023). Atomic force microscopy probing and analysis of polyimide supramolecular systems for sensor devices. *Sensors, 23*(9), 4489.
21. Kim, J., Jang, Y. J., & Jang, Y. H. (2023). Electrodeposition of stable noble-metal-free Co-P electrocatalysts for hydrogen evolution reaction. *Materials, 16*(2), 593.
22. Sattarahmady, N., & Heli, H. (2012). A non-enzymatic amperometric sensor for glucose based on cobalt oxide nanoparticles. *Journal of Experimental Nanoscience, 7*, 529–546.
23. Yonas, S., Gicha, B. B., Adhikari, S., Sabir, F. K., Tran, V. T., Nwaji, N., & Tufa, L. T. (2024). Electric-field-assisted synthesis of Cu/MoS2 nanostructures for efficient hydrogen evolution reaction. *Micromachines, 15*(4), 495.
24. Madden, J., et al. (2021). On-chip glucose detection based on glucose oxidase immobilized on a platinum-modified. *Gold Microband Electrode. Biosensors, 11*(8), 249.
25. Potejanasak, P. (2021). A study on fabrication process of gold microdisk arrays by the direct imprinting method using a PET film mold. *Crystals, 11*(12), 1452.
26. Abermann, R., & Koch, R. (1980). In situ determination of the structure of thin metal films by internal stress measurements: Structure dependence of silver and copper films on oxygen pressure during deposition. *Thin Solid Film, 66*, 217–232.
27. Hassaninejad-Darzi, K., & Yousefi, F. (2015). Electrocatalytic oxidation of glucose on the modified carbon paste electrode with sodalite nanozeolite for fuel cell. *Iran. Journal of Hydrogolgy Fuel Cell, 1*, 47–58.
28. Liu, X., et al. (2021). One-pot synthesis of uniform Cu nanowires and their enhanced non-enzymatic glucose sensor performance. *Journal of Materials Science, 56*, 5520–5531.
29. Medhi, A., & Mohanta, D. (2022). Deciphering highly sensitive non-enzymatic glucose sensor based on nanoscale CuO/PEDOT-MoS2 electrodes in chronoamperometry. *ECS Advces, 1*, Article 046504.
30. Reitz, E., et al. (2008). CuO nanospheres based nonenzymatic glucose sensor. *Electroanalysis, 20*, 2482–2486.

Chapter 5
Nanocoatings: Design, Development, and Application

"Nanocoatings: Design, Development, and Application" focuses on creating coatings with improved performance through the use of nanoparticles. Nanoporous coatings, which are ultra-thin layers of nanomaterials, offer advantages over conventional coatings, including improved wear and corrosion resistance, antibacterial properties, and improved functional characteristics of surfaces. Design: Consists of selecting the appropriate nanomaterials and formulations to ensure the desired results. This requires a thorough understanding of how particle shape and size affect overall performance. Development: Involves setting up efficient manufacturing procedures to consistently create high-volume nanocoatings. The focus here is on coating quality assurance, stability, and preparation methods. Application: Nanoporous coatings improve product performance and extend product life in a variety of industries such as electronics, energy, construction, and healthcare. Nanoporous coatings are becoming increasingly important in technological and industrial advances due to their special properties that open up new possibilities for improving the performance and protection of materials.

5.1 Introduction

A layer of one substance applied on top of another, called a substrate, is a coating. They act as an intermediary between the substrate and the environment and are mainly used to protect materials from deterioration caused by external influences. Coatings also offer a decorative component [1].

Nanocoatings are coatings that have at least one dimension where the particle sizes range from 1 to 1000 nm. These coatings are more resistant to wear as they are stronger and more durable than conventional coatings. The bulk properties of substrate are enhanced by their additional characteristics, which include antibacterial effects, stain and wrinkle resistance, hydrophobic and hydrophilic behavior, UV protection, and antistatic properties [2, 3].

R. A. Al-Samarai and Y. Al-Douri, *Engineering of Nanomaterials*,
https://doi.org/10.1007/978-981-95-5499-7_5

The liquid-phase method and vapour-phase method are the two main methods used to create nanocoatings. Chemical vapor deposition (CVD), laser ablation, vapor condensation, plasma arc, and flame fusion are examples of phase approaches. Sol–gel, precipitation, electrolysis, microemulsion, and hydrothermal processes are used in liquid-phase methods.

It is hypothesized that other capabilities, including electronics or anti-infective applications, can be incorporated into composite nanomaterials through bio-inspired production of soft nanocoatings. This is focused on two methods for producing protein-based and wax-based nanocoatings made of retinal or retinal proteins, metalloproteins with antiviral and antibacterial properties [1]. CVD, laser ablation, vapor condensation, plasma arc, and flame welding are examples of vapor-phase processes [4–7]. Furthermore, it is postulated that our bio-inspired synthesis of soft nanocoatings can serve as a basis for adding numerous other capabilities, such as electronics or anti-infective applications to composite nanomaterials (Fig. 5.1).

To stop corrosion and erosion problems, nano-coatings are used in industrial rolls, turbines, hydraulic shafts, boiler tubes, pumps, landing gear, and aircraft engines. They are used for decorative purposes on cosmetics, watches, pencils, and cars. Banknotes have a nano-coating that prevents counterfeiting. This chapter discusses tannins in detail. In addition, an attempt has been made to combine different processing methods used in their manufacture. Impact coatings are defined as a coating of one material over another, commonly referred to as an underlay. They are mainly used on materials to protect them from degradation caused by external factors. They serve as a mediator between the environment and the environment substrate. In addition, they serve as a decorative element.

Sol–gel, precipitation, electrolysis, microemulsion, and hydrothermal processes are examples of liquid-phase methods. To stop corrosion and erosion problems, nanocoatings are used in industrial rolls, turbines, hydraulic shafts, boiler tubes, pumps, landing gear, and aircraft engines. Also, they are used for decorative purposes on cosmetics, watches, pencils, and cars. Banknotes have a nano-coating that prevents counterfeiting.

5.2 Nanocoatings and Coatings

Starting with the basics of nanotechnology, it covers several methods for creating and manipulating nanostructured films, with an emphasis on size effects at the nanometer scale. This chapter has reviewed methods for fabricating nanostructured films and provided detailed reviews of several methods with ample examples. This is divided into sections that include topics such as nanotechnology, thin films, size effect, nanoparticles, sol–gel method, and applications of nanostructured coatings [8–10]. Next, the strategy of encapsulating metal nanoparticles inside retinal nanocoatings is tested before their self-assembly.

Compared to traditional coatings, nano-coatings provide better wear resistance due to lower porosity concerning the substrate and increased impact strength and

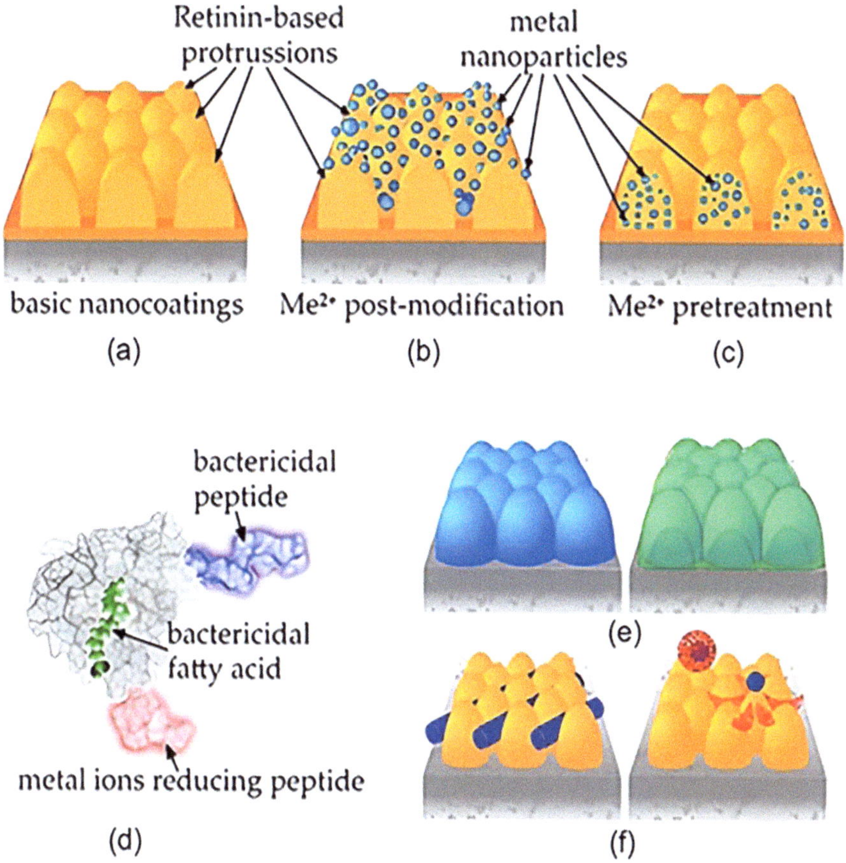

Fig. 5.1 Illustrates the functionalities of bioactive nanocoatings. Metal formation in network nanocoatings (**a**) can occur through bioconjugation before the self-assembly process (**b**) or through the incorporation of metal nanoparticles after the nano-coating has been created (**c**). Modifications of fatty acids and key protein components are examples of potential functionalities that can be used to create nanocoatings with improved anti-infective capabilities (**d**). It is also possible to use single or multiple layers of metals or other compounds (**e**) and incorporate other nanostructures such as carbon nanotubes or complex multicomponent nanoparticles (**f**) in applications ranging from electronics to medicine [1] [Open access]

hardness [11]. In addition, they have antibacterial properties, stain and wrinkle resistance, hydrophobic and hydrophilic properties, UV protection, and antistatic properties, which influence the bulk properties of the substrate material [12]. Examples of materials that can be coated with nanoparticles include nano-TiO_2, TiAlN, Ni-ZrO_2, nanosized layered magnesium–aluminum double hydroxides (nano-ALDs), nanostructured Ni-Mo alloys, silver nanoparticles, nanostructured silver-based materials, nano oxide particles, zinc, copper nanoparticles, and carbon nanotubes (CNTs) [13].

Furthermore, it is argued that since the metals would form biocomplexes and be encapsulated in the protein, this strategy could be more environmentally friendly (Fig. 5.2a). So, to choose copper and silver as metals for this method [1].

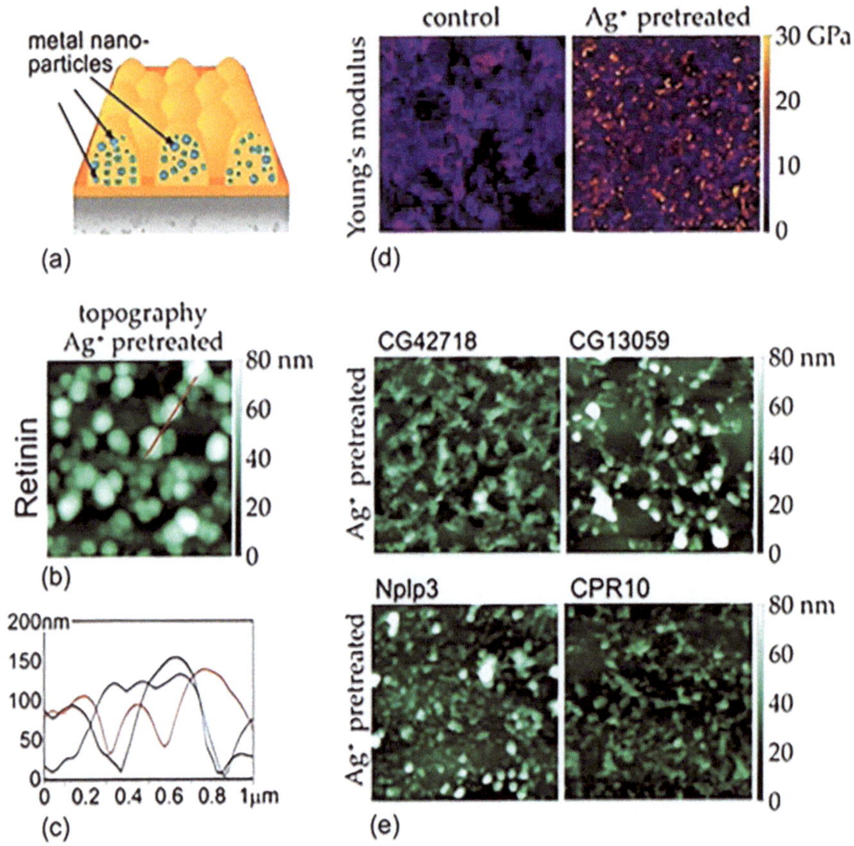

Fig. 5.2 Retinin-based nanocoatings with distinct topologies incorporating silver nanoparticles are fabricated by reinforcing metallic nanoparticles that have evolved before the self-assembly process (**a**). The cross section containing the red line is used to plot the heights. (**c**) Representative heights of the nanostructures are shown through cross sections of the pre-treated silver nanocoatings; three different cross sections are shown in different colors; black and blue are obtained from supplementary images, and red is from the image in (**b**). **d** The pre-treated silver nanocoatings have an intermediate Young's modulus, which is between those of the metallic and protein surfaces. **e** Retinin-like proteins CG42728, CG13059, Nplp3, and CPR10 underlie the topology of the silver-incorporating nanocoatings. The scan sizes in (b, d, and e) are 2 × 2 μm. Color scales showing surface height and Young's modulus are provided next to the images [1] [Open access]

5.3 Methods for Producing Nanocoatings

Biotextile applications such as biofiltration materials, medical clothing, medical textiles, textile implants, textile substrates for cell development, and other textiles for medical systems are increasingly using nanocoatings [4–7, 14]. For example, thin films, nanocapsules, and nanoparticles can form nanocoatings on textile surfaces. Their sizes usually vary from several tens of nanometers to hundreds of nanometers. This supports the theory that metal nanoparticles are hidden inside the retinal nanostructures, and the size of silver nanoparticles in such a composite material is estimated by the area of the regions with the highest Young's modulus, which is 100 nm, smaller than the characteristic size of the nanostructures, about 300 nm. Therefore, the materials are examined using energy-dispersive X-ray spectroscopy (EDS) and scanning electron microscopy (SEM) to confirm these results (Fig. 5.3) [1].

The textile industry is interested in various nanomaterials such as electrospun nanofibers, colloidal suspensions or solutions of nanoparticles, encapsulated nanosystems with controlled delivery (nanocapsules), metal-based nanoparticles,

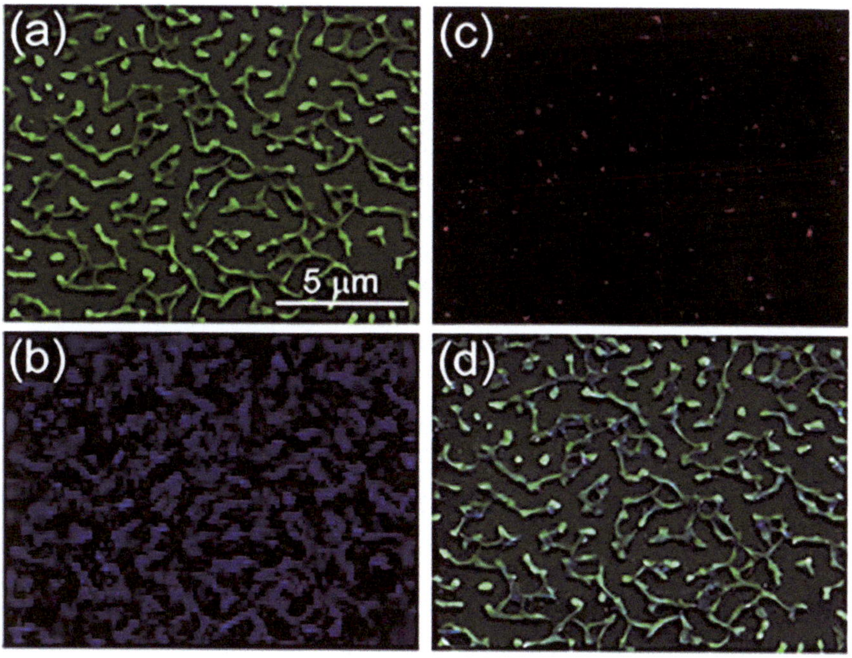

Fig. 5.3 Energy dispersive X-ray spectroscopy (**a**) and SEM of carbon (**c**) and silver (**b**) nanocoatings based on retinal nanoparticles containing silver nanoparticles. **d** Merged panels (**a**–**c**). Note that silver spots are also present in the overlap of carbon signals with the nanostructures. Note the larger scale (15 × 10 μm) compared to the previous SEM images [1] [Open access]

plasma-sprayed coatings, and cold plasma polymerized coatings. For example, electrospinning [15] has been used to deposit polymer nanofibers as a thin, invisible, flexible coating layer on membranes, foams, and nonwovens. Commercial products such as multifunctional filters (for germs, viruses, and ultrafine particles), as well as medical gadgets and fabrics.

5.3.1 Method of Vapor Phase

Surfaces grown on sapphire with M misorientation under our conditions generally showed lower RMS roughness values over 2 × 2 μm square areas compared to A, as can be seen from AFM surface examination in Fig. 5.4. However, as noted previously [16, 17], samples with M misorientation exhibit large-scale waviness, resulting in higher roughness values when measured over larger areas (e.g., 8–9 nm for 1C at 465 μm × 350 μm compared to 2–3 nm for 2C). Metal alloys, amorphous or crystalline materials with different stoichiometries, and elementary and complex semiconductors can be grown into thin films using chemical vapor deposition (CVD), a flexible deposition method. Figure 5.5 illustrates a flowchart of the Vapor Phase Technique. Volatile compounds of deposited material typically undergo a chemical reaction with other gases in CVD techniques to form a non-volatile solid that is deposited at the atomic level on a suitable substrate [2].

In this experiment, the effect of growth conditions, especially ammonia flow, on the surface properties of gallium nitride (GaN) is studied. Two different cases are tested: one with a sapphire substrate at 1150 °C with a nominal growth rate of 0.45 nm/s, and another at 1180 °C with a significantly higher Trimethylgallium (TMGa) TMGa flow. The goal of these experiments is to achieve smoother surfaces, especially under low-misorientation conditions. The ammonia flow is changed for each run, while other parameters are kept constant. The North Pole GaN surfaces at 0.3 M ammonia flow showed significant changes that are visible even to the naked eye. The results showed two clear trends: the HHL (high hexagonal lines) decreased and became shallower at high V/III ratios (greater than 500). This suggests that the ammonia flow plays a major role in determining the surface morphology and crystallinity of GaN, especially under different growth temperatures and flow conditions. These results provide valuable information for optimizing growth conditions to obtain smoother and more uniform GaN surfaces.

The best sample in this series is obtained at the stability edge, even for N-polarized GaN, deep in the absorption-limited growth regime. (Ga-polarized GaN layers are completely absorbed within a few minutes.) Although this is three to four times larger than the average GaN surface roughness of 4–6 nm, the large 13 nm RMS roughness region may still be acceptable for devices. In addition, the surface of this sample shows some evidence of atomic steps. It is observed that only step edges on the low-misorientation N-polarized GaN surface are present. Surfaces such as the one in Fig. 5.6 are typically featureless and have holes about 1 nm deep [2].

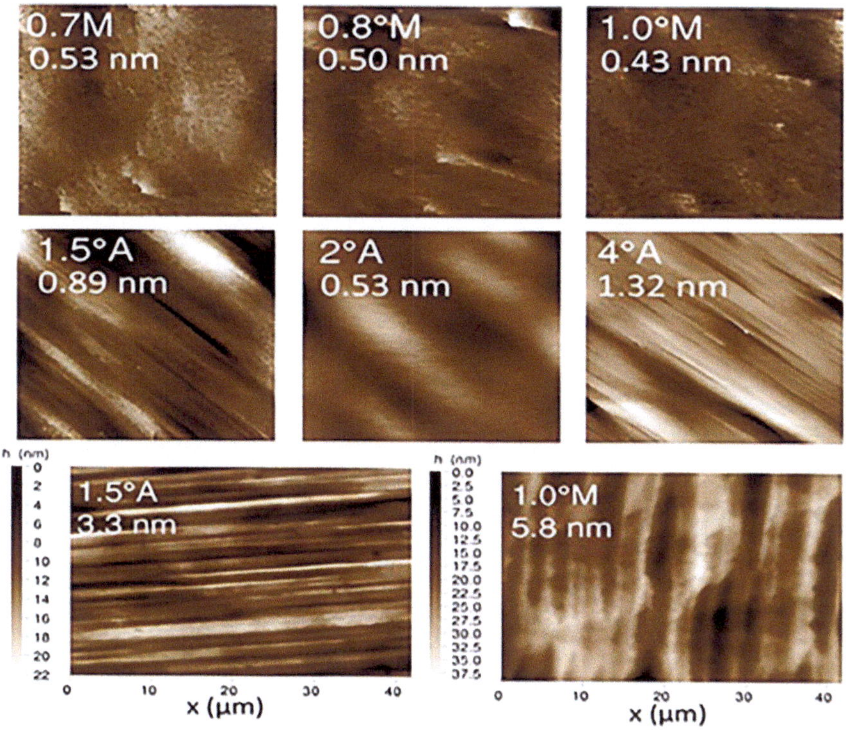

Fig. 5.4 AFM images of GaN growth on sapphire with different misorientations, each measuring 2 μm × 2 μm. The plot shows square roughness values and directional difference. For all images except when the substrate tilt angle used in the experiment is 4 Å (corresponding to 10 nm), the z-scale is set to 4.5 nm. In typical experiments, an angle of 45° is used for flat wafers. The bottom row shows samples with misorientation, 1.5°, and values used to characterize growth (1°m). The 1° sample shows higher square roughness and increased waviness around the step edges [2] [Open access]

The variety of reagents and precursors that can be used to deposit a particular film demonstrates the chemically adaptable nature of CVD procedures. Like physical vapor deposition (PVD), CVD depends on the substrate temperature to influence the film growth rate, while vapor supersaturation affects the film nucleation rate. Together, these two variables influence the degree of epitaxy, grain size, grain shape, and feel. The development of single-crystal film on substrates is facilitated by high substrate temperatures and low gas supersaturation. Low substrate temperatures combined with high gas supersaturation have led to the formation of polycrystalline, nanocrystalline, and even amorphous films. The five main processes in the vapour-phase approach are flame synthesis, laser ablation, plasma arc, chemical and physical vapor deposition, and plasma arc (Fig. 5.6) [9–12].

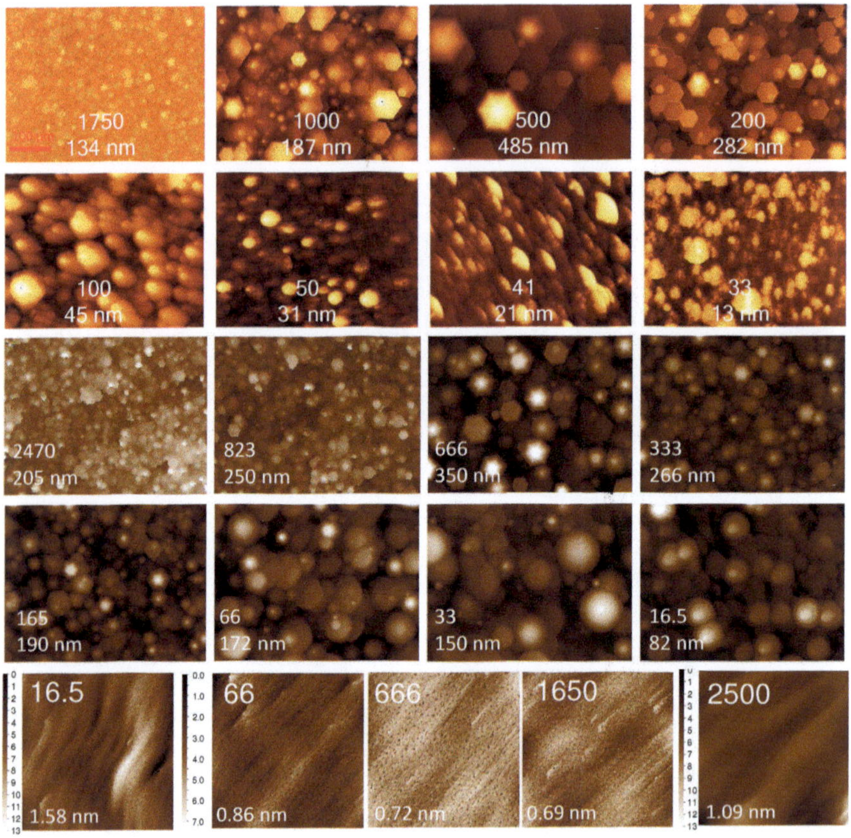

Fig. 5.5 White light holographic surface maps and microscopic images of North Pole GaN grown under different conditions. Top two rows: surface maps of 0.3 M GaN North Pole grown for 30 min at 1180 °C with constant TMGa flux and decreasing NH_3 flux, covering an area of 465 μm × 350 μm. V/III ratio is listed as the first value, and the average RMS roughness under these conditions is shown as the second value. Middle two rows: surface maps of 0.2 M GaN North Pole grown for 45 min at 1150 °C with constant TMGa flux and decreasing NH_3 flux, covering the same area (465 μm × 350 μm). Again, the V/III ratio is listed first, followed by the average RMS roughness. Bottom row: AEM images of 2 μm × 2 μm sample grown for 45 min at 1150 °C on 2° misoriented sapphire at very low, medium, and maximum NH_3 flux levels [2] [Open access]

5.3.1.1 Physical Vapor Deposition

The atomic-level material transfer is a component of the evaporative coating process known as physical vapor deposition, or PVD. The stages that make up this process can be outlined as follows. First, the deposited material is physically converted into vapor (high-temperature vacuum or gas plasma); second, the vapor moves from the source to the substrate into a low-pressure region, and third, the vapor condenses on the substrate, forming a thin layer. As described in several previous examples [18–20], the resulting films are very rough (roughness > 10 nm) with a large number of

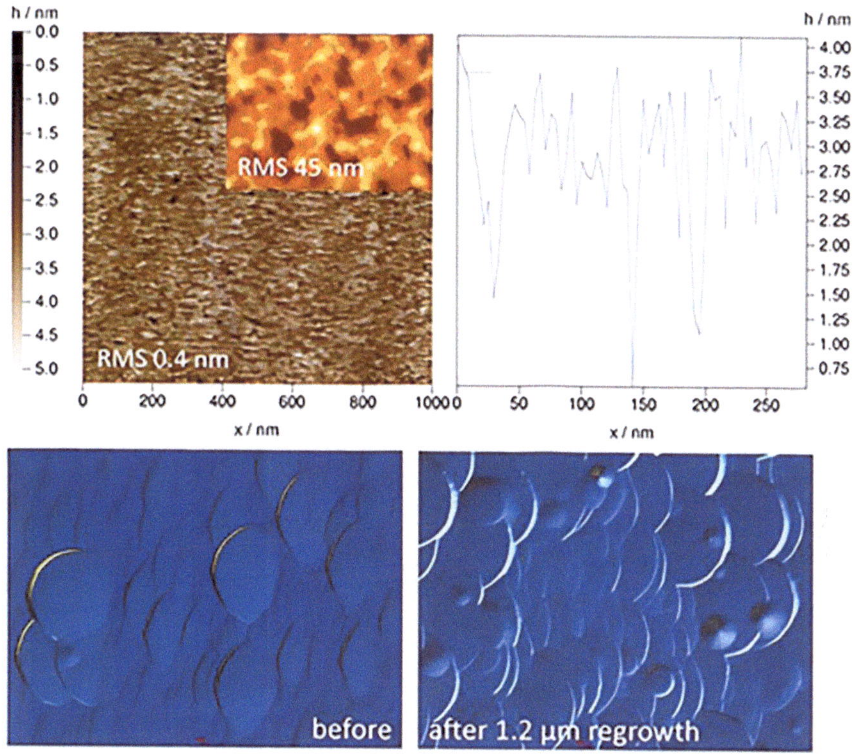

Fig. 5.6 Top row shows an atomic mass micrograph of the surface after two growth steps at V/III 500 and 12 at 1150 °C at 0.2 °C. The inset contains a 3D white light image of the surface measuring 930 × 700 µm. Bottom row: Nomarski images obtained at V/III 1000, before and after re-growth at 1180 °C with 1 µm [2] [Open access]

aggregated particles (Fig. 5.7a–c). This is true regardless of the deposition period. The predominance of adsorbed molecular aggregates in this case indicates the presence of aggregated particles [3].

Typically, films ranging in thickness from a few nanometers to thousands of nanometers are deposited using PVD. However, they can also be used to create free-standing structures, gradient deposits, layered coatings, and particularly thick deposits [21]. Reactive evaporation, ion deposition, electron beam, hot bath evaporation, and other deposition methods are included in the broad category of deposition processes covered by PVD. Processes based on ion beam or plasma sputtering are included in PVD. Other organic molecules (Alq3, NPD, and PCBM) are tested to evaluate the adaptability of the flow-through chemical vapor deposition (FCPVD) method; all molecules showed comparable results (Fig. 5.8). The membranes prepared by FCPVD exhibited ultrasmooth surfaces (RMS roughness less than 0.4 nm) free of noticeable particles or defects, in contrast to the extremely rough surfaces (RMS roughness greater than 10 nm) of the substrate chemical vapor

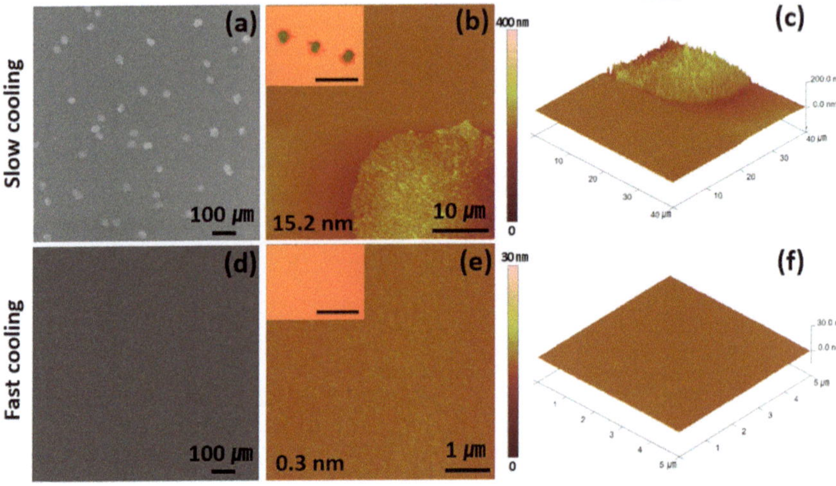

Fig. 5.7 SEM and AFM images of rubrene films as a function of cooling rate. When the deposition time is 100 s, the top row images (**a–c**) and bottom row images (**d–f**) are obtained using SCPVD and FCPVD, respectively. SEM images of each film are included as insets (scale bar: 50 μm) [3] [Open access]

deposition (SCPVD) membranes, which contained a large number of aggregated particles (Fig. 5.8). Furthermore, several substrates, including silica, quartz, mica, and indium oxide, are subjected to the FCPVD process (Fig. 5.8). These results indicate that by suppressing nucleation, the FCPVD approach can be used to create ultrasmooth membranes on a variety of organic compounds and substrates [3].

According to these results, the smooth films are amorphous while the rough films are crystalline (Fig. 5.9). The in-plane scattering vectors are qxy, and the out-of-plane scattering vectors are qz. The rubrene film prepared by SCPVD, a mixture of orthorhombic and triclinic phases, exhibits strong and anisotropic scattering peaks; from large aggregated particles, thermodynamically favorable crystalline forms (Fig. 5.9a) [3]. On the other hand, the rubrene film prepared by FCPVD exhibits broad peaks with anisotropy at q ~ 0.68 Å − 1, indicating the presence of a kinetically favorable amorphous phase (Fig. 5.9b) [3]. The Alq3, NPD, and PCBM films prepared by FCPVD exhibit similar nucleation suppression as rubrene films (Fig. 5.9d–h), whereas the films prepared by SCPVD exhibit sharp and anisotropic crystalline phases as evidenced by scattering peaks (Fig. 5.9c–g). The as-deposited Alq3 and NPD films are reported as thermodynamically favorable α phase and triclinic phase, respectively [4, 22]. In contrast, large scattering peaks are observed in Alq3, NPD, and PCBM films prepared by FCPVD, indicating that the amorphous phase is kinetically favorable (Fig. 5.9d–h) [3]. These results indicate that rapid cooling of the substrate effectively suppresses the nucleation of adsorbed molecules, promoting the formation of ultrasmooth organic thin films [3].

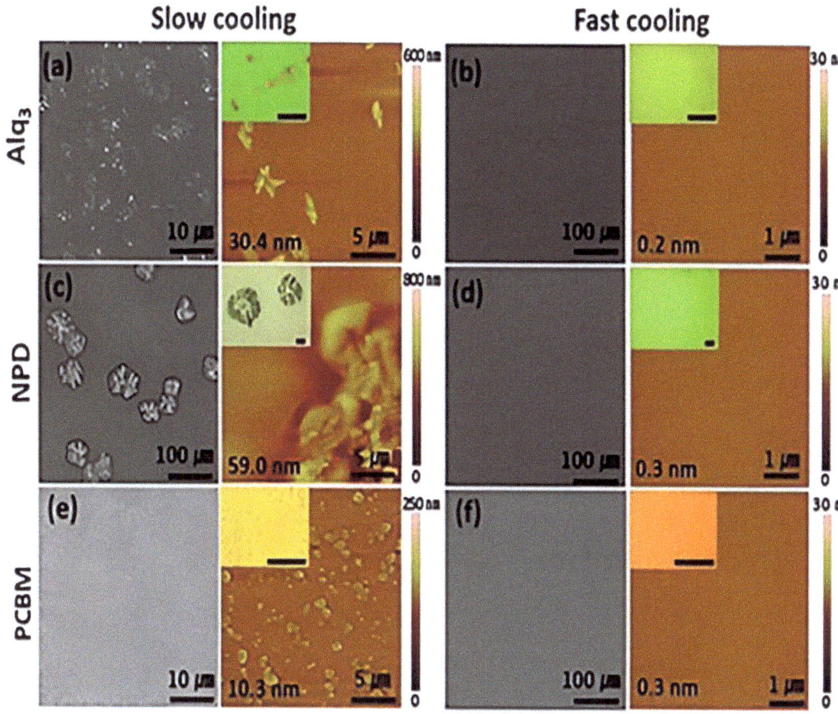

Fig. 5.8 Differential organic films as a function of substrate cooling rate. The following atomic and scanning electron microscope images show Alq3 films (**a**, **b**), NPD films (**c**, **d**), and PCBM films (**e**, **f**) as a function of cooling rate. SCPVD and FCPVD procedures are used to obtain the films on the left side (**a**–**d**) and on the right side (**b**–**f**). The images included with each film are optical microscope images at a scale of 10 µm [3] [Open access]

PVD can also refer to deposition from arc sources, whether filtered or not. This process can be divided into two groups: atomization and evaporation. While sputtering involves the action of gaseous ions (plasma) to dislodge atoms or molecules from a solid target, evaporation describes the heating process used to form thin layers. Both approaches have been expanded to include several specialized strategies [8, 14]. PVD is used to coat a steel substrate with TiN, and it is found that the results are better with nanosized particles than with micron-sized particles.

5.3.1.2 Chemical Vapor Deposition

CVD reactors must be built with the ability to control several aspects of the film, including composition, surface morphology, thickness, and crystal structure. The direction of gas flow to the substrate holder (susceptor) can be used to distinguish between horizontal and vertical systems. Reactors designed for air and low-pressure deposition have been specially designed. Using radio frequency, which is conductive,

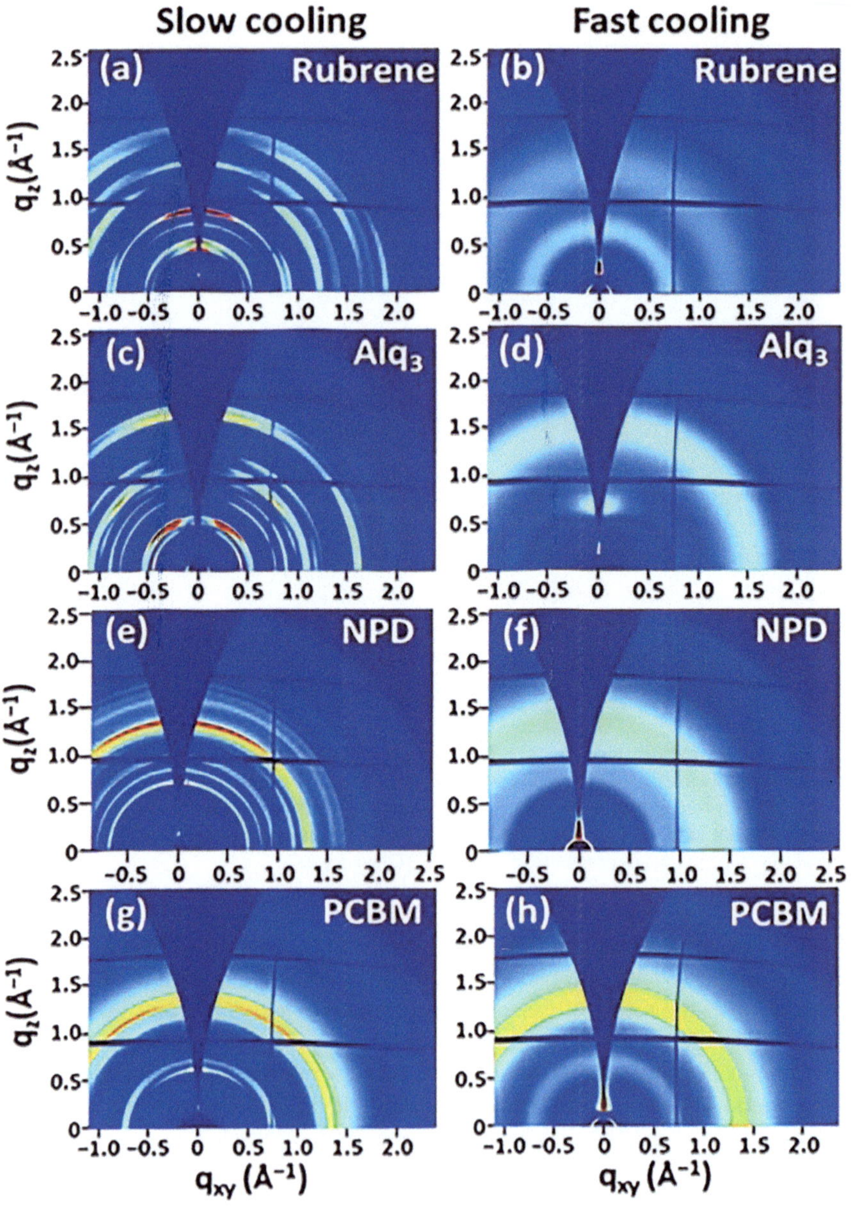

Fig. 5.9 Structural characterization of organic thin films. GIWAXS patterns of films prepared by SCPVD (left column) and FCPVD (right column), including (**a**, **b**) rubrene films, (**c**, **d**) Alq3 films, (**e**, **f**) NPD films, and (**g**, **h**) PCBM films [3] [Open access]

the substrate holder is heated by radiation. Cold- or hot-wall reactors can heat up from the inside out. Thus, the phase composition changes are evident in the layer morphology as confirmed by field emission scanning electron microscope (FESEM) characteristics. The film formed at 800 °C, as shown in Fig. 5.10a, shows the creation of a nanolayer with uniformly distributed grains of hundreds of nanometers in size between rod/platelet-shaped grains of a few microns in size. The inset in Fig. 5.10a illustrates film thickness, 875 ± 25 nm. The sample obtained at 850 °C, as shown in Fig. 5.10b, has a surface characterized by irregular grains. Numerous micron-sized triangular grains can be seen. These triangular helical nanoplatelets, in particular, represent a helical growth [2], most likely associated with the growth of CeO_2 and consistent with the strong CeO_2 reflectance with the same pattern recorded at 28.85°. The cross section (inset in Fig. 5.10b), where the thickness can be estimated as between 700 and 930 nm, shows this rather unusual shape. For the sample obtained at 900 °C, Fig. 5.10c depicts smaller and more uniformly distributed triangular grains [16].

To determine the composition and purity of the $BaCeO_3$ thin films, EDX analysis is performed. The results are shown in Fig. 5.11a, where the spectrum shows the characteristic peaks of oxygen, cerium, and barium elements. In particular, the L lines of Ba show peaks in the range 4.4–5.5 keV, Mα lines of Ce show peaks at 0.88 keV in the range 4.7–5.9 keV, and Kα lines of O show peaks at 0.53 keV. The correct chemical measurements of pure $BaCeO_3$ films with a Ba to Ce ratio of about 1:1 over the entire surface are confirmed by quantitative EDX analysis [16].

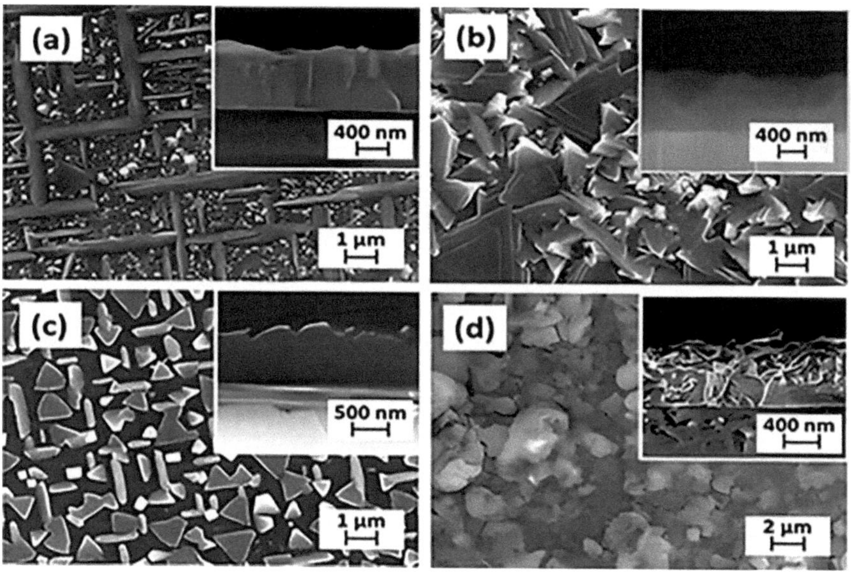

Fig. 5.10 SEM images of Ba-Ce-based Fe-Z thin films, more suitable for extreme temperature ranges: 800 °C, 850 °C, 900 °C, and 950 °C for liquid magnesium (100). This is the hottest thing in the world, according to SEM images of Fe-Zn [16] [Open access]

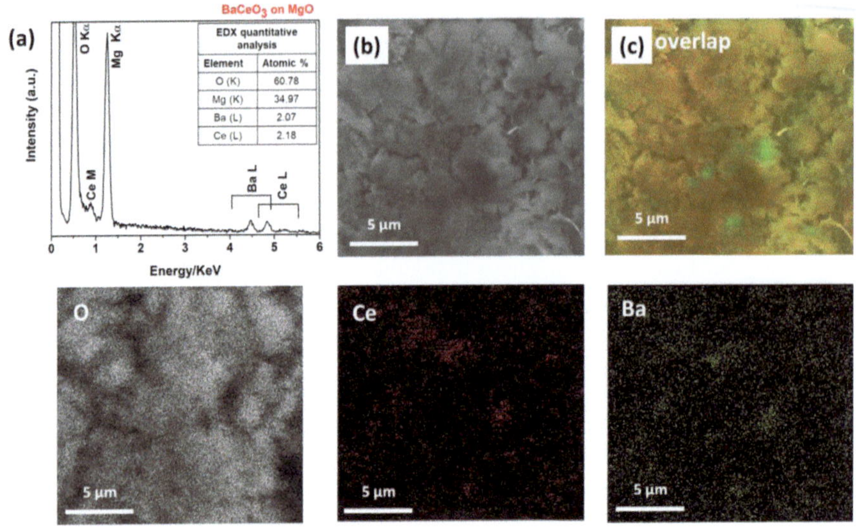

Fig. 5.11 $BaCeO_3$ film deposited on MgO at 950 °C is shown in **a** EDX spectrum, **b** relative FESEM image, EDX maps of O, Ce, and Ba elements, and **c** overlap map image [16] [Open access]

Rotating substrate holders are designed to improve film uniformity and flow dynamics. Large-scale CVD production is possible, so reactors such as barrel reactors, pancake reactors, and spinning disk reactors have been built that allow simultaneous deposition onto multiple substrates. The selected deposit is created on the surface of the substrate as a result of the reaction or destruction of one or more volatile precursors when the substrate is exposed to them by CVD (Fig. 5.11). In addition, volatile by-products are often formed and are ejected from the reaction chamber by the gas stream [8]. System conditions play an important role in determining the process's success. During this procedure, the temperature in the area varies from 800 to 2000 °C. An excellent adhesion and coating thickness of up to 20 μm are achieved [10, 15]. Characterization techniques such as SEM and XRD have concluded that the pore surface characteristics changed as a result of CVD nano-coating.

These results, which show larger grains for the pure film formed on MgO, are consistent with what is seen in FESEM images and can be attributed to the special nature of the substrate. Since the former may be the result of the coalescence of several crystals, it is important to note that FESEM grain sizes are much larger than crystal sizes. The pattern shown in Fig. 5.12a, b shows the formation of a uniform layer over a large area with rectangular grains of 50–100 nm in size and visible nanoscale features. The film grows with well-defined and regular grains on the YSZ(100) substrate, in contrast to the similar undoped film deposited on MgO(100) (see Fig. 5.12d) [16].

The elements Ba, Ce, and O are uniformly distributed in the relative maps in Fig. 5.13, and dopant Y is uniformly distributed throughout the Y-doped $BaCeO_3$ thin film (Fig. 5.13b). This evidence is a crucial element in determining whether the described procedure is suitable for the fabrication of large-area doped $BaCeO_3$

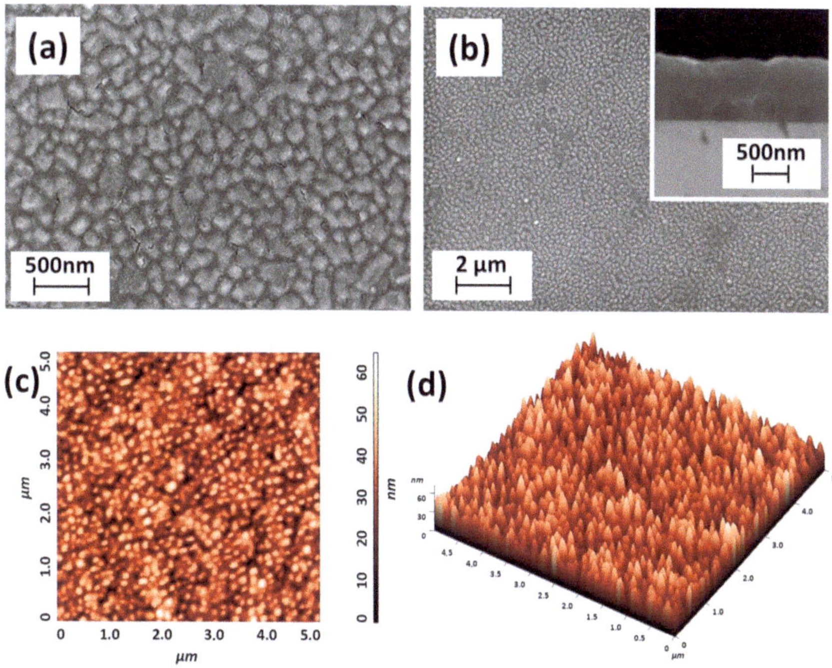

Fig. 5.12 $BaCe_{0.8}Y_{0.2}O_3$ thin film formed on YSZ(100) substrate at deposition temperature of 950 °C. **a**, **b** FESEM images; **c** 2D and **d** 3D AFM topography (RMS roughness of 10.2 nm over an area of 2.5 × 2.5 μm). The cross-sectional image of the film is shown in Appendix (A) [16] [Open access]

systems. The final functional properties of the material largely depend on the uniform distribution of the dopant ion [16].

5.3.1.3 Plasma Arc

The electrically conductive material is heated, melted, and cut into unique patterns and designs using a plasma torch. This manufacturing technology is known as plasma cutting, sometimes known as fusion plasma cutting or plasma art cutting. Alternatives to plasma cutting include oxy-fuel, waterjet, and laser cutting. The size of features is the main difference in this case. Roughness focuses on small deviations from a perfectly smooth surface, whereas topography includes all surface features at different scales. For example, AFM images in Fig. 5.14a–d show that the surface roughness of polytetrafluoroethylene (PTFE) samples increases with different exposure times to the RF plasma torch [9]. The degree of deviation from a smooth surface without regard to any directional structure or regularity is called roughness. In contrast, texture refers to some regularity or structural features that are arranged.

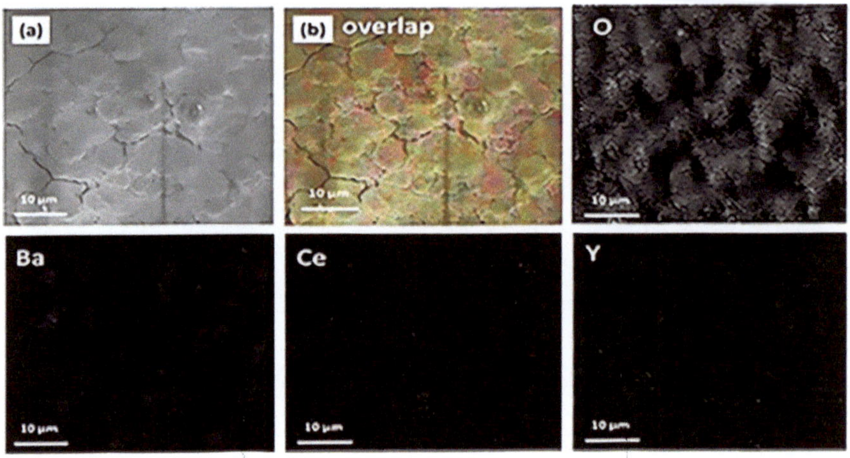

Fig. 5.13 EDX distribution maps of O, Ba, Ce, and Y elements, FESEM images, and interference map images of $BaCe_{0.8}Y_{0.2}O$ film formed on MgO substrate at a deposition temperature of 950 °C [16] [Open access]

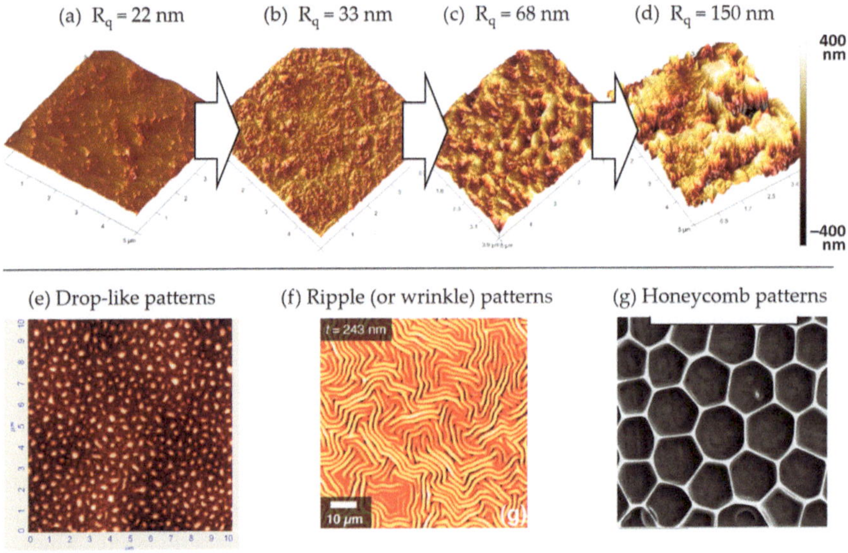

Fig. 5.14 Panels **a**–**e** show AFM images of PTFE surfaces exposed to a radiofrequency plasma torch for different exposure times: 0 min, 2 min, 26 min, and 50 min. The root mean square roughness is denoted as Rq. **e** AFM image of plasma-treated polypropylene showing a droplet-like pattern. **f** AFM image showing the waviness of plasma-treated polydimethylsiloxane (PDMS). **g** SEM image showing the honeycomb structure of plasma-treated FEP. All rights reserved. 2009 American Chemical Society [17] [Open access]

While texture describes specific structures or directional aspects, topography is a broader term that includes all surface features and qualities of these features. Figure 5.14e–g is a representation of the teardrop, ripple, and honeycomb structures, respectively, as seen in AFM or SEM images [17].

Plasma cutting can cut a wide range of metal materials, such as copper, aluminum, alloy steel, and structural steel, which can cut materials from 0.5 mm to 180 mm thick. Rubber, plastic, and other non-conductive materials are not suitable for plasma cutting. There are different types of plasma cutting, such as water injection plasma arc cutting, secondary plasma arc cutting, and standard (or traditional) plasma arc cutting. The technique, known as plasma cutting, uses a jet of hot plasma that is accelerated and cuts electrically conductive materials. When an electrically conductive gas, such as oxygen, nitrogen, argon, or a combination of gases, is ionized and forms a plasma arc, it binds to the material being cut, transferring energy and melting it. This is exactly how plasma cutting works. When a gas is heated to such a high temperature that electrons are stripped from the atoms, creating a mixture of free electrons and ions (atoms that have gained or lost electrons), a unique state of matter known as plasma is formed. Ionization is the process by which atoms lose their electrons [11].

Three states of matter are most often mentioned: solid, liquid, and gas. But plasma, which is found both inside and outside the galaxy (for example, in the centers of stars and the form of lightning), is sometimes called the fourth state of matter. From a technical point of view, plasma is the most common state of matter, both by volume and mass. When more energy is given to gas, the molecules can move faster and collide with greater force, resulting in plasma [12].

Molecules can be separated into their constituent atoms by collisions and these individual atoms can be separated from their outer shell electrons to form ions. The gas reaches a critical point where the number of atoms gaining free electrons and the number of atoms losing electrons balance out as more atoms lose their electron [13]. Comparing substrates with low Critical Process Spray Parameter (CPSP) and high CPSP coatings, the latter is found to exhibit the least wear resistance [23].

5.3.1.4 Flame Synthesis

An industrially scalable method for creating nanomaterials is called flame spray pyrolysis (FSP), which can produce a wide range of metal or metal oxide nanoparticles with specific physicochemical properties [24]. The complex process architecture, including precursor atomization, combustion, and nanoparticle production, is the basis of FSP technology [25]. This enables rapid, one-step synthesis and eliminates the need for post-production modifications that are often necessary with other approaches. So, the importance of FSP has increased in the creation of nanodevices, in which nanoparticles can be inserted in situ into complex functional arrays [26]. Consequently, FSP is not only transforming the manufacturing of nanomaterials, but with recent advances, it is opening the door to the development of next-generation nanodevices [18]. Therefore, a revised summary of the current state of the art in nanodevice development, new materials synthesis, and innovative reactor and process projects in FSP technology is provided.

Figure 5.15 shows transmission electron microscope (TEM) and high-resolution TEM images of carbon films formed on the copper foil for different growth times (the copper foil is above the flame, and the ethanol flow rate is 2 mL/min). TEM grid images (no sample is embedded) are shown in Fig. 5.15a, b. These figures show that each TEM grid has an independent carbon layer supporting it, and the middle of each grid is a space without any physical connection. As a result, the carbon films in the middle of the grids are obtained during the synthesis experiments [22].

In the field of nanotechnology, the synthesis of nanostructures is an important area of research that includes a wide range of techniques in addition to FSP. These various methods, including electrospinning, sol–gel processing, and CVD, provide specific qualities in terms of shape, chemical composition, and particle size control. The choice of an appropriate synthesis technique depends on several factors directly related to the intended use of the nanostructures.

When assessing the suitability of a method for a variety of applications, from solar devices to drug delivery systems, it is important to consider factors such as temperature range, synthesis time, environmental impact, and material adaptability. Several synthesis methods are used to create nanomaterials (including FSP) and describe the specific standards applicable to their use. The carbon film surface showed nanometer-scale island features in the overall morphology of amorphous carbon film.

5.3.1.5 Laser Ablation

One method for creating several types of nanoparticles is laser ablation. These include core–shell nanoparticles, semiconductor quantum dots, carbon nanotubes, and nanowires. Using this method, particles subjected to laser evaporation in a background gas are nucleated and multiplied to form nanoparticles. It is beneficial to produce high-purity nanoparticles in the quantum size range, <10 nm by extremely rapid steam cooling. Figure 5.16 shows a diagrammatic representation of the nanocoatings laser ablation process. This review provides an overview of the process of creating nanoparticles using laser ablation. Next, it reviews recent developments in nanoparticle size control and the challenges of synthesizing functional nanoparticles using modern laser ablation technologies.

Figure 5.16 illustrates an overview of the structures formed at different irradiance levels. AFM height profile, TEM cross section, and SEM top view are shown for each irradiance level. The sinusoidal irradiance profiles have maximum irradiances that are more than twice the average recorded irradiance. The maximum irradiance, 54 mJ/cm^2 with an average irradiance, 27 mJ/cm^2 is well below the measured multi-pulse ablation threshold, 150 mJ/cm^2, using the large-area ablation method. The ablation threshold for a single pulse is difficult to estimate because, using conventional procedures, the small changes that occur after a single pulse approaches this threshold make it impossible to distinguish between material modification and ablation [4].

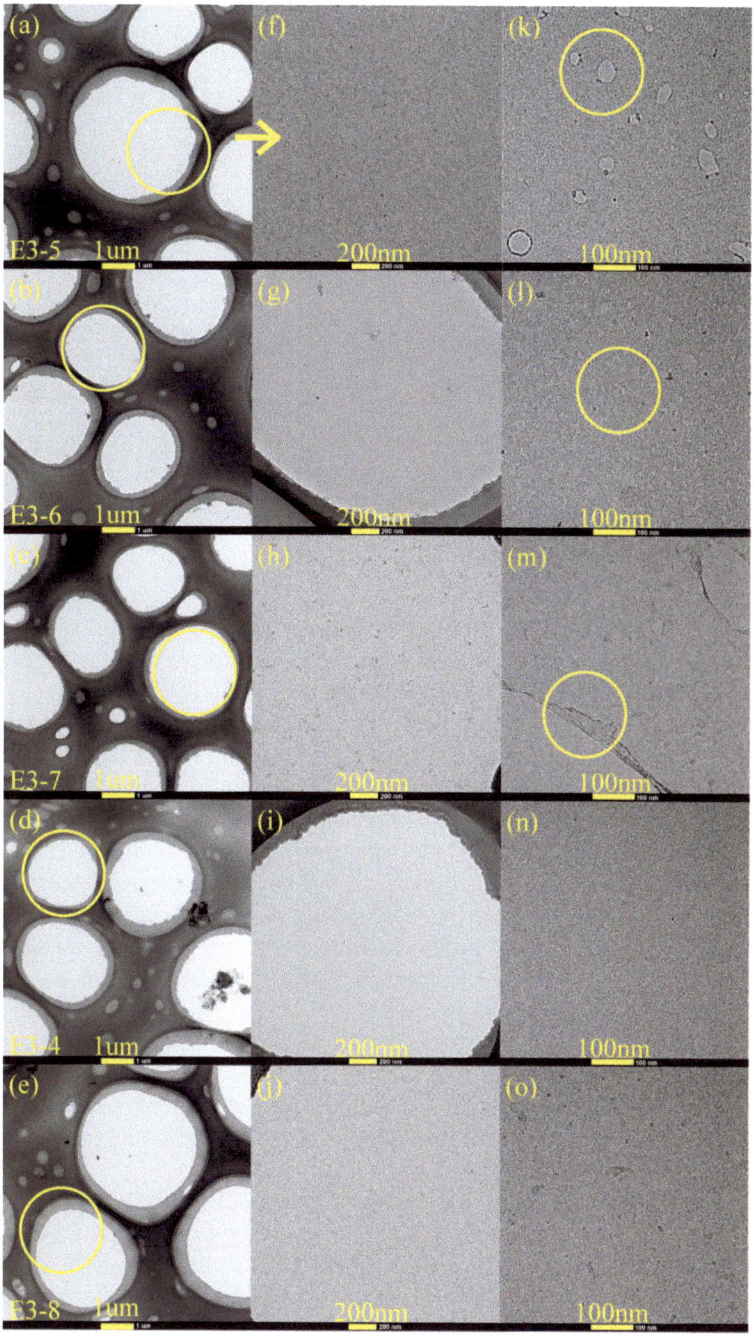

Fig. 5.15 TEM images of carbon films with different growth times at a height of 14.0 mm and an ethanol flow rate of 2 mL/min. 10 s (E3-5) for (**a**, **f**); 30 s (E3-6) for (**b**, **g**); 2 min (E3-7); 5 min (E3-4); and 10 min (E3-8) for (**c**, **h**). High-resolution TEM images of carbon films grown at different growth rates at a height of 14.0 mm and an ethanol flow rate of 2 mL/min: **k** 10 s; **l** 30 s; **m** 2 min; **n** 5 min; and **f** 10 min [22] [Open access]

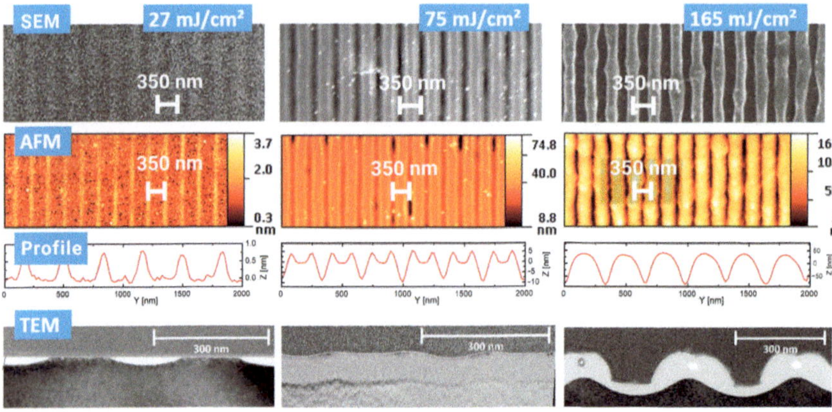

Fig. 5.16 How the structure is formed on silicon by changing the laser beam speed (middle). SEM (top), AFM (middle), and TEM (bottom) records are shown [4] [Open access]

5.3.2 Method of Liquid Phase

Another method for creating phosphorus compounds based on silicates is liquid-phase synthesis. This procedure can produce approximately 90% more transformation than the high-pressure process, which produces only 20%. Solvents are widely used in liquid-phase synthesis as reaction media. The response speed and system can vary significantly depending on the range of solvents. The effect of solvation is the term for this process, which can be analyzed quantitatively using the free energy of solvation. Understanding the factors influencing appearance during liquid-phase synthesis has received much less attention than gas-phase synthesis methods. A change in SthK (potassium (K^+) ion channel) that is cyclic nucleotide-gated, specifically responding to the presence of cyclic guanosine monophosphate (cGMP). It is found in some bacteria and is considered a model for studying the mechanisms of cyclic nucleotide-gated channels. Channels from an active/open state to a resting/closed state upon the addition of excess cGMP using HS-AFM. cGMP is added in increasing doses to a final concentration of 7 mM, starting with an initial concentration of 0.1 mM cAMP. The image showed a clear change in 2D crystal packing and surface texture. As seen in Fig. 5.17a, this change has started with molecules on the outside of the crystal patch and slowly moved toward the center, from left (0.1 mM cAMP) to right (7 mM cGMP) [6].

Using the method of liquid-phase precursors, a green-emitting phosphor (Ba1−x, Sr1−x)SiO_4:2xEu^{2+}, where x = 0.03, 0.05, 0.1, and 0.15 is created [4]. When phosphors are produced using a liquid-phase precursor, their emission intensity is reported to be significantly higher than that of other synthesis methods. The most widely used film is SiO_2, which is easily hydrolyzed by metal, boric acid, or aluminum. In addition to basic solutions, extremely acidic solutions are also formed when creating a thin film.

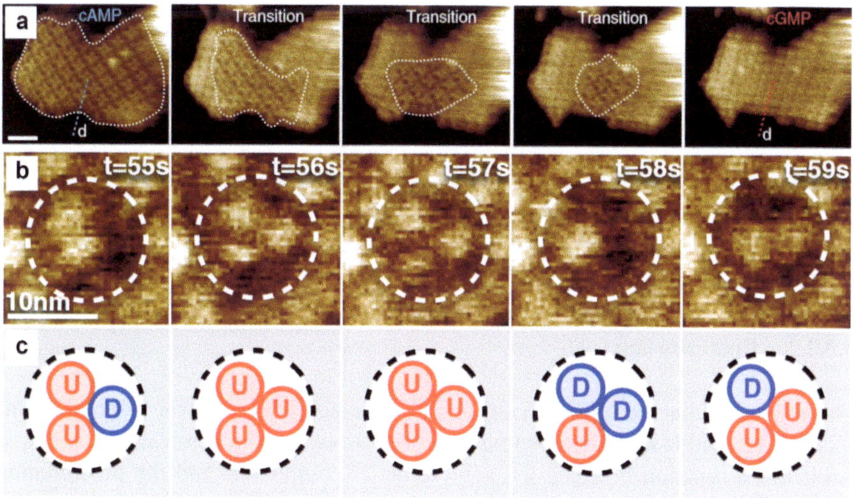

Fig. 5.17 **a** Dynamic ligand-induced conformational changes in SthK revealed by real-time high-resolution HS-AFM microscopy from 0.1 mM initial cAMP to 7 mM final cGMP. The dotted line denotes conformational changes in SthK channels. The scale bar is 30 nm. **b** Sequential frames of high-resolution HS-AFM images illustrating conformational changes in GltPh trimers. **c** Schematic representation of the up (U) or down (D) state of each protomer for the AFM images shown in (**b**) [6] [Open access]

5.3.2.1 Electrochemical

Electrochemistry is the study of chemical processes involving the separation of charges, usually in liquid media such as solutions. Charge transfer, which can occur heterogeneously on the surface of electrodes or homogeneously in solution between different chemical compounds, is often associated with charge separation. Thus, it is immediately obvious that its application is very wide and will be discussed later.

At least two charge transfer half-reactions: oxidation (loss of electrons or increase in oxidation state) and reduction (gain of electrons or decrease in oxidation state) occur simultaneously and in opposite directions to ensure electrical neutrality.

With increasing volume percentage of body-centered cubic (BCC) phase and the amount of aluminum in the $Al_{0.7}CoCrFeNi$ alloy, no pitting corrosion is observed. On the other hand, as shown in Fig. 5.18, selective decomposition of the (Al, Ni)-rich organized BCC phase and localized corrosion along the dendritic/interdendritic boundaries are observed. Therefore, the author considered that the degradation of the passivation coating shifted from pitting to decomposition at the phase boundaries when the microstructure changed from a single solid solution to a multi-phase, which reduced the corrosion resistance [7].

Oxidation and reduction half-reactions are spatially separated at electrode surfaces and often occur at different electrodes immersed in solution within the same cell or in different compartments of the cell. The electrodes are connected by conductive paths that allow the transfer of charge and completion of an electrical circuit, both externally (via electrical wires, etc.) and internally (via ion transport).

Batteries and fuel cells can be used to capture the electrical energy released when the total number of Gibbs energy changes at both electrodes is negative. If the difference is positive, external electrical energy can be used to counteract the positive Gibbs energy and drive electrode reactions that convert chemicals (electrolysis). Therefore, as in industrial electrolytic reactions, the products of two-electrode reactions can be separated if the cell design allows it [5].

5.3.2.2 Precipitation

The most popular method for extracting dissolved (ionic) metals from solutions such as hazardous metal-containing process wastewater is chemical precipitation. A chemical reaction between the soluble metal compounds and the precipitating reagent converts the ionic metals into an insoluble form (particle). By settling and/or filtration, the particles resulting from this process are removed from the solution. In this method, the typically required unit steps are neutralization, sedimentation, coagulation/flocculation, separation of particles from liquids and dewatering.

The efficiency of the chemical precipitation process depends on several variables, such as the type and number of ionic metals in the solution, the precipitant used, reaction conditions (especially the pH of the solution), and the presence of additional substances that may interfere with the precipitation reaction. The most widely used chemical precipitation method is hydroxide precipitation, known as pH precipitation, which uses precipitants such as sodium hydroxide (caustic) or calcium hydroxide (lime) to create metal hydroxides. The ideal pH for the precipitation of hydroxides of each dissolved metal ranges from 7.5 for chromium to 11.0 for cadmium. Because metal hydroxides are amphoteric, they become more soluble at both high and low pH levels. Thus, the metal may dissolve or begin to reenter the solution depending on the ideal pH for that metal to precipitate. Precipitation of these various metals as hydroxides can be a complex procedure since most process wastewaters contain mixed metals [23].

5.3.2.3 Sol–Gel

One of the common wet chemical methods for applying nanocomposite films is the sol–gel procedure. Material sources that are first prepared in this process have included metal oxide and alkoxide precursors, metallic nanoparticles, tetraethoxysilane for the silica matrix, stabilizers, catalysts and other porosity additives. After this, the sol undergoes processes of polycondensation and hydrolysis, which gradually lead to the development of a network resembling a gel. The fundamental morphology or structure of the solid phase can take several forms, such as continuous polymer networks or discrete colloidal particles [20].

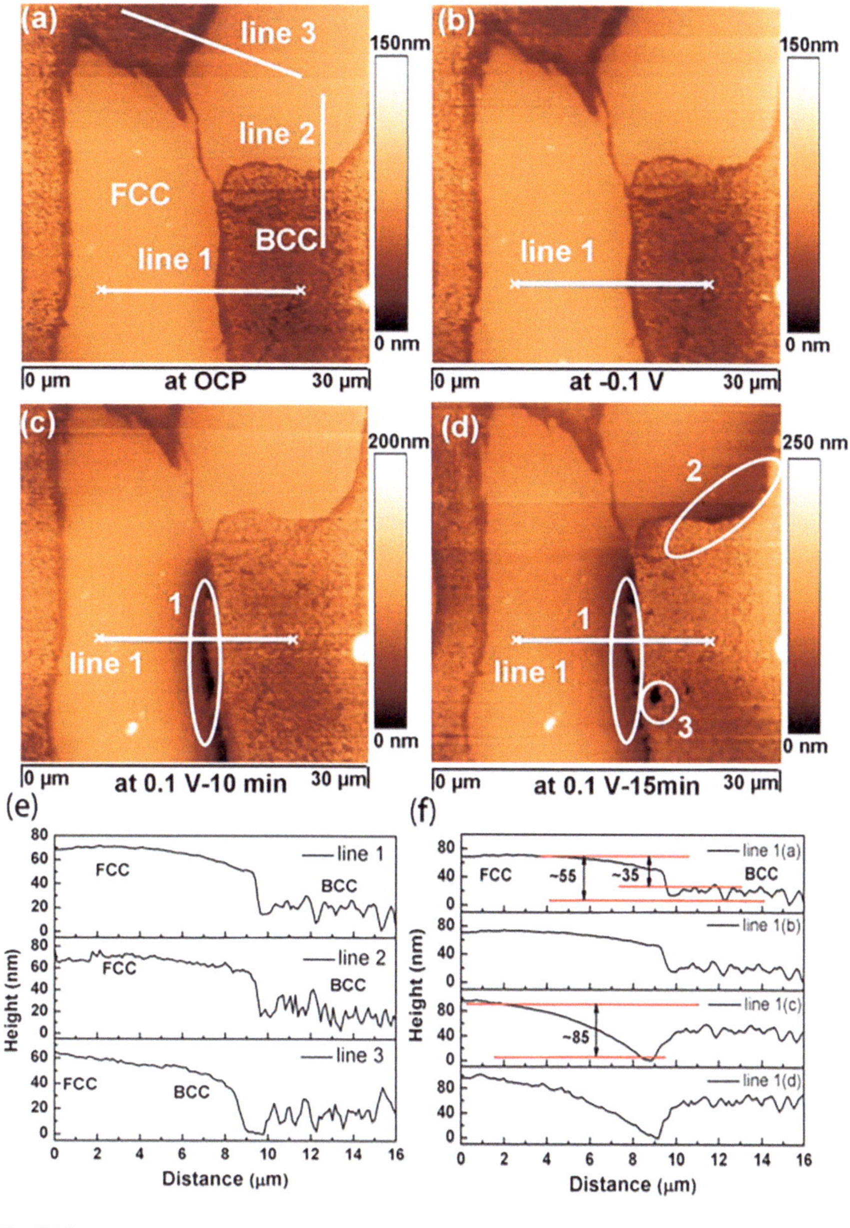

Fig. 5.18 In situ EC-AFM topography images of $Al_{0.7}CoCrFeNi$ HEA at **a** OCP, **b** 0.1 V, **c** 0.1 V, and **d** after 15 min soak at 0.1 V compared to SCE after 10 min soak in 3.5 wt% NaCl solution. **e** The phase boundaries shown in (**a**) are crossed by trilinear profiles. **f** Vertical profiles in (**a**–**d**) along surface line 1[7] [Open access]

The sol–gel precursor is converted into a nanocomposite film by spin-coating or dip-coating onto a substrate, followed by pyrolysis and decomposition of compounds and extraction of organics and water from the resulting network, after which crystallites nucleate and develop. The crystallite size and film porosity are significantly influenced by the thermal decomposition behavior of gel precursor [27]. In contrast, in ash, the dispersed phase is solid, and the continuous phase is liquid. The fact that solid particles are scattered throughout the liquid in the ash distinguishes it from a non-colloidal liquid [13].

Using atomic force electron microscopy (AFM), the textural properties of enzyme complexes are investigated. For the native horseradish peroxidase (HRP) and lignin peroxidase (LiP) enzymes deposited on the surface, aggregated enzyme molecules are visible in AFM images (Fig. 5.19a, b). When magnetite is immobilized, silica aggregates are visible on the surface, indicating that the enzymes are encapsulated (Fig. 5.19c). Figure 5.19e, f shows pristine magnetite nanoparticles and silica-capped magnetite nanoparticles, respectively [15].

Scattered particles in a liquid can be precipitated by centrifuging the sol. A non-colloidal liquid solution, such as salt dissolved in water, does not experience this [12]. An overview of various sol–gel methods is shown in Fig. 5.19. (A) Sol–gel creation. The chemical processes of sol–gel formation are depicted in Fig. 5.19 and include the stages of hydrolysis, condensation, gelation, aging, drying, and compaction. Then, the addition of an acid (e.g., hydrochloric acid) or basic substance (e.g., potassium hydroxide) activates the precursor. Subsequently, the activated precursors combine to form a network. With time and temperature, the network expands and ages until it reaches the size of a container. From this point, the viscosity of the liquid increases exponentially until gelation occurs [21]. The structure of gels obtained in two different modes is shown in Fig. 5.20. Several variables, including pH, solvent, temperature, time, catalyst, and agitation, affect sol–gel chemistry. The morphology (scan) and thickness (full cavity) of composite brushes are investigated using atomic force microscopy (AFM). Figure 5.20 depicts the surface topography and phase images of composites at different pH values [20].

5.3.2.4 Hydrothermal

Hydroxyapatite (HAp) is a naturally occurring form of calcium apatite that is an inorganic component of bones. About 80% of it is minerals, 22% is protein and 8% is water. Hexagonal crystal structure (Hap) is a form of calcium phosphate with the chemical formula $Ca_{10}(PO_4)6(OH)_2$. Hydroxyapatite, a needle-like material 60 nm long and 5–20 nm wide, imparts rigidity and strength to bones and accounts for about 65% of bone mass [25].

The wide range of possible applications of nanostructured HAp has attracted the interest of investors, industry, and researchers. The osteoconductivity and biocompatibility qualities of hydroxyapatite attract attention. The substance is widely and effectively used in drug delivery systems and regenerative medicine [18]. Hydroxyapatite particles with a nanostructure can be used in caries therapy as building blocks

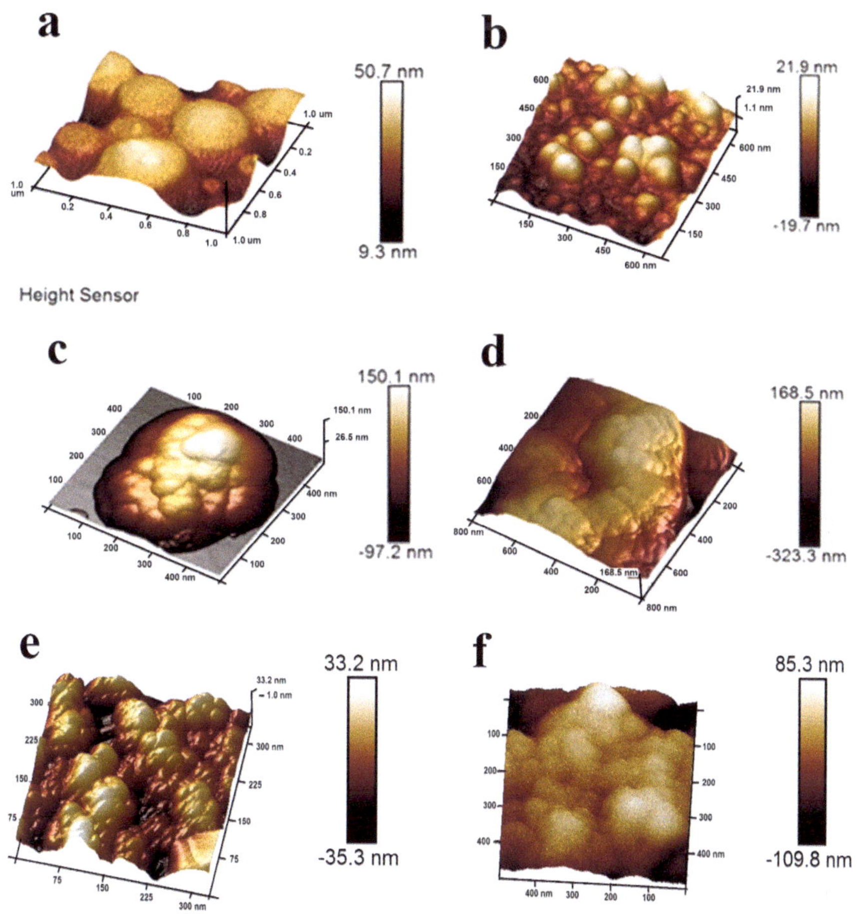

Fig. 5.19 AFM images of native magnetite (**e**), silicon-coated magnetite (**f**), sol–gel-coated HRP (**c**) and LiP (**d**) complexes, and native HRP (**a**) and LiP (**b**) enzymes [15] [Open access]

for damaged enamel [19]. Numerous studies show that nanosized HAp can improve its mechanical properties due to its high surface energy, which allows for faster surface renewal of implants [20]. Hydroxyapatite materials are used as sunscreen, anti-wrinkle agents, and bone replacement. They help speed up wound healing and are used in toothpaste formulations. Polish scientists [9] have created patches and ointments that use the special properties of HAp [23]. There is a great interest in hydroxyapatite to restore bone loss in animals. Typically, solid hyaluronic acid particles (Linhai Hyaluronic Acid Co., Ltd., Linhai, China) are added to deionized water and quickly sink to the bottom of the container due to their insolubility (left in Fig. 5.21a). After being placed in a Teflon-lined stainless-steel vessel, the solid–liquid mixture of hyaluronic acid and deionized water is subjected to hydrothermal treatment for ten hours at 190 °C. After 30 min of sonication and 10 min of centrifugation at 9000

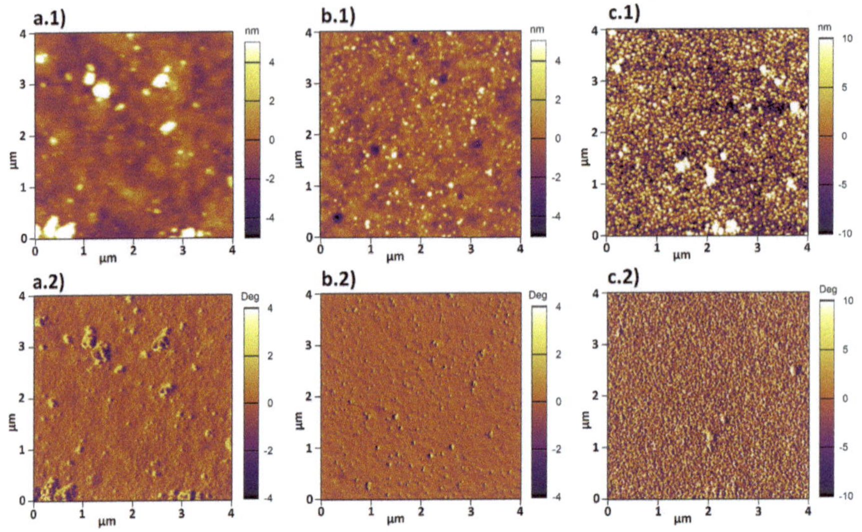

Fig. 5.20 AFM is used to record height (a.1–c.1) and phase (a.2–c.2) images of compounds (a.1/a.2) at pH 4, b.1/b.2) at pH 6 and c.1/c.2) at pH 8 [20] [Open access]

rpm to eliminate insoluble by-products, the homogeneous mixture yielded a brown GOH solution (as shown on the right of Fig. 5.21a). Using the freeze-drying method, the solid GOH could be collected, and GOH could be converted to reduced GOH (r-GOH) by thermal reduction in N_2 environment for 40 s at 900 °C [9].

It has been demonstrated that HAp nanoparticles (nano-HAp) rather than micrometer-sized hydroxyapatite (micro-HAp) are more suitable to function as a bone apatite substitute in biomedical applications [23]. The effects of differently shaped nano-HAp particles on extremely aggressive melanoma cells are investigated. The results obtained showed that the appearance of a nanoscale effect, in contrast to the morphology of HAp particles, more effectively suppresses the development of such malignant cells [15]. A further study examined the effect of different sizes of nano-HAp, from 20 to 80 nm, on the growth of bone-associated cells, specifically bone marrow mesenchymal stem cells and osteosarcoma cells [8]. A cell culture experiment confirmed that the hydrophilicity of the nanophase mineral is increased by using nano-HAp compared to traditional HAp. In addition, nano-HAp is found to enhance stem cell viability and proliferation, especially for the smallest particles of 20 nm. Similarly, the smallest 20 nm particles are found to be the most effective inhibitor of osteosarcoma cell development, while each nano-HAp slowed down the growth rate. The anticancer activity of nano-HAp is also found in another study [9].

Fig. 5.21 The obtained solution of humic acid (HA) and graphene oxide (GOG) is shown in **a** a photograph; **b** an SEM image; **c** an SEM image of GOG; and **d** an AFM image of GOG [9] [Open access]

5.3.2.5 Microemulsion

Nanotechnology is a branch of research that originated in the physical, chemical, biological, and engineering sciences. These domains are known for their innovative methods of manipulating individual atoms and molecules. In the context of nanotechnology, a nanoparticle is a small object with transport characteristics and possible applications that behaves as a whole. They involve intriguing modifications in their optical, magnetic, catalytic, electrical, and drug delivery properties, coupled with improved physical properties such as mechanical hardness, thermal stability, or chemical passivity. They are one of the most challenging areas of modern scientific and technological research.

Simple or compound nanoparticles are useful building blocks for creating innovative nanodevices with a wide range of physical, biological, biomedical, and pharmacological applications. For many years, scientists have been fascinated by nanoparticles and nanostructured materials due to their applications in Raman scattering, catalysis, photonics, electronics, labeling, imaging, and sensing. When comparing the optical characteristics of single crystals of the same chemical composition with nanoparticles, significant differences are often found. The properties of nanomaterials differ from their bulk phases due to the influence of a significant number of surface atoms and the effect of quantum confinement of electronic states on nanosized crystallites.

The surface morphology of different bulk-heterojunction (BHJ) films is analyzed by tap-mode AFM. The Au/PEDOT: PSS film layers are peeled off to expose the BHJ layers (PTB7:EH-IDTBR and PTB7:ITIC) after J-V measurements. Then, the background electrolyte solution (BGE) is used to remove the acceptor materials from the BHJ layers. AFM images of BHJ films after BGE washing have revealed obvious differences in the molecular assembly of PTB7, in contrast to AFM images of BHJ films without BGE washing (Fig. 5.22a–d). PTB7 has created distinct granular aggregates in ITIC-containing BHJ films, while it has created an amorphous-like texture in the EH-IDTBR-containing BHJ films [10].

Compared to bulk materials and individual atoms or molecules, the properties of matter at this scale are very different. These nanoparticles, both organic and inorganic, have been prepared using a range of physical and chemical methods such as thermal and irradiation methods. One of the many preparation methods that can control many particle properties, including size, geometry, morphology, uniformity, and surface area, is the microemulsion method, which covers various aspects of microemulsions. Colloidal metals, colloidal AgCl, nanocrystalline oxides Fe_2O_3, TiO_2, Al_2O_3, and high-temperature oxides $YBa_2Cu_3O_7$ are synthesized using the microemulsion method. In addition to summarizing recent studies on the synthesis of organic and inorganic nanoparticles using this technique, this work aims to provide a clear overview of the use of microemulsions to produce nanoparticles and control their particle size and shape [12, 13, 21, 23].

5.4 Advancements in Nanocoating Technology

In every field, including mechanical, electrical, electronic devices, and computer applications are mainly used to improve performance, prevent corrosion, prevent weathering, and prevent water damage, so nanotechnology is the fastest-growing field. Thus, it is used everywhere nowadays. In addition, as they become more accessible, we are trying to use nanotechnology in turning and other machining operations. In this case, the nanoparticles will be coated using traditional methods, and hard, difficult-to-cut material will be turned. Before going deep into this project, we will first learn about nanotechnology, the materials that we will use, the place of our work, and the types of tools that will be used, as we can see one by one. A nanometer is approximately 100,000 times smaller than the diameter of a human hair, 1000 times smaller than a red blood cell, and half the size of a DNA molecule. Nanotechnology is defined as the creation and application of structures, devices and systems that, due to their small size, have unique properties and functions, the ability to control or manipulate matter on an atomic scale, and research and development of technologies at the atomic, molecular, or macromolecular levels using a length

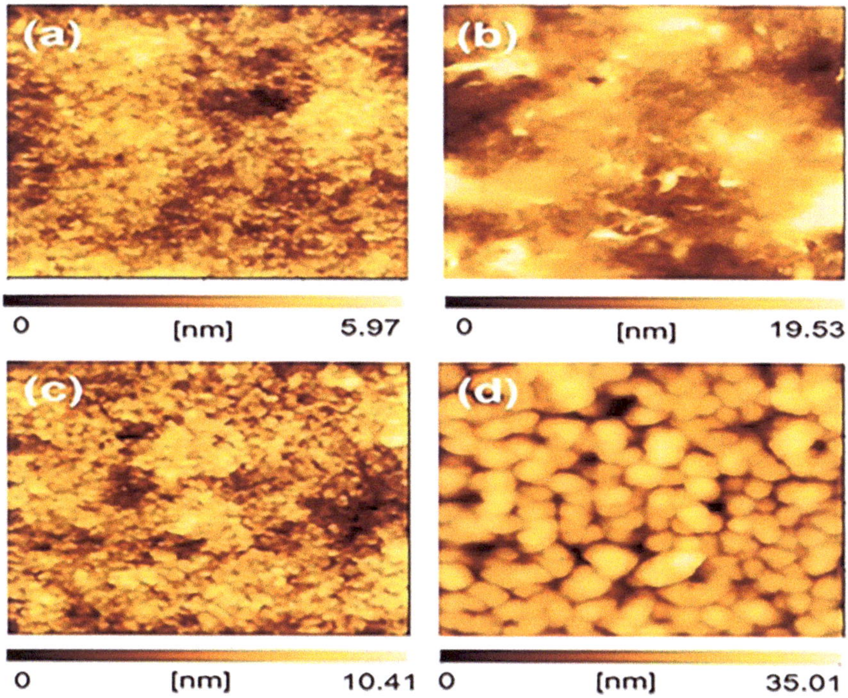

Fig. 5.22 AFM (2 μm × 2 μm) before (**a**, **b**) and after (**c**) BGE washing. PTB7/ITIC films (**b**, **d**) and PTB7/EH-IDTBR films (**a**, **c**). [10] [Open access]

scale of approximately 1 to 100 nm in any dimension. Engineering at the molecular level (groups of atoms) is known as nanotechnology. It is a group as a whole word used to describe a variety of methods, instruments and procedures involving the manipulation of matter at the smallest scale, 1 to 100 nm^2. On such an extremely small scale, the classical rules of physics and chemistry are not easily applied for two reasons. First, compared to their larger counterparts, very small particles can have completely different electrical characteristics. Second, the material's characteristics change because the surface atoms are often the most reactive, resulting in a much larger surface area-to-volume ratio [18, 19, 24–27].

5.5 Application of Nanocoatings

Typically, single- or multi-phase solid structures called nanocoatings, about 100 nm thick or less, are deposited onto an immediate surface to impart specific properties or purposes to the substrate material. Different materials have different characteristics at the nanoscale. The particle size of material has a significant impact on overall properties such as hydrophobicity, electrical conductivity, magnetic permeability, melting point and reactivity. Intermolecular forces such as hydrogen bonds, van der Waals interactions and electrostatic forces control material properties at the nanoscale. The length scales over which these intermolecular forces act vary from a few angstroms to several nanometers.

When the dimensions of material reach the nanoscale, or same order of magnitude as intermolecular force distances, the properties of material change significantly, thus, many new applications can be explored by improving the physical and chemical properties of nanocoatings made from nanoparticles or multiple nanoscale layers. The main industrial applications of nanocoatings are wear-resistant coatings, self-cleaning (non-stick) surfaces, corrosion prevention, and thermal protection. These applications are particularly important to the aviation sector as they allow aircraft engine, cabin, and frame components to be modified to improve performance, reduce operating costs, and reduce fuel efficiency [27].

It is reported that the solar absorber used in the collector panel determines how well solar radiation is converted into thermal energy. A solar absorber with low emissivity and high absorption coefficient is ideal. It has been found that electrochemical deposition of selective nickel-based coatings can provide both cost-effective and good optical performance required for solar energy absorbers [28]. However, scientists are moving toward physical vapor deposition technology to be used to coat medium- and high-temperature solar photovoltaic panels due to the poor thermal stability of the CVD method and environmental concerns. Sputtering is one of the PVD processes that has been successfully used to coat receiver tubes and flat plate solar collectors [20].

The study "Nanocoatings: Design, Development and Application" highlights how the use of nanoparticles to create superior coatings with improved performance properties is revolutionary. Ultra-thin layers of nanoparticles define nanoporous coatings, which offer some advantages over conventional coatings, such as improved surface characteristics, higher wear and corrosion resistance, and antibacterial properties. Design: Selecting the right nanomaterials and formulations to achieve specific performance goals is essential to effectively design nanocoatings. It is important to have a thorough understanding of how particle size and shape affect the overall performance of the coating. Development: The goal of the development phase is to establish efficient manufacturing procedures that will reliably produce high-quality nanocoatings. Optimizing preparation methods, ensuring coating stability, and quality control are important factors. Applications: Nanoporous coatings have proven their effectiveness in many industries, including electronics, energy, construction, and medicine. These coatings extend the service life of materials and improve the performance properties of the product.

Problems

1. What are hydrothermal vents?
2. What are the different methods of laser ablation of nanomaterials?

3. How is nanomaterial electrochemistry applied in engineering?
4. List the materials for applying nano-coatings made of nanomaterials.
5. Describe advances in nano-coating technology in engineering applications.

References

1. Kryuchkov, M., Adamcik, J., & Katanaev, V. L. (2022). Bactericidal and antiviral bionic metalized nanocoatings. *Nanomaterials, 12*(11), 1868.
2. Pristovsek, M., Furuhashi, I., & Pampili, P. (2023). Growth of N-polar (000 1-) GaN in metal-organic vapour phase epitaxy on sapphire. *Crystals, 13*(7), 1072.
3. Yoon, Y., Lee, J., Lee, S., Kim, S., & Choi, H. C. (2021). Ultrasmooth organic films via efficient aggregation suppression by a low-vacuum physical vapor deposition. *Materials, 14*(23), 7247.
4. Blumenstein, A., Simon, P., & Ihlemann, J. (2023). High-resolution laser interference ablation and amorphization of silicon. *Nanomaterials, 13*(15), 2240.
5. Ruan, Y., Miyagi, A., Wang, X., Chami, M., Boudker, O., & Scheuring, S. (2017). Direct visualization of glutamate transporter elevator mechanism by HIGH-SPEED AFM. *Proceedings of the National academy of Sciences of the United States of America, 114*, 1584–1588.
6. Rotondi, S. M. C., Ailuno, G., Mattioli, S. L., Pesce, A., Cavalleri, O., & Canepa, P. (2023). Morphological investigation of protein crystals by atomic force microscopy. *Crystals, 13*(7), 1149.
7. Chen, H., Qin, Z., He, M., Liu, Y., & Wu, Z. (2020). Application of electrochemical atomic force microscopy (EC-AFM) in the corrosion study of metallic materials. *Materials, 13*(3), 668.
8. Xie, R., Li, D., Hou, B., Wang, J., Jia, L., & Sun, Y. (2011). Solvothermally derived Co_3O_4@m-SiO_2 nano-composites for fischer-tropsch synthesis. *Catalysis Communications, 12*, 380–383.
9. Huang, G., Kang, W., Geng, Q., Xing, B., Liu, Q., Jia, J., & Zhang, C. (2018). One-step green hydrothermal synthesis of few-layer graphene oxide from humic acid. *Nanomaterials, 8*(4), 215.
10. Nakano, M., Takahara, A., Genda, K., Shahiduzzaman, M., Karakawa, M., Taima, T., & Takahashi, K. (2021). Selective extraction of nonfullerene acceptors from bulk-heterojunction layer in organic solar cells for detailed analysis of microstructure. *Materials, 14*(9), 2107.
11. Ghosh Chaudhuri, R., & Paria, S. (2012). Core/shell nanoparticles: classes, properties, synthesis mechanisms, characterization, and applications. *Chemical Review, 112*, 2373–2433; Montero, D. A., Arellano, C., Pardo, M., Vera, R., Gálvez, R., Cifuentes, M., Berasain, M. A., Gómez, M., Ramírez, C., & Vidal, R. M. (2019). Antimicrobial properties of a novel copper-based composite coating with potential for use in healthcare facilities. *Antimicrobial Resistance and Infection Control, 8*, 3.
12. Arendsen Linda, P., Thakar, R., & Sultan Abdul, H. (2019). The use of copper as an antimicrobial agent in health care, including obstetrics and gynecology. *Clinical Microbiology Reviews, 32*, e00125-e218.
13. Urnukhsaikhan, E., Bold, B.-E., Gunbileg, A., Sukhbaatar, N., & Mishig-Ochir, T. (2021). Antibacterial activity and characteristics of silver nanoparticles biosynthesized from Carduus crispus. *Science and Reports, 11*, 21047.
14. Hövelmann, J., Putnis, C. V., & Benning, L. G. (2018). Metal sequestration through coupled dissolution–precipitation at the brucite–water interface. *Minerals, 8*(8), 346.
15. Pylypchuk, I. V., Daniel, G., Kessler, V. G., & Seisenbaeva, G. A. (2020). Removal of diclofenac, paracetamol, and carbamazepine from model aqueous solutions by magnetic sol–gel encapsulated horseradish peroxidase and lignin peroxidase composites. *Nanomaterials, 10*(2), 282.

16. Pellegrino, A. L., Lo Presti, F., & Malandrino, G. (2023). Metalorganic chemical vapor deposition approach to the synthesis of perovskite $BaCeO_3$ and $BaCe0.8Y0.2O_3$ thin films. *Molecules, 28*(8), 3303.
17. Dufour, T. (2023). From basics to frontiers: A comprehensive review of plasma-modified and plasma-synthesized polymer films. *Polymers, 15*(17), 3607.
18. Mas-Moruno, C., Su, B., & Dalby, M. J. (2019). Multifunctional coatings and nanotopographies: Toward cell instructive and antibacterial implants. *Advances Healthcare Materials, 8*, 1801103.
19. Staneva, A. D., Dimitrov, D. K., Gospodinova, D. N., & Vladkova, T. G. (2021). Antibiofouling activity of graphene materials and graphene-based antimicrobial coatings. *Microorganisms, 9*, 1839.
20. Kesal, D., Christau, S., Krause, P., Möller, T., & Von Klitzing, R. (2016). Uptake of ph-sensitive gold nanoparticles in strong polyelectrolyte brushes. *Polymers, 8*(4), 134.
21. McNeilly, O., Mann, R., Hamidian, M., & Gunawan, C. (2021). Emerging concern for silver nanoparticle resistance in Acinetobacter baumannii and other bacteria. *Frontiers in Microbiology, 12*, Article 652863.
22. Zhu, J., Li, F., Liu, G., Liu, D., Li, Q., & Kan, E. (2018). Synthesis of amorphous carbon film in ethanol inverse diffusion flames. *Nanomaterials, 8*(9), 656.
23. Román-Kustas, J., Hoffman, J. B., Reed, J. H., Gonsalves, A. E., Oh, J., Li, L., Hong, S., Jo, K. D., Dana, C. E., Miljkovic, N., et al. (2020). Molecular and topographical organization: Influence on cicada wing wettability and bactericidal properties. *Advanced Materials Interfaces, 7*, 2000112.
24. Imani, S. M., Ladouceur, L., Marshall, T., Maclachlan, R., Soleymani, L., & Didar, T. F. (2020). Antimicrobial nanomaterials and coatings: Current mechanisms and future perspectives to control the spread of viruses including SARS-CoV-2. *ACS Nano, 14*, 12341–12369.
25. Cloutier, M., Mantovani, D., & Rosei, F. (2015). Antibacterial coatings: Challenges, perspectives, and opportunities. *Trends in Biotechnology, 33*, 637–652.
26. Escobar, A., Muzzio, N., & Moya, S. E. (2021). Antibacterial layer-by-layer coatings for medical implants. *Pharmaceutics, 13*, 16.
27. Li, W., Thian, E. S., Wang, M., Wang, Z., & Ren, L. (2021). Surface design for antibacterial materials: From fundamentals to advanced strategies. *Advancement of Science, 8*, 2100368.

Chapter 6
Nanocomposites-Based Sensors and Medical Devices

Many beneficial properties such as high mechanical strength, elasticity, tunable optical transparency, photodynamic and photothermal activity, nanoscale nature, and high absorption, provide the basis for progress in the field of nanocomposites. Because of flexible temperature sensors provide conformal attachment to human skin and provide consistent and continuous measurements, they have received a lot of attention recently. However, there is still room for improvement in strain sensitivity and stability. These include a limited measuring range, slow response in dynamic situations, high hysteresis, and low accuracy under high mechanical loads. Additional benefits include the separation of individual biomolecules and the electrical stimulation of affected tissues (e.g., heart, nerves, bones). Modern wound dressings include, for example, dressings with antimicrobial, antivascular, tissue engineering and anticancer activities. However, it is important to consider the potential cellular and immunotoxicity of chemicals and their components.

6.1 Introduction

To maintain the contact properties of the hydrogel network, nanoparticles can be used to cross-link hydrogels or microspheres [1]. Due to electrochemical, mechanical, optical, thermal, dielectric, acoustic, magnetic, and electrocatalytic abilities possessed by most nanoparticles and composite materials [2, 3]. These features enable the use of these nanomaterials in various biotechnological applications, including antibacterial activity and biofilm inhibition studies, drug delivery, separation devices, and catalysis [4–6]. They can create appropriate nanostructures such as clay minerals in combination with various polymers to create an intercalated nanocomposite. Clay minerals are represented by polymer chains that connect monomer units, forming well-organized multilayer structures. However, it should be emphasized that polymer nanocomposites usually contain a composite

R. A. Al-Samarai and Y. Al-Douri, *Engineering of Nanomaterials*,
https://doi.org/10.1007/978-981-95-5499-7_6

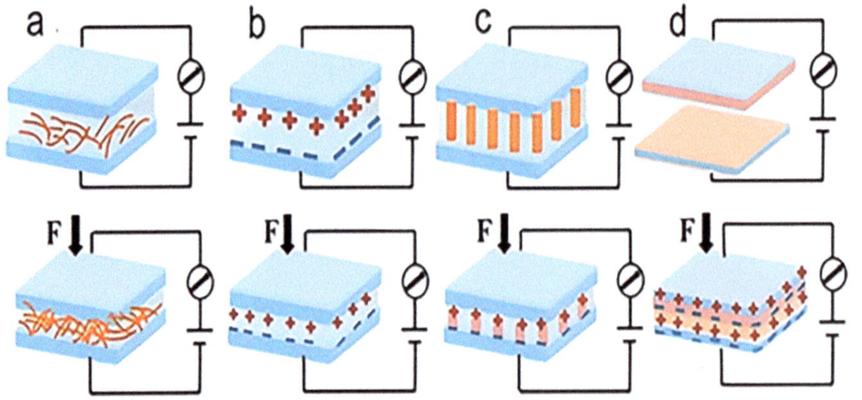

Fig. 6.1 Schematic representations of four common transduction mechanisms: **a** piezoelectric, **b** capacitive, **c** piezoelectric, **d** triboelectric [1] [Open access]

(partially exfoliated) As(III) nanogroup (Fig. 6.1). Due to their exceptional properties, nanocomposites, a special class of materials created by combining nanoparticles with a matrix material have attracted much interest in recent years. These materials enhance mechanical, electrical, optical, and thermal capabilities by combining the beneficial properties of the matrix and nanoscale components. Nanocomposites are very promising in various applications such as sensors and medical devices due to their exceptional properties.

Nanocomposites have been used in the sensor industry to create extremely responsive, selective, and sensitive devices. Incorporating nanoparticles such as metal oxides, graphene, or carbon nanotubes into a polymer matrix allows for improved signal detection and reduced response times. Numerous chemical, biological, and environmental analytes can be detected by these sensors at low concentrations, making them promising for enhancing the functionality, performance, and biocompatibility of implants and diagnostic tools in the medical device industry. Nanocomposite materials are being explored for use in drug delivery systems, tissue engineering, and prosthetics. Medical devices can be made to more closely resemble natural tissue, integrate more successfully with the body, and improve drug delivery by altering the strength, elasticity, and conductivity of nanocomposites. Additionally, biosensors, wound dressings, and implant coatings that suppress infection or promote healing can benefit from the use of nanocomposites due to their ability to be tailored to specific purposes [1–3].

By combining modern nanotechnology with traditional materials science, nanocomposite-based sensors and medical devices have the potential to deliver smarter, more efficient, and customized medical solutions. By offering faster, more accurate and more affordable solutions, their use in therapeutic procedures, medical diagnostics and health monitoring systems have the potential to completely transform modern medicine. Nanocomposites, which combine some remarkable properties not found in conventional materials are the result of recent integration of nanomaterials with traditional materials. Nanocomposites, which are created by combining nanoparticles in a matrix of metals, polymers, or ceramics, have exceptional mechanical, electrical, thermal, optical, and chemical properties that make them ideal for a variety of advanced applications. These substances have shown particularly promising results. The detection of several biological, chemical, and environmental factors has been transformed by using nanocomposites in sensors. Graphene, carbon nanotubes, quantum dots, and metal oxide nanoparticles are examples of nanoparticles that enhance the detection capabilities of sensors by increasing their surface area, conductivity, and functionalization. Even at lower analyte concentrations, nanocomposites can demonstrate improved selectivity, sensitivity, and response time in sensors. For example, ammonia, carbon dioxide and nitrogen dioxide can be detected using gas sensors based on nanocomposite technology, which is especially useful for industrial and environmental monitoring. Another important application of nanocomposite technology is biosensors, where the increased surface area and functionalization of nanoparticles allow the detection of certain biomarkers such as glucose, proteins, or DNA [4–6].

6.1.1 Processing Methodologies

Clay-based polymeric nanocomposites have been produced using three different techniques: melt intercalation, solution exfoliation, and in situ polymerization. Each method starts with organic clay and sometimes with virgin clay before producing polymer nanocomposites in a series of steps. One of the most important parts of flexible and wearable electronics is the pressure sensor, which is capable of converting mechanical stimuli or external deformation into electrical data. The four transduction mechanisms that we primarily discuss are piezoresistive, capacitive [1, 2], piezoelectric, and triboelectric (Fig. 6.1). The change in the distribution and contact states of internal sensing material causes a change in the corresponding electrical signal when an external force is applied to the piezoresistive pressure sensor (Fig. 6.1a). These sensors have a simple device design, which is easy to manufacture and requires little signal processing. However, they are limited in certain applications due to their high-power consumption and response hysteresis. As shown in Fig. 6.1b, capacitive pressure sensors change the capacitance of the device by changing the area of the dielectric layer or the distance between electrodes on each side through the application of an external mechanical force. Charge polarization is created when piezoelectric pressure sensors are subjected to external forces that cause unequal charges within the material (Fig. 6.1c). Depending on the friction material and different characteristics of friction electrodes, triboelectric pressure sensors can generate electrostatic charges as a result of the periodic mechanical energy of positive and negative electrodes touching and separating (Fig. 6.1d) [1].

Clay nanocomposites improve physical properties such as barrier properties, abrasion resistance, fire resistance, thermal stability, chemical stability, and increased flexibility and hardness. Several types of polymers have been used to create polymer nanocomposites. Understanding the nanomechanical properties of versatile base materials or polymers such as polylactic acid (PLA) is essential, especially when they are used in various devices or sensors. These characteristics are investigated using the advanced AM–FM mode of amplitude–frequency modulation atomic force microscopy (AFM) in combination with topographic and morphological studies. This technique provides insight into the mechanical behavior of polymer surfaces by accurately analyzing the surface properties at the nanoscale.

Thin films created by spin-coating a suspension of poly (3-hexylthiophene (P3HT)) and styrene-butadiene-styrene (SBS) in combination with anisole and chloroform are investigated using the scanning force microscopy (SFM) technique. To create nanocomposites with tailored properties, the obtained P3HT nanofibers have a fingerprint pattern that shows the microphase separation of styrene–butadiene–styrene (SBS) in a solution containing anisole and chloroform. To create nanocomposites with specific properties for use in areas such as sensors or medical devices, the obtained P3HT nanofibers have a fingerprint pattern that shows the microphase separation of SBS.

In conclusion, this work investigates the morphology and nanomechanical performance of nanocomposite materials by combining multiple techniques including AM–FM, AFM, and SFM. This helps to clarify how the special structure of these polymers can improve their usefulness in specific applications. This finding suggests that the production of P3HT nanofibers is not affected by SBS polymers present in the suspension. Moreover, a large number of P3HT nanofibers visible in the SFM image are embedded in the matrix due to microdomain rupture at the interface of P3HT nanofibers [1] (Fig 6.2).

Due to the ease of fabrication of these polymers and relatively low mechanical requirements, they are commonly used in low-cost applications. One way to increase the value of these goods is to create nanocomposites. The conductivity increased with increasing P3HT ratio, but the mechanical strength seems to decrease. Therefore, after the elongation treatment, the conductivity of free-standing film containing P3HT (0.05 wt%) and SBS (5 wt%), which is the ratio that produces a free-standing film with sufficient mechanical strength. After stretching the film twice, SFM is performed. Figure 6.3 shows the SFM phase images of stretched and unstretched portions. P3HT nanofibers are consistently observed in the unstretched portion, but not observed at all in the stretched portion [1, 2].

6.1.2 Nanocomposites' Characteristics

Comparison of polymer nanocomposites with other polymers shows that they have much better properties and characteristics. Poly methyl methacrylate (PMMA) doped with halloysite nanotubes (HNTs) or titanium dioxide (TiO_2) nanoparticles at two different concentrations (1–3 wt%) is characterized [4] using atomic force microscopy (AFM). For example, A Clay makes up 2–8% of nanopolymer

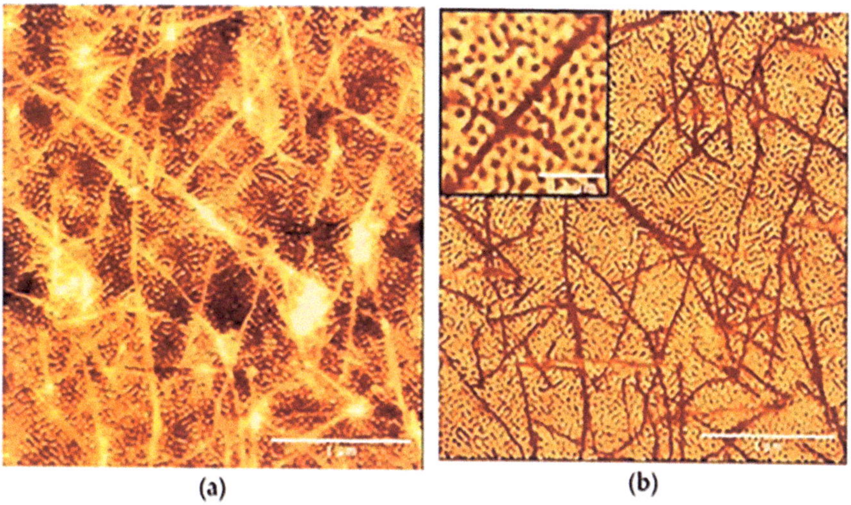

Fig. 6.2 **a** SFM topography and **b** phase pictures of the thin film produced by spin-casting a suspension of SBS (0.5 wt%) and poly(3-hexylthiophene) (P3HT) (0.05 wt%) in a 70:30 (v/v) chloroform and anisole (3 × 3 μm). SBS's microphase separation and P3HT's nanofiber production are seen in both pictures. The phase picture's magnified image is displayed in the upper left inset of (b) [2] [Open access]

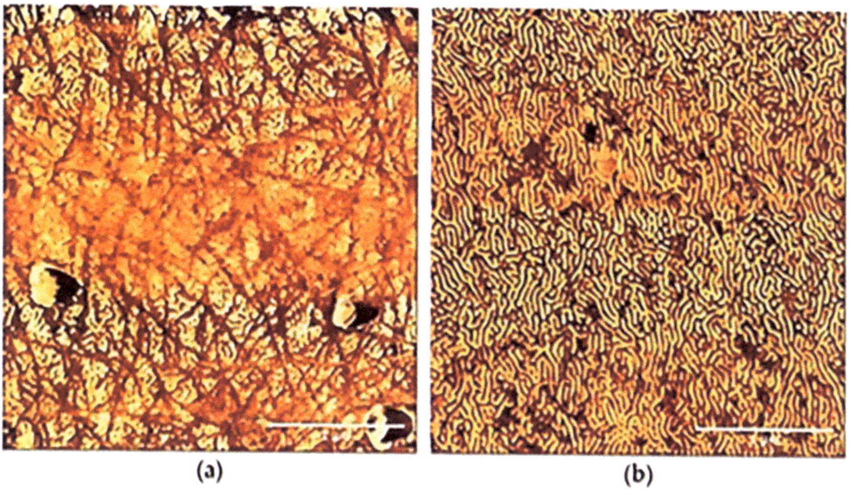

Fig. 6.3 SFM phase image of thin film obtained by spin-casting a solution of P3HT (0.05 wt%) and SBS (0.5 wt%) in a 70:30 (v/v) mixture of chloroform and anisole (3 × 3 μm) in **a** unstretched and **b** stretched sections after elongation treatment [2] [Open access]

composite and improves mechanical (stress and deformation) and thermal (dimensional) stability while lowering gas and liquid permeability. Additionally, the optical clarity of pure polymer is maintained while the fire resistance is increased. Last but not least, it might exhibit intriguing conductivity and enhanced biodegradability [2, 5]. Examples of these fillings include silicon carbide, aluminum nitride, and boron nitride. Materials with high thermal conductivity can have faster response times, which may be particularly useful when detecting dynamic temperature or heat [2, 6]. Thermal conductivity determines the amount and rate of heat extraction from a contacting object by mechanically mixing a chitosan-CuMnO nanocomposite, which is sensitive to temperature in the range of 21°C–80 °C.

6.1.2.1 Mechanical Characteristics

The high hardness and strong affinity between polymer materials and organoclay are key factors in the improvement of the mechanical characteristics of polymer nanocomposites. Stronger edge connections, for instance, dramatically reduce the stress absorption point of recurrent deformation, which happens naturally in ordinary glass fiber reinforced composites and leads to subpar fatigue performance. The addition of organoclay to polymer nanocomposites significantly improves their mechanical properties. The high hardness of the polymer matrix and the strong affinity between the polymer components and organoclay are two important elements that have contributed to this development.

The composite is stronger and able to withstand mechanical loads due to its high hardness, which increases its rigidity and resistance to deformation. However, a better distribution of clay particles within the polymer matrix is made possible by the strong interaction between the polymer and organoclay, which increases strength, load-bearing capacity, and overall structural integrity. Together, these elements improve the performance of material under mechanical loads, allowing the use of polymer nanocomposites in a variety of complex applications. Figure 6.4 illustrates the relatively low-magnification photographs of polystyrene (PS) and styrene–butadiene–styrene (SBS) blend. In the literature, it is suggested that PS forms the matrix, while the islands are believed to exhibit SBS phase [7]. Since the phase is based on a wide range of physical properties, it is usually not possible to estimate a specific mechanical property from the phase difference. As discussed earlier, the phase image (Fig. 6.4c) can be qualitatively, not quantitatively, distinguished between different polymers. According to some studies, the softer material is found in the darker phase regions (i.e., the smaller or more negative phase) [8–10], while other studies show the opposite [1, 2, 4]. In this case, the SBS thermoplastic elastomer is softer than the nearby brittle PS matrix; hence, in this case, the harder material is represented by the darker (smaller) phase. It should be noted that the topographic image (Fig. 6.4a) is the only one that shows the actual differences in height, but the amplitude image (Fig. 6.4b) emphasizes the edges, making it appear much sharper [3].

Since the concentration of nanofillers in the above formulations significantly exceeds the leakage current threshold [3, 5, 6], it is decided to investigate the

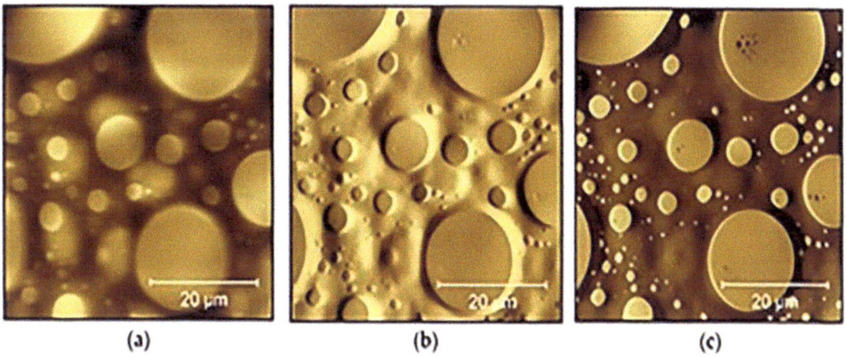

Fig. 6.4 AFM images of **a** a combination of polystyrene and styrene–butadiene–styrene block copolymer (PS/SBS), **b** topography (height range 2.4 µm) amplitude, and **c** phase image (phase range 137°) [4] [Open access]

morphology of these samples. This serves as a useful reference point for assessing the dispersion efficiency of nanofillers in the resin. To ensure consistent results, the entire surface of epoxy nanocomposites is analyzed. In particular, the multiple points of the sample surface are scanned to ensure the accuracy of the images. Since the etched samples are used for AFM examination, the carbon nanofillers are visible and distinguishable from the surrounding epoxy matrix [3]. Interestingly, as shown in Fig. 6.5, PS remnants can also be seen in SBS islands at higher magnification. In this case, the amplitude image (Fig. 6.5b) gives a clearer picture of the distribution of islands within the islands, the phase image (Fig. 6.5c) again shows clear differences, and topographic differences (Fig. 6.5a) are much smaller.

ABS is investigated as an additional sample whose composition is known after this extensive test of the phase imaging approach. The results are shown in Fig. 6.6, which

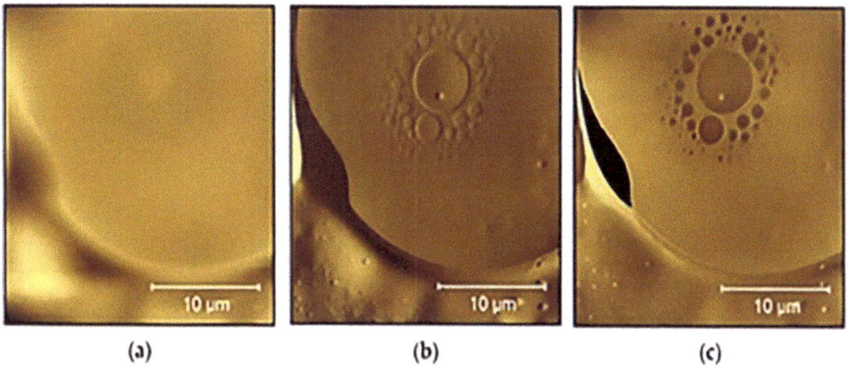

Fig. 6.5 HR-AFM images of polystyrene and styrene–butadiene–styrene (PS/SBS): **a** topography (height range: 1.0 µm); **b** amplitude; phase image (phase range: 55°); **c** except for the black area that is not accurately measured [4] [Open access]

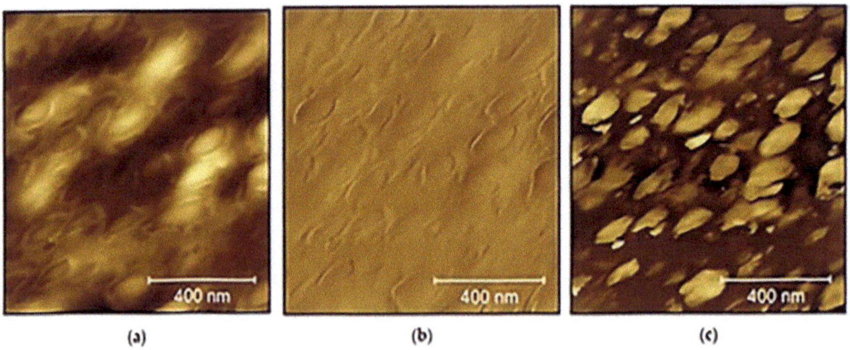

Fig. 6.6 AFM images of acrylonitrile butadiene styrene (ABS) surface: **a** topography (height range: 16 nm); **b** amplitude; **c** phase image (phase range: 90°) [4] [Open access]

is captured at a higher resolution than the previous images with an edge length of 1 μm. Although the amplitude (Fig. 6.6b) does not allow a clear distinction between the islands and matrix, and very small differences in height (the height range in Fig. 6.6a is 16 nm) are not very noticeable, the amplitude shows a strong contrast (phase range, 90°), allowing the two different polymers to be distinguished.

In terms of strength and modulus, it begins by describing a particular nanocomposite material and demonstrating its remarkable mechanical properties, including 40% tensile strength, 60% flexural strength, 68% elongation strength, and 126% bending strength. It then cites a study by rapid thermal processing (RTP) that demonstrates comparable properties in nylon-6 clay-based nanocomposites. The passage explains that the increase in modulus (stiffness) is likely due to the final nanostructure and the higher aspect ratio of the clay layers, which help strengthen the material. In addition, significant progress has been made in creating exfoliated (separated) nanostructures, such as amine-treated thermoset montmorillonite (MMT) epoxy nanocomposites, which further improve the material properties.

The text moves on to a discussion of flexible temperature sensors, explaining that these sensors must work effectively on complex or uneven surfaces. Thinner materials with low elastic modulus and high stretchability are ideal for this purpose. Due to their flexibility, polymeric materials PI (polyimide), PDMS (polydimethylsiloxane), PES (polyethersulfone), PET (polyethylene terephthalate), and PU (polyurethane) are preferred for use in these sensors.

Finally, the paragraph links the improved mechanical properties of nanocomposites with their possible use in stretchable, flexible sensors that can operate on a variety of surfaces.

6.1.2.2 Properties of Thermal

In general, the easily thermoset nature of polymer nanocomposites results in the formation of a higher volume of volatile compounds [11]. Enhanced thermal stability of nanocomposites composed of multiple organic layers and polymer matrix [9]. Increased thermal stability, for example, explains the presence of intercalated nanocomposites [10]. Figures 6.7, 6.8, and 6.9 show the high-resolution micrographs (height and amplitude profiles) of the fracture surface of epoxy resin samples heat-treated to TBD + 3%EG, TBD + 0.64% multi-walled carbon nanotubes (MWCNTs), and TBD + 1%CNFs, as well as Transmission electron microscopy (TEM) micrographs obtained for each of the heat-treated carbon nanofillers, EG, MWCNTs, and Microfibrillated or nanofibrillated cellulose (CNFs). TEM images are a reliable means of visualizing the morphological properties of nanoparticles and help distinguish them from each other when incorporated into epoxy resin. It is worth noting that the morphological changes in carbon nanofibers resulting from heat treatment are visible in TEM images of the CNF sample (see Fig. 6.7) [3].

Nanoparticles such as carbon nanotubes, nanoclay and graphene are often used to modify the chemical, mechanical, electrical, optical and thermal properties of polymer nanocomposites. But before the full potential of polymer nanocomposites can be fully realized, several important obstacles must be overcome. Various parameters, including processing method, interfacial contact of nanoparticles with the host polymer, and dispersion state of nanoparticles, influence how well the

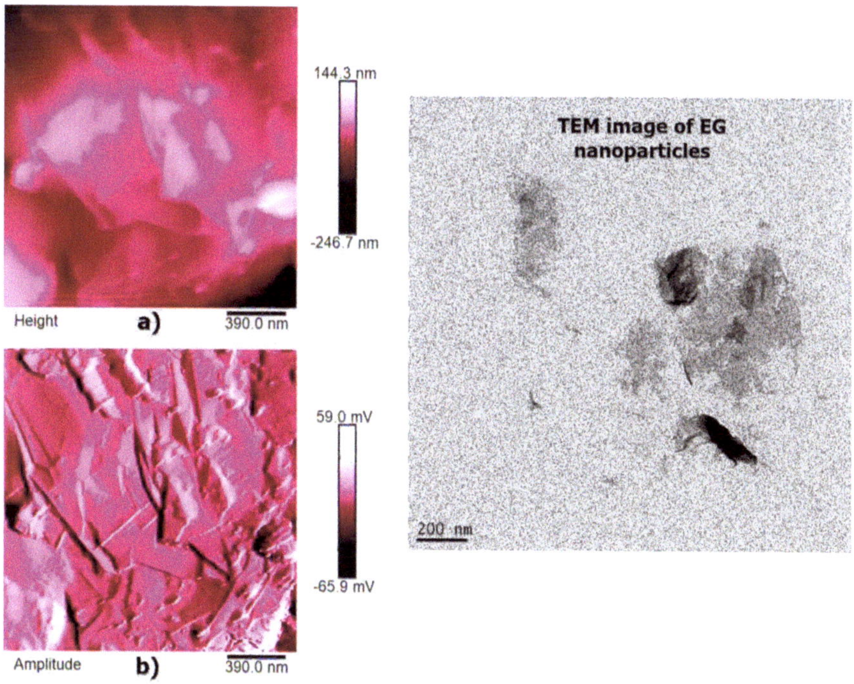

Fig. 6.7 TEM of EG nanoparticles (see right) and high-resolution atomic electron micrograph of the fracture surface of TBD + 3%EG sample in tapping mode (see left: **a** height profile, **b** amplitude profile) [3] [Open access]

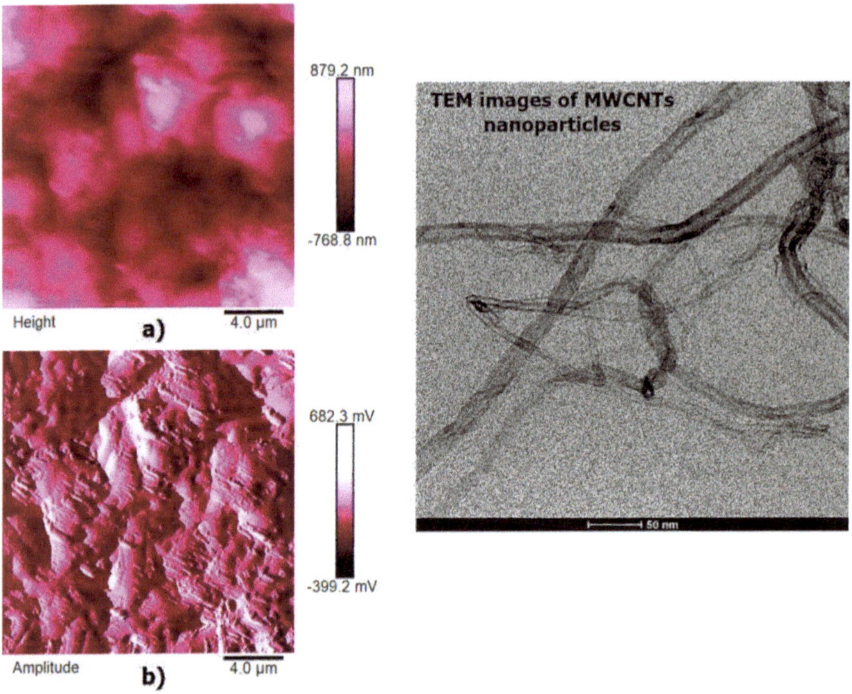

Fig. 6.8 HR-AFM micrographs of the fracture surface of the MWCNT sample in tapping mode: **a** height profile, **b** amplitude profile, and (right) transmission electron micrograph of nanoparticles made from MWCNTs [3] [Open access]

addition of nanoparticles improves the mechanical, thermal, electrical, and rheological properties of polymers. Despite many recent advances in the field of polymer nanocomposites, there is a lack of research on how processing parameters influence the morphology, structure, and functional properties of these materials. There is a need to establish a connection between structural optimization of distributions and final properties of polymer nanocomposites. Improved characterization methods are needed to measure the distribution and concentration of nanoparticles,, and evaluate the type of contact that exists between the polymer and nanoparticles. In addition, there is an urgent need to develop more accurate models that can predict the mechanical properties of polymer nanocomposites as a function of relevant variables, including nanoparticle orientation, type of functional groups and molecular weight of polymer chain. Figure 6.10a shows a flexible temperature sensor with two curved electrodes; (b) substrate with waves, and (c) the final, optimized substrate design [11]. Thermal resistance under external load, which can be quantified by thermal distortion temperature (HDT) is another important thermal behavior. Improving HDT in typical microparticle reinforced polymer composites is extremely difficult. (a) A temperature sensor fastened to the yarn in Fig. 6.10 .

Finally, the mechanical and fire-retardant properties, while the mechanical properties of polymer composites with traditional fire retardants deteriorate, are better

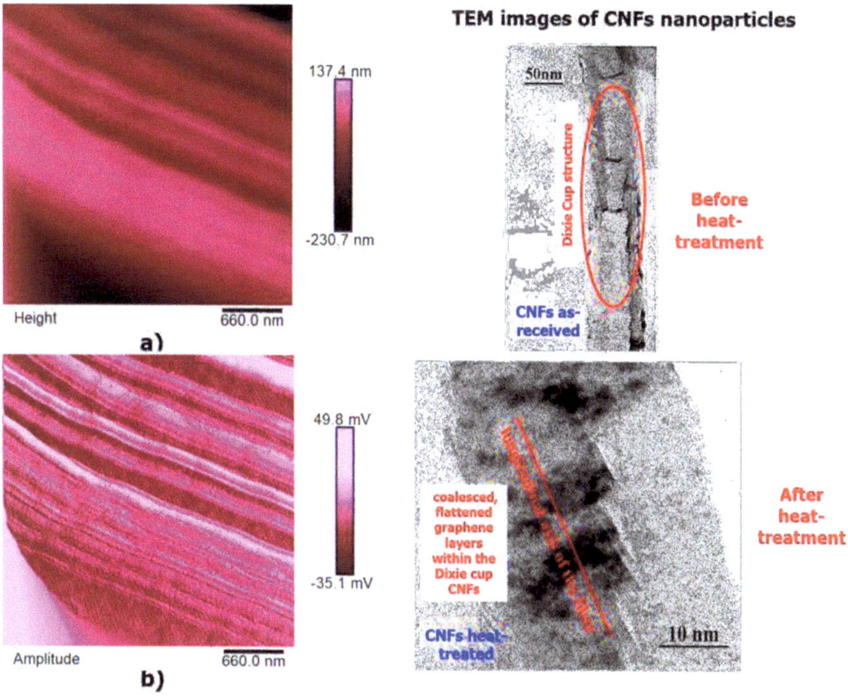

Fig. 6.9 HR-AFM micrographs of CNF nanoparticles in near-contact mode before and after heat treatment (left–amplitude profile, right–height profile), as well as TEM micrographs of the fracture surface of the sample (right—CNF nanoparticles) [3] [Open access]

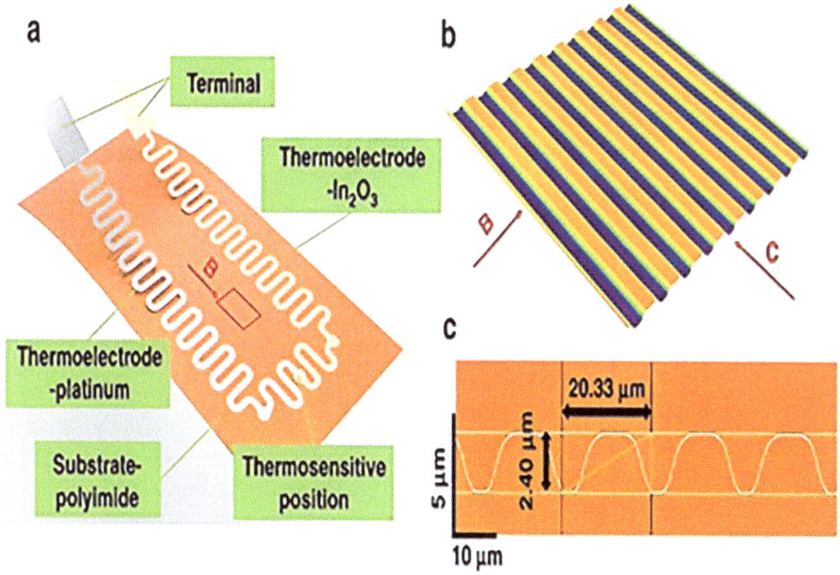

Fig. 6.10 **a** Flexible sensor with two curved electrodes for temperature sensing **b** Substrate with waves **c** Final optimized substrate design [39] [Open access]

for clay-based polymer nanocomposites. The carbon layers formed during combustion and the structure of polymer nanocomposites give them fire-resistant minerals contained in clay. The multilayer structure of clay makes it an excellent mass transfer barrier and insulator. Degraded volatiles cannot be released from the polymer matrix due to the formation of charcoal and clay structure [12]. The conductive path built into a thermal and tactile sensor using extra graphene and carbon nanotube tube (CNT).

6.1.2.3 Properties of Nanocomposites' Barriers

In terms of gases such as oxygen, nitrogen, and carbon dioxide, water, and hydrocarbons, polymer nanocomposites are superior. This is demonstrated by creating electrospun nanofibrous mats from polyvinyl acetate and chitosan [12]. Fish protein extracts isolated from Clupeonella Cultriventris caspia are used as antioxidants in the partner. It is found that increased concentration of protein extracts increases fiber diameter (Fig. 6.11). Meanwhile, the materials containing chitosan and protein hydrolysate are found to have a higher surface roughness index compared to the electrospun nanofibrous mat prepared from polyvinyl alcohol (PVA) alone. This is due to the addition of these materials, which increases the viscosity of the spinning solution [6].

There has been increased interest in polymer nanocomposites reinforced with chips. Numerous studies have shown that the aspect ratio of clay platelets has a significant effect on reducing permeability, with high ratios significantly increasing gas barrier performance. For polymer nanocomposites filled with 2% organ clay and having a completely exfoliated coefficient of relative permeability of polyimide (PI), the greatest gas barrier characteristics are noted. Fillers such as layered silicate clay, graphene and graphene oxide, boron nitride, and molybdenum disulfide [2, 13]. This interest can be explained by the excellent properties of polymer matrices (low processing temperatures, low weight, low cost, versatility, and ease of manufacture) as well as the significant improvement in their mechanical and physical properties as a result of adding only a small percentage of polymer. Filling factor. In particular, the results show that a polymer nanocomposite with the addition of silicate clay [2], graphene, graphene oxide [14], boron nitride [2], and molybdenum disulfide [14] significantly improves the barrier ability. Local crystallinity contours are shown in the center of the simulation area at about 15 ns (higher crystallinity with added nanoparticles) and at about 56 ns (lower crystallinity with added nanoparticles) for neat polymers (top row) and synthesized nanopolymers (bottom row, $D = 4.5$ nm, $\varphi = 2.31\%$). Crystallinity greater than 90% is found in the red areas (color online) (see also movies in supplementary material showing sections above simulation area). The boxes of sections show the corresponding molecular orientations of neat and synthesized polymers at the end of crystallization; for clarity, the molecules are shown in different colors. Movies of the molecular crystallinity of neat and synthesized polymers can be seen in the supplementary material and are shown in Fig. 6.12 [31].

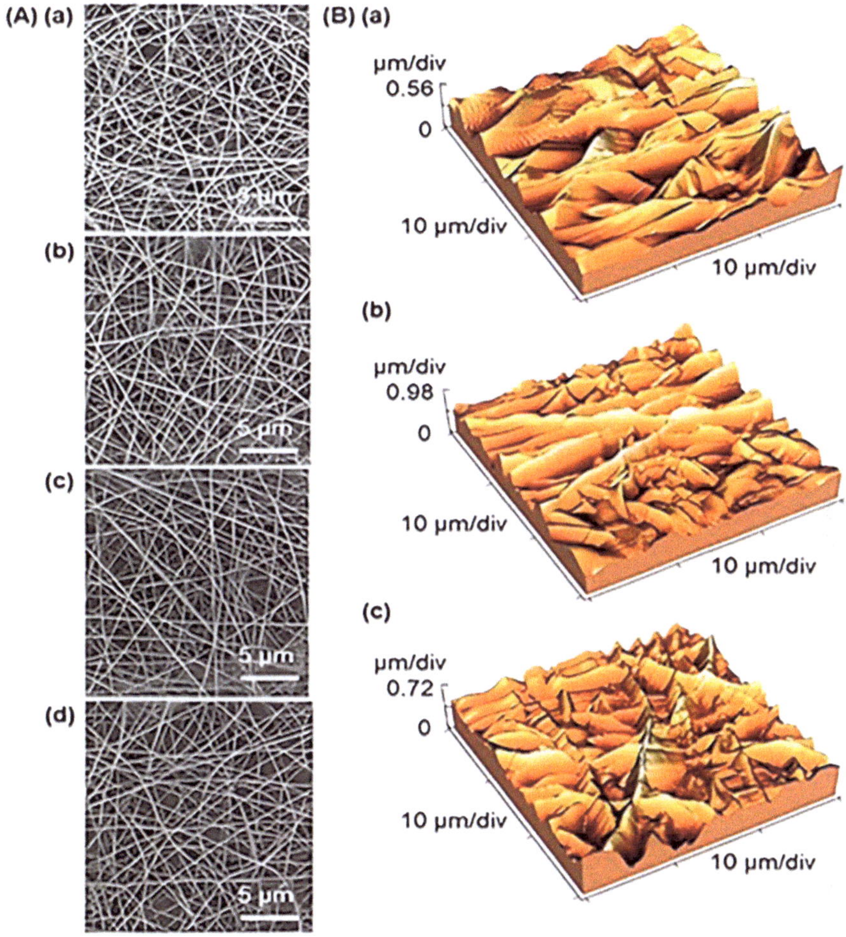

Fig. 6.11 **A** SEM images of electrospun polyvinyl acetate chitosan nanofibrous mats with four different protein extract concentrations: **a** 0 mg/mL, **b** 1 mg/mL, **c** 3 mg/mL, and **d** 5 mg/mL. **B** 3D AFM images of electrospun nanofibrous mats made from **a** polyvinyl acetate alone, **b** polyvinyl acetate mixed with chitosan, and **c** polyvinyl acetate and chitosan mixed with protein solutions [6] [Open access]

6.1.2.4 Nanocomposites' Electrical Conductivity

Although the clay layers can be considered as insulators, the protection and mobility of water layers provide a greater ionic conductivity. Natural clay resources exhibit only electrical properties, especially ionic conductivity. The water shells of interlayer signatures can be affected by the intercalation of neutral particles, which can significantly change ion mobility, electrical conductivity, and other electrical properties. Crown ether clay has ionic conductivity several orders of magnitude higher than that of equivalent clay. The approach is simple and effective and demonstrates

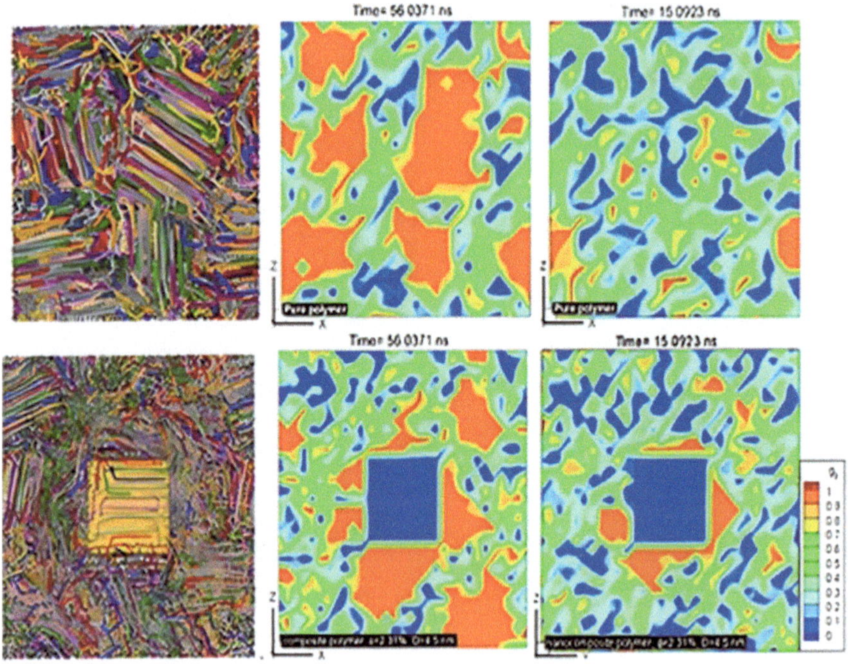

(b)

Fig. 6.12 Molecular crystallinity of neat and synthesized polymers can be seen in supplementary material [31] [Open access]

the presence of leakage paths in small samples [13]. Using energy-dispersive X-ray (EDX) analysis, the chemical composition of film surfaces is studied at the microscopic spatial domain level. The distribution of C, S, Ca, and Ti on the surface of the 01GNPs sample is shown in EDX photographs in Fig. 6.13. In particular, the elements Ca (see Fig. 6.13d) and Ti (see Fig. 6.13e) are present in the forms of $CaCO_3$ and TiO_2, respectively, while the element S (see Fig. 6.13c) is present in the form of $BaSO_4$ [13].

It is possible to predict the further development of conductivity. Polyethylene oxide (PEO) and ionic conductive clay nanocomposites have higher ionic conductivity than clay, with a maximum temperature of 600 K, which prevents the migration of anions (negatively charged clay layers) and interaction of complex cations. In addition, similar to the nanocomposite, the highest conductivity is recorded when PEO (40%) is directly dissolved in Li-MMT, in the range 105–104 S/cm in the direction parallel to the clay [2].

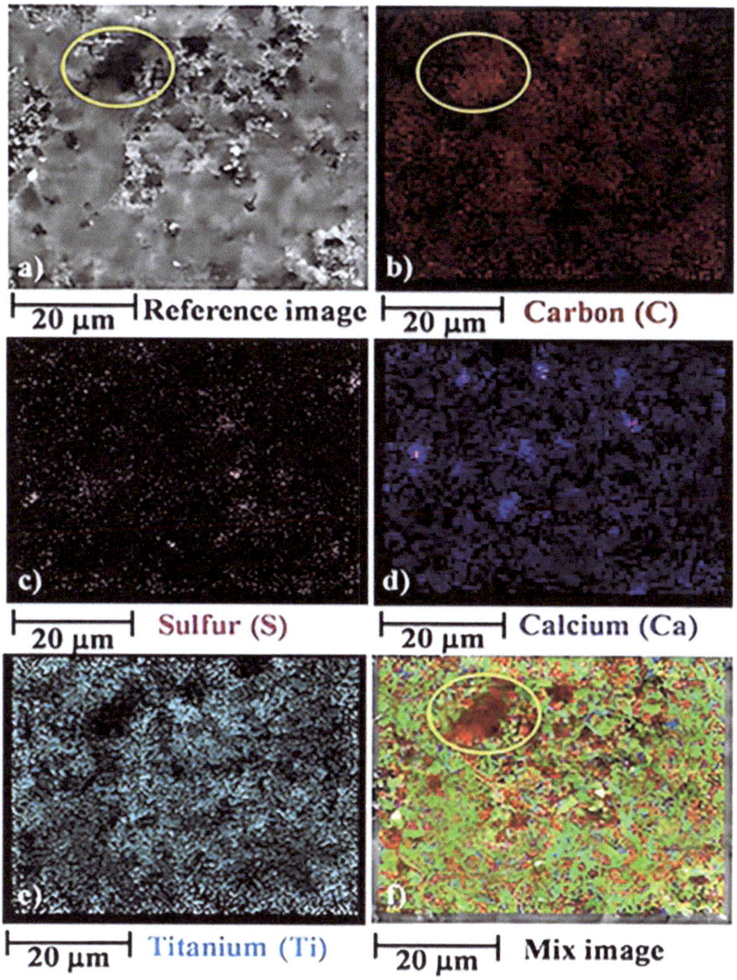

Fig. 6.13 Fracture surface of the 01GNPs sample is subjected to energy-dispersive X-ray spectroscopy (EDX) as follows: **a** control image; **b** C distribution map; **c** S distribution map; **d** Ca distribution map; **e** Ti distribution map; and **f** mixed image [13] [Open access]

6.1.2.5 Nanocomposites of Nanocellulose and Graphene

Graphene consists of a single layer of sp^2-hybridized carbon atoms arranged in a two-dimensional honeycomb lattice. Scientists have been working hard to combine graphene with nanocellulose to create flexible, high-performance nanoelectronic composites. The unique properties of nanocellulose, such as its renewable nature, biodegradability, and biocompatibility, have attracted much attention. However, its use in flexible electronics is limited by its inherent lack of electrical conductivity.

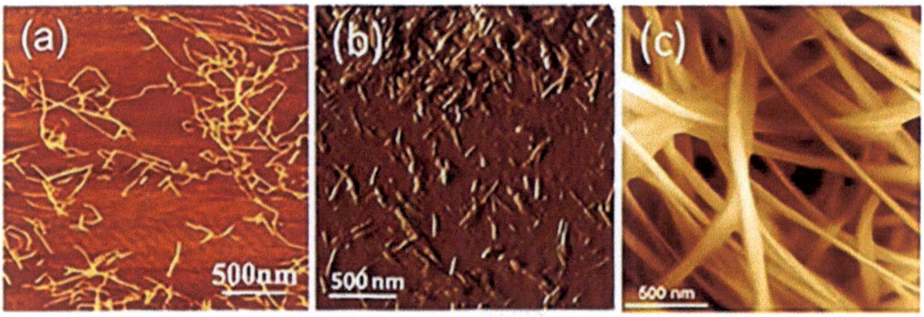

Fig. 6.14 AFM images of **a** CNF, **b** CNC, and **c** BC [14] [Open access]

On the other hand, graphene, a promising carbon-based nanomaterial, has remarkable mechanical strength, a large specific surface area, and extraordinary electrical conductivity. As a result, there is a growing interest in using graphene to improve the electrical performance of nanocellulose-based materials for flexible electronics. The synergistic effects of graphene and nanocellulose in electronic composites are still not fully understood despite continuous research [2, 13].

Petroleum-based products have historically served as flexible substrates for electrical devices [2, 5]. However, because they are polluting and difficult to dispose of in an environmentally acceptable manner, these materials pose serious environmental problems. In addition, they often have less-than-ideal mechanical properties. Amorphous regions are removed from cellulose nanocrystals (CNC) by acid hydrolysis, forming rigid, rod-like, and highly crystalline structures [9, 10].

Figure 6.14a, b shows AFM photographs of CNCs and CNFs, illustrating their structural differences [14]. Furthermore, glucose polymerization is a step in the synthesis of bacterial cellulose (BC), leading to the formation of β-1 $\rightarrow$ 4 glucan chains, which are then extruded extracellularly through the pores of the bacterial cell wall. These chains combine to create aggregates, 1.5 nm in width. As these aggregates form protofibrils and then flat ribbons, as shown in Fig. 6.14c [14], the BC network is further strengthened.

The researchers want to create next-generation flexible electronic materials with improved electrical conductivity, mechanical strength, and environmental sustainability by combining graphene and nanocellulose. However, further research is needed to fully understand the interactions between them and thermal stability. This demonstrates the urgent need for conductive materials with exceptional mechanical and electrical properties, as well as being renewable, biodegradable, and environmentally friendly.

The single-layer honeycomb lattice structure of graphene, a two-dimensional nanomaterial, consists of closely packed carbon atoms connected by sp^2 hybridization [12]. Due to its remarkable electrical, mechanical, and thermal properties as well as its large specific surface area and excellent biocompatibility, graphene has been extensively studied in several fields, including energy storage, biomedicine, sensors, catalysis, and optoelectronics [2, 15]. In addition, its exceptional electrical and mechanical strength are ideal for flexible electronic devices such as electron shells and flexible energy storage systems due to its conductivity and flexibility [13, 14].

Despite these advantages, graphene does have its drawbacks. Its performance can be affected by a noticeable increase in granule size upon high-temperature magnetization. To address this, scientists have considered combining graphene with nanocellulose, which is classified into two main types depending on how it is processed. CNF, which is produced through mechanical processing, has a high aspect ratio and is flexible. In contrast, CNCs, also known as nanowhiskers, are created through acid hydrolysis and have higher crystallinity but a lower aspect ratio. While the more amorphous regions of fiber structure are preserved through the gentle chemical and mechanical processes used to separate CNF, these substances increase their functionality in nanoelectronic applications.

6.1.3 Metallic Nanocomposites

Nanocomposite materials with one or more dimensions less than 100 nm are new materials. With the constant need to reduce vehicle weight, fuel consumption, and exhaust emissions—all of which have a direct impact on environmental pollution—their use in the automotive industry has increased. New lightweight and durable materials are needed to reduce the weight of vehicles. These materials can be created by varying combinations of base, reinforcement, and the amount of reinforcement. If more than one reinforcement is used, hybrid nanocomposites can be created by adding more diverse reinforcements if the base material is unsuitable. Nanocomposites have many advantages. These include increased productivity, improved thermal and environmental stability, fewer wear parts, lighter vehicles, and reduced emissions of pollutants into the environment [16]. However, the quantum effect causes metal nanoparticles to have completely different properties depending on size compared to their equivalent size [29]. Found that a very low dispersion, or 0.25 vol.%, of the most realistic functional filler content for an effective temperature sensor is 10 S m [2, 4]. The ability of poly-n-isopropylacrylamide/Polyethyleneimine (PNIPAm/PEI) to interact with both pH and temperature stimuli is one of its special properties. Certain PNIPAm chains coexist with PEI chains in the shell of particles, as mentioned previously by XPS data. It is observed that the morphology of microgels in aqueous solution using liquid-state AFM confirms the morphological changes of microgels above and below the bulk phase transition temperature (VPTT) of 32 °C. The image of microgel particles with smooth surface morphology and particle sizes between 150 and 200 nm at 29 °C is shown in Fig. 6.15a. Once the temperature reaches 45 °C, which is higher, the microgels shrank to 100–150 nm in size and formed a porous surface (Fig. 6.15b) [17].

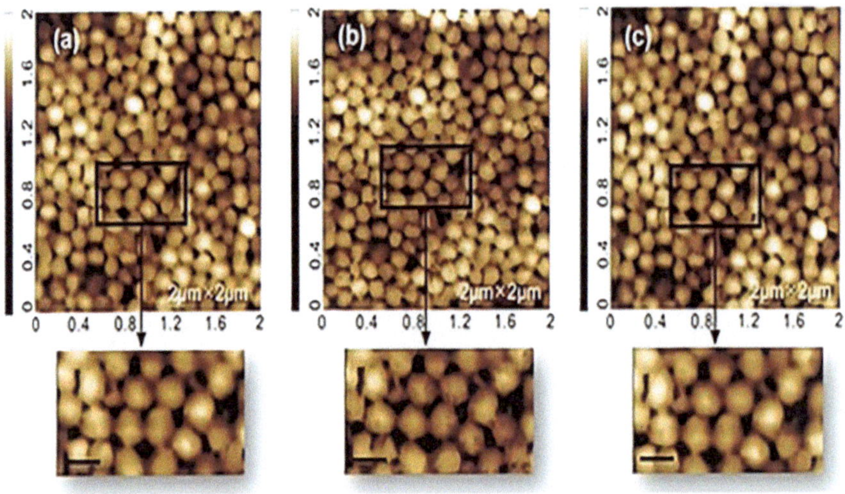

Fig. 6.15 PNIPAM/PEI microgel particles are recorded in liquid mode at three different temperatures: **a** 29 °C, **b** 45 °C, and **c** cooled from 45 to 29 °C. (Scale bar: 200 nm) [17] [Open access]

6.1.3.1 Nanocomposites of Ceramic

Thermal conductivity refers to the ability of a material to transfer heat. In the context of chitosan-CuMnO nanocomposites, a study found that as temperature increases (ranging from 21 °C to 80 °C), the thermal transfer resistance decreases, which in turn increases the thermal conductivity. This is found at pressure, 312 MPa, during fabrication, with the highest temperature sensitivity recorded at 347.5 K [2, 5].

Furthermore, a study using a nickel oxide matrix on a PI (polyimide) substrate to measure temperatures up to 200 °C using inkjet printing technology found that bending did not significantly affect the performance [2, 15]. Furthermore, an ultra-sensitive thermomechanical dual-parameter monitoring sensor is developed.

The spatial distribution of nanocomposites in Fig. 2.16 shows the distribution of nanoparticles (NPs) in SiO_2, MgO, and Al_2O_3 nanocomposites at 3 and 5 wt%. Using the SEM–EDS mapping approach, the distribution is clearly shown. The epoxy nanocomposites show a semi-uniform dispersion of NPs, with SEM–EDS maps of epoxy-containing 3 wt% SiO_2, MgO, and Al_2O_3 shown in Fig. 6.16a–c, respectively [18].

6.1.3.2 Materials for Flexible Substrates

Any flexible sensor must include substrates, since substrates greatly influence the flexibility of the sensor. Without affecting the sensing mechanism, it gives the sensing material flexibility and stability. There has been extensive research into the potential use of flexible electronics in low-cost electronic products for several interesting applications, including electronic skin, health monitoring patches, and other lighting

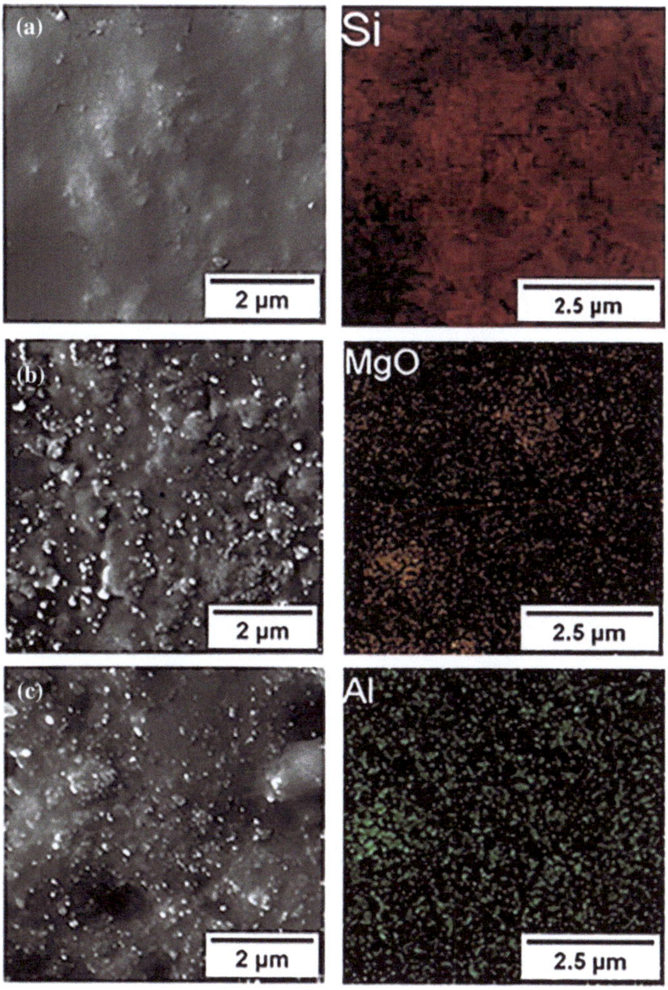

Fig. 6.16 SEM mapping &EDS image **a** Epoxy/3 wt% SiO_2, **b** Epoxy (3 wt% MgO), **c** 3 wt% Al_2O_3 epoxy [18] [Open access]

and sensing applications. Because organic field-effect transistors are a fundamental component of most of these systems and can be processed at low temperatures and with inherent flexibility, they are being extensively explored for use in wearable electronics. These flexible gadgets are demonstrated using a variety of new materials. However, before this technology can be applied in practical and credible ways, several challenges need to be addressed. In addition, to reduce the environmental impact of e-waste, it is necessary to develop and optimize manufacturing procedures suitable for green electronics [2, 4]. Compared to rigid glass substrates, flexible substrates have many advantages, including low weight, ease of coil fabrication, and new applications such as building-integrated photovoltaic systems. Copper indium gallium selenide

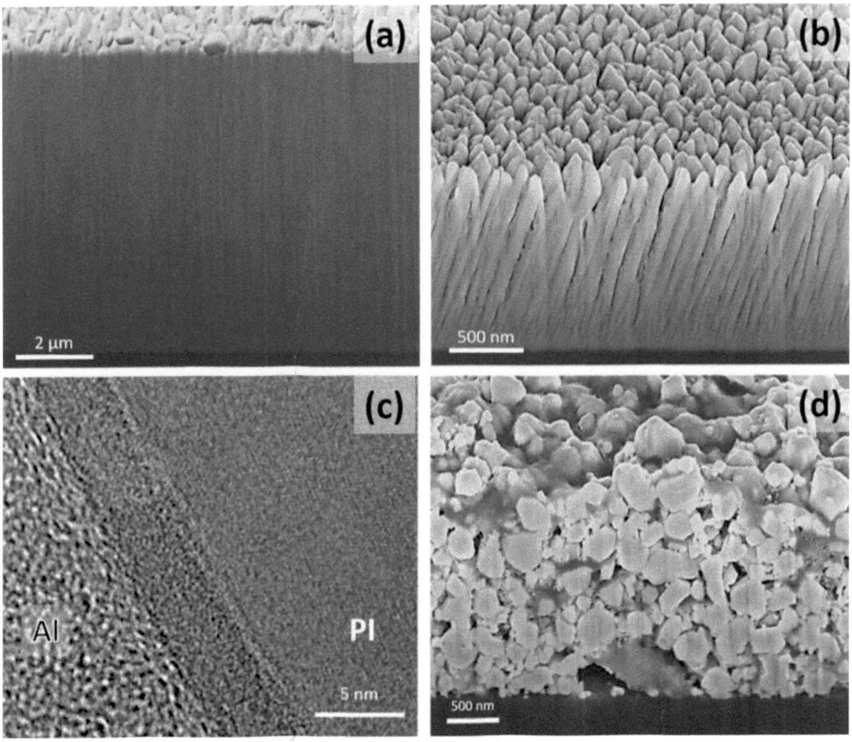

Fig. 6.17 Cross sections of thin films. A rotating molybdenum-aluminum-titanium target is placed in front of a molybdenum-aluminum-titanium composite film, and an SEM image **a** shows thin vertical film growth with vertical columns on the substrate surface; **b** vertical growth of an open-air film grown from the same target and using the same deposition parameters under tilted conditions, resulting in a molybdenum-aluminum-titanium composite [7]. **c** TEM cross section of an aluminum film thermally deposited on a polyimide (PI) substrate, clearly showing the variable region near the surface that formed during the film deposition process **d** SEM image of copper film deposited [19] [Open access]

(CIGSe) solar cells use a variety of metal substrates, including industrial steel and titanium foil. Steel and stainless steel are used. When selecting suitable metal substrates, it is important to consider factors such as cost, impurity content, surface roughness and coefficient of thermal expansion. Exceptionally high efficiency (about 18%) is found using stainless steel and (expensive) titanium foil. For industrial steel, a reasonable efficiency of about 13% has also been achieved using suitable insulating layers to stop the spread of impurities, especially the spread of iron from the substrate [2, 6]. This growth mode results in the formation of columnar V-shaped grains, as shown in the SEM cross section obtained by focused ion beam (FIB) milling in Fig. 6.17a [19]. In this case, the column boundaries typically intersect the entire film. Before continuous films can completely cover the substrate surface, a certain thickness is required for the formation of islands, which depends on the mixture of the substrate material and the thin film, as well as on the growth conditions [19].

6.1.3.3 Pigments and Nanocomposite Coatings

The process of creating hybrid nanomaterials involves combining layers of artificial or natural clay with copolymers or surfactants to create new nano-additives. The exceptional adhesion of these materials, which improves their mechanical and thermal properties allows them to form a strong bond with the polymer matrix.

Since their molecules are smaller than the wavelength of light, these materials can be used as nano pigments, making materials colored with plant dyes completely transparent. In addition to their high oxygen content, which increases their long-term stability, these compounds have exceptional qualities such as bright fluorescence, resistance to UV radiation, and high temperatures.

An important factor in increasing the effectiveness of nanocoatings is vegetable oil, which improves the ability to withstand water (hydrophobicity) and serves as a natural binding agent between nanoparticles, preventing cracking and contributing to the creation of uniform, heat-resistant coating layers. While the coatings containing hydrogenated oils exhibited multi-level roughness with open porosity, improving the water contact angle to 128° in the steady state and 138° in the advanced state, the coatings based on unsaturated oils generally demonstrated higher water resistance. The use of this technology with plant oil-coated nanoparticles deposited on paper surfaces is shown in Fig. 6.18, demonstrating the high efficiency of these materials in improving the performance of smart coatings and materials [20].

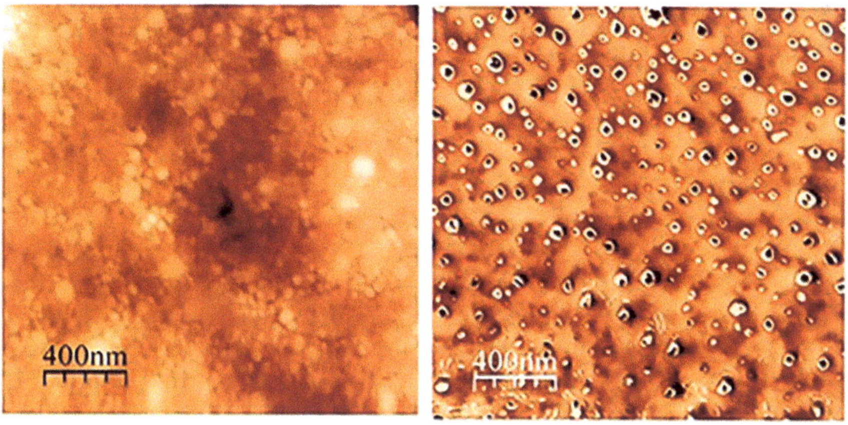

(a) (b)

Fig. 6.18 SMI/oil nanoparticles are applied using encapsulated vegetable oils such as **a** 50% soybean oil and **b** 70% palm oil, creating a continuous coating layer [20] [Open access]

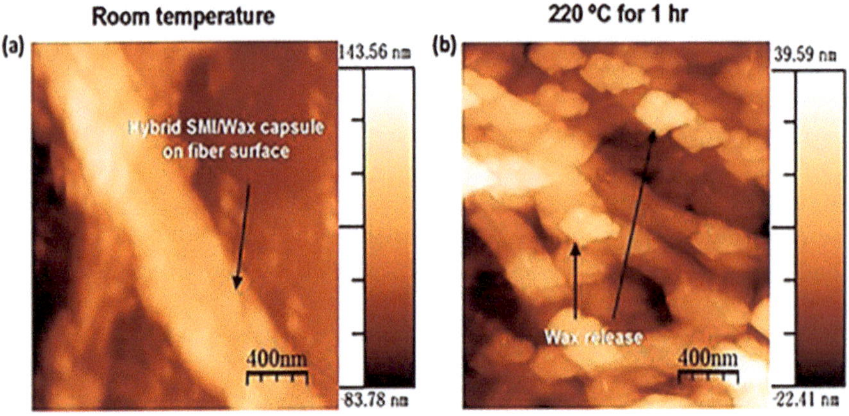

Fig. 6.19 Morphology of MFC obtained using AFM; **a** untreated, **b** treated at temperatures up to 220 °C

Modified cellulose microsheets (MFC) fiber structure with multi-scale morphology formed thin wax spots (Fig. 6.19) as a result of combined wax release and migration, which determined the resulting surface hydrophobicity. As mentioned earlier, the barrier properties of MFC/NFC films are significantly reduced in the presence of moisture, but the current design allows for a combination of hydrophobicity and oxygen barrier resistance. Alternatively, modified MFC fibers can be blended as a nanofiller in a biopolymer matrix to create a barrier coating that is completely protective. Thermal wax release and increased surface hydrophobicity during processing (e.g., extrusion) ensure interface compatibility with the non-polymer matrix [20].

It is expected that hazardous dyes such as cadmium (Cd) and palladium (Pd) will be replaced by plant dyes or nanoparticles composed of clay and organic dyes. In bulk polymers and coatings, plant pigments are common at the nanoscale. By selecting appropriate pigments from a large variety of organic pigments, pigments of different colors can be created.

6.1.3.4 Hybrid Nanocomposites

Combining two or more types of fillers to capitalize on their unique advantages results in hybrid fillers. MoS-BN nanospheres and TiO-silver are a few examples. In studies, these hybrid materials typically exhibit lower filtration thresholds than individual nanofillers. So, a temperature-sensitive nanocomposite is fabricated by physically combining chitosan and FeO-doped copper oxide nanopowder as the temperature increases, the resistance of the nanocomposite decreases markedly [21].

In addition, the effect of adding Ag-coated Ni particles and Ni nanocomposites to Spurr B polymer composites (a type of epoxy resin), as well as introducing CeO-CoO hybrid nanoparticles into a polymer adhesive, is investigated. Compared to composites with a single Ni filler, these composites show markedly lower stability in the temperature range of 60–110 °C. SilkyCombining two or more types of fillers to take advantage of their unique advantages results in hybrid fillers. MoS-BN nanospheres and TiO-silver are a few examples. In studies, these hybrid materials typically exhibit lower filtration thresholds than individual nanofillers. Therefore, a temperature-responsive nanocomposite is made by physically combining chitosan and FeO-doped copper oxide nanopowder; with increased temperature, the resistance of the nanocomposite decreases markedly [22].

In addition, the effect of adding Ag-coated Ni particles and Ni nanocomposites to Spurr B polymer composites (a type of epoxy resin), as well as the incorporation of CeO-CoO hybrid nanoparticles into a polymer adhesive, is investigated. Compared to composites with a single Ni filler, these composites show markedly lower stability in the temperature range, 60–110 °C [7, 23–25].

6.1.3.5 Materials for Flexible Electrodes

In addition to a flexible substrate, flexible temperature sensors require flexible electrodes with high conductivity. The resistance of these materials to cracking and diffusion during bending is limited by their hardness and brittleness [8]. By using carbon and metal nanoparticles as an alternative, an efficient channel for electrons to move through a network structure of connected nodes [9]. Since nanowires have a high aspect ratio, the contact resistance is reduced and, as a result, electrons can travel faster [10]. Electrical stability under bending, however, is hampered by contact resistance at the nanowire interface. Reducing the number of contacts points through alignment and nanowire welding can reduce contact resistance, although this method is useful for specific resources [26]. With improved electronic and optomechanical capabilities, various types of nanomaterials, such as nanogrids and nanoelectrodes, are available. It has been established that the grain boundaries of nanolithography are electrically conductive, optically transparent, and have good extensibility [26]. Another common material for flexible electrodes is conductive ink, which is preferred because it is easy to apply through printing or coating [11]. To create conductive inks, metal nanoparticles of silver, gold, copper, or brass, carbon nanotubes, and other carbon-based materials are introduced into an aqueous environment [29] in various printed electrode designs that can withstand normal bending and stress [16]. The most commonly used carbon compounds to improve adhesion to textile fabrics are r-GO and CNTs [29].

6.1.3.6 Sensor and Medical Device Applications

When the polymer is combined with xenograft platelets, it may be possible to improve hydrogel performance beyond drug delivery. Clay-based nanocomposites are expected to find use in a variety of biological applications, including fast motors and artificial muscles, especially with regard to thermal sensitivity, swelling, swelling rate, and molecular diffusion. A photoacoustic sensing (PAS) system is described, designed to detect carbon dioxide (CO_2) in the blood. At the center of the development is a PA sensor, which consists of a detection chamber and a measurement cell, according to the so-called "dual-chamber principle." The goal is to continuously and reliably monitor transdermal carbon dioxide levels. This is extremely important, for example, for monitoring patients in intensive care units. The light source is an infrared light-emitting diode (LED) with a peak wavelength of 4.3 μm. To selectively detect the target gas, the hermetically sealed detection chamber contains a microelectromechanical system (MEMS) microphone and target carbon dioxide. The absorption path length of a 2.0 miniature PA CO_2 sensor in laboratory conditions is determined based on calculations and measurement data. A commercial infrared CO sensor is compared with a similar meter to study cross-sensitivity, detection limits, and signal stability. With a 3.0 mm diameter sensor. In laboratory testing, the detection limit of the CO_2 PA sensor shown is 1 vol.% CO_2. Our PA sensor demonstrated faster response and recovery time compared to a commercial sensor, as well as reduced exposure to moisture and oxygen on the recorded signal. When the created sensor system is finally applied to the skin of the test subject, it is possible to calculate the arterial time of 181 min.

6.1.3.7 Nanocomposite Hydrogels

Stimulation-responsive smart polymers have been pioneered as sensors, actuators, tissue engineering materials for regenerative medicine, and drug delivery. Evaluated several polymers systems, their applications, and operating procedures. Natural and artificial polymer materials form the basis for the creation of polymer hydrogels. Collagen, hyaluronan, shitosan, alginate, chondrocyte sulfate, fibrin, and silk are examples of natural materials used. Both chondroitin sulfate (CS) and hyaluronan (HA), which have viscoelastic properties, have been used in cartilage engineering. Glycidyl methacrylate is included in CS to modify it. It has been demonstrated to be harmless for cartilage replacement and can be photopolymerized into a homogeneous hydrogel.

6.1.3.8 Gaps in Technology and Potential Fixes

Better sensitivity, longer range, and quicker response time can be created by careful selection of materials, microstructures, and logical production procedures. But there are still issues with multifunctionality and limitations related to uniform and homogeneous dispersion of nanofillers to prevent agglomeration, as well as using sustainable, disposable, and recyclable materials in the production of flexible temperature sensors. Temperature sensors may require extensive signal processing to separate mixed information, as well as more complicated manufacturing and design if more than one sensor is created that reacts to a particular stimulus in order to directly acquire individual information [11].

6.1.3.9 Functional Uses

Foil coating agent used in various food processing applications, especially when moisture resistance or protection is required. PVA protects active chemicals from moisture, oxygen, and other external elements present in the formulations covering striped food tablets and dietary supplement tablets, while destroying their taste and aroma. This makes the final product easy to use and delicious. PVA coatings can be applied to tablets, capsules, and other dosage forms that typically operate under fairly high solid loads due to the thickness (viscosity) of PVA. Use food category apps to maintain the taste, texture, and quality of foods with high moisture content. Polyvinyl alcohol is used. PVA can be used in confectionery products to keep moisture-sensitive compounds intact. In addition, TiO_2 is applied to make it stiffer without affecting the adhesion of the PVA layer to the paper surface. Its antibacterial properties can also be used in the production of disposable covers and wipes using UV-activated photocatalysis. Incorporating uniformly dispersed TiO_2 nanoparticles into PVA nanofiber coating is an interesting option used in conventional bulk coating. PVA/nano-TiO_2 fibers, especially when electrospinning, allow the formation of very thin layers when the solvent evaporates during the spinning process, and under the correct process conditions, a thin dry mat can be applied to the substrate [16].

6.1.3.10 Loading of Drug

Drugs incorporated into nanoparticles offer several attractive advantages over free drugs, such as increased bioavailability, reduced toxic side effects, improved drug delivery, and the ability to include additional features such as controlled release, imaging agents for imaging, targeted delivery, and administration of several drugs for combination therapy. Drug loading, or the mass ratio of drug to drug-loaded nanoparticles, is one of the important characteristics. Most nanoparticle systems currently have relatively low drug loading (<10 wt%), and it remains difficult to develop strategies to improve drug loading. An overview of current research on the development of nanoparticles with high drug loading (>10 wt%) is presented in this mini-review, which focuses on synthesis methods such as post-loading, co-loading, and pre-loading [10].

These three different approaches allow the creation of different nanoparticle systems with different synthesis methods, drug loading, encapsulation efficiency, release profiles, stability, and drug delivery applications that have been described and studied for both materials and pharmaceuticals. The purpose of presenting the advantages and disadvantages of these methods is to offer practical guidance for the development of the next generation of highly drug-loaded nanoparticles.

Its mechanism of action is to inhibit bacterial protein synthesis. Biologically active tetracyclines are slowly released after binding to the root surface.

6.1.3.11 Assessment of Nanocomposite

Tetracyclines have a broad spectrum of antimicrobial activity against both Gram-positive and Gram-negative bacterial strains. Nanocomposites are endowed with properties that are not only a synergistic combination of those inherent in the bulk components but also a set of new ones arising from using the substance in its molecular state, due to their hierarchical structure spanning different time and length scales.

In addition to their exceptional properties, one of the most intriguing aspects of using nanofillers is that, compared to traditional (micro-sized) fillers, they provide significant improvements at much lower concentrations, helping to achieve high levels of performance in a variety of engineering applications. This is a huge advantage because lighter composites, which are preferred in many applications, are produced with less filler [11]. One possible explanation for the remarkable physical and chemical characteristics of nanocomposites is the "nano effect," the effect of the hierarchical structure of this type of material. Regardless of filler design, uniform spraying of nano reinforcement results in the formation of a huge interfacial area per unit volume [26]. This drug specifically binds to 30S ribosomes and appears to interfere with protein synthesis by preventing aminoacyl from reaching the active transport machinery, which may require the presence of a peripheral carrier protein at the acceptor site of the mRNA-ribosome complex. The inhibitory properties of tetracycline can be reversed by ash, suggesting that the antibacterial effect of the antibiotic is due to the reversibly bound antibiotic rather than to a limited amount of permanently bound drug [29]. Because tetracycline easily binds to magnesium, aluminum, iron, and calcium, interfering with the body's ability to fully absorb it, it is usually taken after a meal, followed by a two-hour break. It is not recommended to consume dairy products or supplements containing iron immediately after taking the medicine.

6.1.3.12 Hydrogel Nanocomposite in Drug Delivery Formulations

Due to their unusual activity, which is determined by their unique structure at nanoscale, nanomaterials have a wide range of applications in the field of nanomedicine. However, nanomaterials have some inevitable disadvantages, including explosive release, rapid removal and poor adhesion. Hydrogels are three-dimensional scaffolds that have good drug release capacity and biocompatibility. In addition, hydrogels have disadvantages for biological applications, including low sensitivity, insufficient bioimaging capabilities, and poor antitumor ability. One potential way to overcome the disadvantages of nanoparticles is to incorporate them into a three-dimensional hydrogel network through physical or chemical covalent action. Multifunctional nanoparticles often serve as the functional core of nanocomposite hydrogel systems, imparting a range of characteristics to the hydrogels such as photothermal conversion, tumor targeting, conductivity, magnetothermal conversion, etc. Hydrogels can enhance the retention effect of nanomaterials and provide good plasticity to nanoparticles, allowing them to be easily adapted to a wide range of biomedical applications, including biosensors. There are several potential applications of nanocomposite hydrogel systems in biomedicine. This summarizes and discusses the most recent developments in composite nanomaterial hydrogels in biomedicine, such as drug and cell delivery, cancer treatment, tissue regeneration, biosensing, and bioimaging. We will also briefly touch on the current state of their commercialization in biomedicine. Although the hydrogel delivery formulation is believed to contain the necessary physical and transport components, the local inflammatory response may prevent it from performing its therapeutic function when present in vivo. Additional diffusion barriers created by the growth of fibrous capsules adjacent to the delivery mechanism can reduce the rate of drug release while increasing drug levels [11, 29]. The rate of matrix and drug degradation may increase in response to proteolytic activity. Therefore, the development of biocompatible controlled-release hydrogel formulations requires careful consideration of material selection, fabrication methods and device surface texture.

6.1.4 Techniques for Hydrogel Nanocomposites

One of two methods can be used to incorporate pharmaceuticals into the nanoporous hydrogel delivery matrix. The polymer precursor allows the creation of hydrogel networks and the encapsulation of a drug. Drug release can be controlled by diffusion and hydrogelation Swelling, reversible drug-polymer interactions or disruption of labile covalent bonds. Drug absorption depends on hydrogel networks, and diffusion is the main factor in drug absorption and release. Spreading and swelling of the gel affect the absorption of the drug. Drug diffusion reflects the release of the drug from the hydrogel network because hydrogels have ligand binding sites that mediate the interaction between the drug and polymer [10, 26, 27].

6.1.5 Nanocomposite Hydrogel Devices

Drug release routes will be determined by the physical properties of the hydrogel network and drug technique.

- Controlled diffusion
- Controlled swelling
- Governed by chemicals

Diffusion control models of hydrogels often use diffusion control, which describes a click law for diffusion with either fixed or variable diffusion coefficients. According to hydrodynamic or obstruction-based theories of drug diffusion on a free volume basis [10]. When drug diffusion is faster than hydrogel swelling, controlled swelling and release occur. Moving boundary conditions are used in the modeling process to simulate the release of molecules at the boundaries between the rubbery and glassy phases of the swollen hydrogels. The swelling rate of gel affects the particle discharge rate in these systems. Hydroxypropyl methyl cellulose (HPMC) is a kind of drug delivery system that is regulated by swelling. HPMC pills with excess drug content, hydrophilic 3D matrix materials, stay dry and glassy. The release of molecules based on the occurrence of a reaction in the delivery matrix is explained by chemically controlled release. Chemically regulated release can be classified based on the type of chemical reaction that occurs during drug release. Encapsulated pharmaceuticals may be released due to harsh conditions of pendant chains, external corrosion, or significant degradation of the polymer backbone. The parameters and standards for the composite preparations [28]. One of the hottest pieces of wearable technology, flexible temperature sensors, has received a lot of attention recently. High sensitivity and stability under mechanical loads are goals that scientists have achieved by applying new materials, designs, and production techniques. With an application domain focus on wearable electronics and a focus on nanocomposite materials, this study provides a comprehensive overview of contemporary flexible temperature sensors. This discusses the different types of nanocomposite sensing materials recently created and used, and carefully explains the sensing mechanism used. In general, the current study can be summarized as follows:

- The selection of appropriate materials and production techniques is essential for the good performance of sensors.
- Compared to using separate fillers, less material is required to get the same effect. However, the compatibility of the fillers must be carefully considered. Textile-based substrates, which can be knitted or woven, are becoming more popular than polymer substrates. Additionally, they are washable and are potential candidates for wearable technology and texting devices due to their biocompatibility.
- Carbon inks and printed metal nanomaterials can overcome the stiffness and brittleness limitations of standard electrode materials. Flexible temperature sensors can now maintain their performance properties under mechanical stress due to advanced materials, rational design, and efficient manufacturing methods. Zigzag, wavy, and curved design features can provide a variety of advantages, including improved sensitivity, pressure isolation, and resistance to wrinkles and tearing. Despite its limited use, printing is a popular technology because it produces less waste and is more affordable than photolithography.
- To address remaining issues, further study is needed, and a complete strategy should be used. Flexible temperature sensors have difficulties that must be overcome in order to become a commercial reality. There can be many fields of study. Seal the sensor to protect against damage caused by high humidity, constant exposure to moisture, or other external factors. Complete separation of temperature and mechanical signals, taking into account that sensor deformation can affect temperature estimation. Issues related to storing, processing, and transmitting huge amounts of real-time data from a variety of sensors that can accurately analyze geographic and temporal information. Temporary discrimination [12, 16, 11.

The field of nanocomposites has witnessed tremendous developments due to their numerous beneficial properties, including high mechanical strength, flexibility, tunable optical transparency, and photodynamic and photothermal activity.

These properties make nanocomposites very attractive for some applications, such as advanced wound dressings and flexible temperature sensors. Flexible Temperature Sensors: These sensors are extremely valuable as they can adhere to human skin in a conformal manner and provide reliable and continuous readings. Even with their advantages, there are still challenges related to strain sensitivity and stability. To improve their performance, issues such as small sensing range, slow response time in dynamic scenarios, high hysteresis, and decreased accuracy under heavy mechanical stress need to be addressed. Due to their antibacterial, angiogenic, and tissue engineering properties, nanocomposites are increasingly used in sophisticated wound dressings in addition to their anticancer properties. They are extremely useful for wound healing and treatment. To ensure the safety and effectiveness of these dressings, the potential cytotoxicity and immunological effects of chemicals and components used in them must be taken into account.

Problems

1. Explain the importance of vacuum technology, thin films, nanoscience, biomaterials, and plasma in surface engineering and science.
2. List several interfaces and surfaces modified by the nanostructure.
3. How would relative wear rates change if the surface geometry changed along a circular path?
4. Why are advanced microspheres useful for hydrogen storage?
5. What are the main causes of problems in nanocrystal-based electronic devices?

References

1. Lai, Q. T., Sun, Q. J., Tang, Z., Tang, X. G., & Zhao, X. H. (2023). Conjugated polymer-based nanocomposites for pressure sensors. *Molecules, 28*(4), 1627.
2. Goto, T., Morita, J., Maekawa, Y., Kanehashi, S., & Shimomura, T. (2020). Semiconducting properties of the hybrid film of elastic poly (styrene-b-butadiene-b-styrene) block copolymer and semiconducting poly (3-hexylthiophene) nanofibers. *Polymers, 12*(9), 2118.
3. Raimondo, M., Naddeo, C., Vertuccio, L., Lafdi, K., Sorrentino, A., & Guadagno, L. (2019). Carbon-based aeronautical epoxy nanocomposites: Effectiveness of atomic force microscopy (AFM) in investigating the dispersion of different carbonaceous nanoparticles. *Polymers, 11*(5), 832.
4. Werner, E., Güth, U., Brockhagen, B., Döpke, C., & Ehrmann, A. (2023). Examination of polymer blends by AFM phase images. *Technologies, 11*(2), 56.
5. Jeanmaire, D. L., & Van Duyne, R. P. (1977). Surface Raman spectroelectrochemistry: Part I. hetero-cyclic, aromatic, and aliphatic amines adsorbed on the anodized silver electrode. *Journal of Electroanalytical Chemistry and Interfacial Electrochemistry, 84*, 1–20.
6. Lai, W. F., & Wong, W. T. (2022). Design and practical considerations for active polymeric films in food packaging. *International Journal of Molecular Sciences, 23*(11), 6295.
7. Pu, J., Shen, Z., Zhong, C., Zhou, Q., Liu, J., Zhu, J., & Zhang, H. (2020). Electrodeposition technologies for Li-based batteries: New frontiers of energy storage. *Advanced Materials, 32*, 1903808.
8. Liu, J., Li, P., Bi, J., Zhu, Q., & Han, B. (2022). Design and preparation of electrocatalysts by electrodeposition for CO_2 reduction. *Chemistry—A European Journal, 28*, Article e202200242.

9. Ji, L., Wang, J., Teng, X., Meyer, T. J., & Chen, Z. (2020). CoP nanoframes as bifunctional electrocatalysts for efficient overall water splitting. *ACS Catalysis, 10*, 412–419.
10. Gupta, S. K., & Misra, R. D. (2023). Experimental pool boiling heat transfer analysis with copper-alumina micro/nanostructured surfaces developed by a novel electrochemical deposition technique. *International Journal of Thermophysics, 44*, 112.
11. Jabbarzadeh, A. (2019). The origins of enhanced and retarded crystallization in nanocomposite polymers. *Nanomaterials, 9*(10), 1472.
12. Huang, Z., Chen, B., Ren, B., Tu, D., Wang, Z., Wang, C., Zheng, Y., Li, X., Wang, D., & Ren, Z. (2023). Smart mechanoluminescent phosphors: A review of strontium-aluminate-based materials, properties, and their advanced application technologies. *Advancement of Science, 10*, 2204925.
13. Guadagno, L., Naddeo, C., Raimondo, M., Speranza, V., Pantani, R., Acquesta, A., & Monetta, T. (2019). UV irradiated graphene-based nanocomposites: Change in the mechanical properties by local harmoniX atomic force microscopy detection. *Materials, 12*(6), 962.
14. Oprea, M., & Panaitescu, D. M. (2020). Nanocellulose hybrids with metal oxides nanoparticles for biomedical applications. *Molecules, 25*(18), 4045.
15. Jensen, T. R., Van Duyne, R. P., Johnson, S. A., & Maroni, V. A. (2000). Surface-enhanced infrared spectroscopy: A comparison of metal island films with discrete and nondiscrete surface plasmons. *Applied Spectroscopy, 54*, 371–377.
16. Nemani, S. K., Annavarapu, R. K., Mohammadian, B., Raiyan, A., Heil, J., Haque, M. A., Abdelaal, A., & Sojoudi, H. (2018). Surface modification of polymers: methods and applications. *Advanced Materials Interfaces, 5*, 1801247.
17. Tan, N. P. B., Lee, C. H., & Li, P. (2016). Green synthesis of smart metal/polymer nanocomposite particles and their tuneable catalytic activities. *Polymers, 8*(4), 105.
18. Alsoud, A., Daradkeh, S. I., Al-Bashaish, S. R., Shaheen, A. A., Jaber, A. M., Abuamr, A. M., & Holcman, V. (2024). Electrical characterization of epoxy nanocomposite under high DC voltage. *Polymers, 16*(7), 963.
19. Cordill, M. J., Kreiml, P., & Mitterer, C. (2022). Materials engineering for flexible metallic thin film applications. *Materials, 15*(3), 926.
20. Mladenović, I. O., Vuksanović, M. M., Dimitrijević, S. P., Vasilić, R., Radojević, V. J., Vasiljević-Radović, D. G., & Nikolić, N. D. (2023). Mechanical properties of electrolytically produced copper coatings reinforced with pigment particles. *Metals, 13*(12), 1979.
21. Mladenovic, I., Lamovec, J., Vasiljevic-Radovic, D., Radojevic, V., & Nikolic, N. (2022). Determination of the absolute hardness of electrolytically produced copper coatings by application of the Chicot-Lesage composite hardness model. *Journal of the Serbian Chemical Society, 87*, 899–910.
22. Marmur, A. (2012). Hydro- Hygro- Oleo- Omni-Phobic? *Terminology of Wettability Classification. Soft Matter, 8*, 6867.
23. Song, S., Guo, M., Zhang, S., Zhan, K., Yan, Y., Yang, J., Zhao, B., & Xu, M. (2020). Plasma-assisted synthesis of hierarchical NiCoxPy nanosheets as robust and stable electrocatalyst for hydrogen evolution reaction in both acidic and alkaline media. *Electrochimica Acta, 331*, Article 135431.
24. Guo, M., Song, S., Zhang, S., Yan, Y., Zhan, K., Yang, J., & Zhao, B. (2020). Fe-doped Ni–Co phosphide nanoplates with planar defects as an efficient bifunctional electrocatalyst for overall water splitting. *ACS Sustainable Chemistry Engineering, 8*, 7436–7444.
25. Sun, H., Yan, Z., Liu, F., Xu, W., Cheng, F., & Chen, J. (2020). Self-supported transition-metal-based electrocatalysts for hydrogen and oxygen evolution. *Advanced Materials, 32*, 1806326.
26. Requena, G. C., & Degischer, H. P. (2022). Effects of particle reinforcement on creep behaviour of AlSi1MgCu. *International Journal of Materials Research, 96*, 807–813.
27. Triantou, K. I., Pantelis, D. I., Guipont, V., & Jeandin, M. (2015). Microstructure and tribological behavior of copper and composite copper+alumina cold sprayed coatings for various alumina contents. *Wear, 336–337*, 96–107.
28. Chen, P., Ye, J., Wang, H., Ouyang, L., & Zhu, M. (2021). Recent progress of transition metal carbides/nitrides for electrocatalytic water splitting. *Journal of Alloys and Compounds, 883*, Article 160833.

Chapter 7
Nanomaterials' Significance in Engineering Applications

Nanomaterials have become more important in technology development due to their superior performance compared to bulk materials and unique properties. These exceptional and distinctive properties of nanoparticles are revolutionizing a wide range of industrial applications. This chapter provides an overview of basic engineering ideas of nanomaterials, with an emphasis on carbon nanomaterials. In addition, this chapter highlights the engineering uses of nanomaterials in a variety of industries, including biomedicine, food processing, biotechnology, environmental remediation, construction, renewable energy, electronics, and energy storage. The possible impact of nanoparticles on human health is discussed. In short, the nanomaterials are expected to significantly advance the field of advanced nanotechnology.

7.1 Introduction

The industrial use of materials at the atomic, molecular, and supramolecular levels is known as nanotechnology or molecular nanotechnology. The precise manipulation of atoms and molecules to produce macroscale goods is the original definition of nanotechnology. But over time, this definition has changed. Later, the National Nanotechnology Initiative expanded the term to include the manipulation of matter in at least one dimension at scale, 1 to 100 nm. This change made nanotechnology a more comprehensive branch of research that studies the special qualities of matter at the quantum level rather than just a specific technological goal. The terms "nanotechnology" and "nanoscale technologies" are now used to refer to a wide range of research and applications that have comparable size-dependent properties as quantum mechanical phenomena become relevant at these quality scales.

With their many advantages, design and manufacturing are among the most promising areas for nanotechnology. The service life of auto parts and other mechanical components can be significantly extended by incorporating nanotechnology.

R. A. Al-Samarai and Y. Al-Douri, *Engineering of Nanomaterials*,
https://doi.org/10.1007/978-981-95-5499-7_7

The distinctive physical and chemical characteristics of nanomaterials improve the quality of the material, which in turn improves mechanical strength, optical quality and magnetic properties. In addition, scientists are constantly creating new precursors that impart specific properties to materials, allowing for the production of more reliable and efficient products.

The ability of nanotechnology to control and regulate the properties of materials at the nanoscale is another important feature. Since individual atoms and molecules make up every material on earth, advances in nanotechnology make it possible to precisely manipulate these important building blocks. In particular, the smooth surface of MoS_2A is revealed by lateral force microscopy (LFM) and atomic force microscopy (AFM) analysis of molybdenum disulfide, as shown in Fig. 7.1a. However, non-uniform lateral current injection is observed when conductive atomic force microscopy (C-AFM) is used to analyze the material (Fig. 7.1b). The boron-doped diamond conductive tip and MoS_2 surface interacted at the nanoscale, revealing a series of dark circular patterns with sizes ranging from 3 to 4 nm. These features indicate reduced tip/MoS_2 contact resistance since their current levels are higher than the surrounding areas. Subsequent investigation has revealed that these high-current structures are nanoscale in nature with an areal density ranging from 10^{10} cm^{-2} to 10^{11} cm^{-2}. Further experiments are conducted by overlaying MoS_2 grid over the revealed circular pits to better understand these high-current regions in scanning tunneling microscopy (STM) maps (Fig. 7.1c). With all this in mind, the application of nanotechnology in materials science and engineering continues to open up new avenues for the development of stronger, more efficient, and incredibly resilient materials. The potential for groundbreaking discoveries in industrial applications will only increase as research in this area advances.

The density of conductive structures visible in the C-AFM images closely matched the surface density of observed pits, which is quantified as $(8 \pm 3) \times 10^{10}$ cm^{-2}. The average lateral size of these pits is up to 5 nm. The researchers have hypothesized that electronic effects rather than topographic variations are the source of the features observed in STM maps based on an experimental study. The absence of these holes in the topographic and lanthanum strontium manganite (LSM) images has led to this conclusion, indicating that the source of anomalies is subsurface defects rather than surface anomalies.

The use of microscopic materials has been around for centuries, despite the relatively young fields of nanoscience and nanotechnology. But it has sophisticated microscopy techniques developed that allow researchers to view objects at a remarkably precise nanoscale. The special properties of nanoparticles, which provide advantages such as increased strength, reduced weight, improved control over the light spectrum, and stronger chemical reactivity compared to their bulk counterparts, are the reason for a growing interest in nanotechnology. Modern scientists and engineers are constantly exploring a wide range of nanomaterials for various applications in order to fully utilize their properties.

The study of electrical conductivity at the nanoscale is one of the most important areas of research in nanotechnology. For example, the conductivity loss associated with grain boundaries (GBs) is assessed by nano-resolution current mapping with

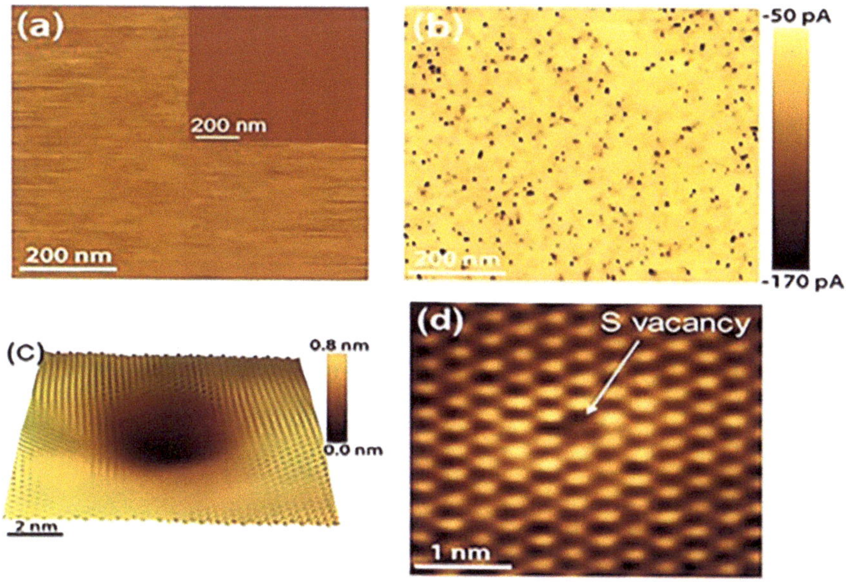

Fig. 7.1 **a** Topography (insert image) and LFM of MoS_2 surface. **b** Simultaneous recording of C-AFM image. **c** STM image shows a subsurface defect corresponding to one of the high-current regions in the panel. **d** Sulfur void in the outer sulfur layer as seen in the STM image [1] [Open access]

C-AFM in a single-layer MoS_2 film prepared by chemical vapor deposition (CVD) on SiO_2/Si substrate [2]. An optical micrograph of synthetic MoS_2 domains, which are typically triangular in shape (Fig. 7.2a), shows them to be up to 50 μm in size. This picture shows that MoS_2 domains are consolidated, leading to the formation of GB.

A platinum-iridium alloy (PtIr) tip is used to form a nanocontact with a microscopic electrode, which partially overlapped several MoS_2 domains in the experimental setup for C-AFM measurements, as shown in Fig. 7.2b. For the current measurements, this contact acted as a critical point. Figure 7.2c, d illustrate the surface morphology and corresponding current map, which are obtained with an applied bias of tip voltage (Vtip) = 1 V. Two MoS_2 domains (labeled D1 and D2) are located in the region of interest, which is indicated by a red rectangle in Fig. 7.2b. Interestingly, a direct contact is observed between the D1 domain and the macroscopic electrode, highlighting the importance of domain connectivity for electrical performance.

This highlights the expanding capabilities of nanotechnology in understanding and modifying electronic characteristics at the atomic level. The possibilities for developing high-performance nanomaterials for a variety of applications, such as electronics, energy storage, and optoelectronic devices, will increase as nanoscale imaging and material synthesis technologies advance, opening the door to new technological developments.

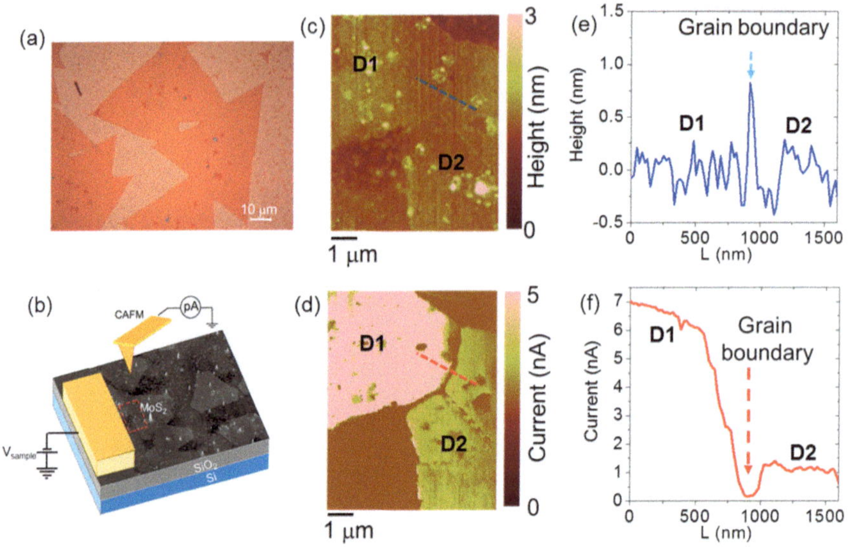

Fig. 7.2 **a** Single-layer MoS_2 domains grown by CVD on a SiO_2/Si substrate as observed by optical microscopy. **b** Schematic illustration of the experimental setup used to detect the local current using C-AFM on MoS_2 domains. **c** C-AFM morphology measurements and **d** current map in a region containing two MoS_2 domains (D1 and D2) separated by a grain boundary. The macroscopic contact is directly connected to the D1 domain. The height scan (**e**) and the current line across the grain boundary (**f**) are shown [1] [Open access]

The two domains are separated by GB. Both in the figure (as a sharp peak of about 1 nm height), as shown in the GB map, can be seen. While the current change is almost constant between the different domains, there is a noticeable drop in conductivity from the D1 domain to the D2 domain, which is separated by GB. The two scan lines in Fig. 7.2e, f, taken from the figure and current map, respectively, illustrate this further. The production of nanomaterials with high functionality and appropriate properties is critical for various applications. In general, there are two ways to synthesize nanomaterials: bottom-up and top-down approaches. The top-down method involves creating devices and structures from bulk materials without precise atomic control. Large-scale production in an industrial context is one of the key advantages of the top-down method [3]. However, the main disadvantages of this method are its high cost and energy consumption, long processing time, and structural deformation of the surface [4]. Mechanical milling, cutting, chemical etching, laser ablation, electrical blasting, and lithography are just a few examples of top-down techniques. In contrast, the bottom-up approach assembles atomic-sized components to create a tunable nanostructure.

7.2 Types of Nanomaterials

The two main groups of nanomaterials are inorganic and organic. Carbon structures, including fullerene, graphene, single-walled carbon nanotubes (CNTs), and buckyballs, are the most common types of organic nanomaterials. The remarkable mechanical, electrical, and thermal properties of these materials have attracted much interest.

The development of affordable electrochemical devices is essential to ensure efficient energy conversion and storage, which is one of the major challenges in sustainable development. Nanomaterials can significantly improve the performance of these devices, but large-scale production still remains a major challenge [2]. The researchers have created a number of synthesis methods, such as solution-based, template-based, and microwave synthesis, to address this issue. High-quality nanomaterials can be produced in large quantities using these methods, which also have the potential to reduce manufacturing costs.

While nanoparticles have several advantages, their incorporation into another significant challenge is electrochemical devices. Since the characteristics of nanomaterials are very different from those of their bulk counterparts, it is essential to carefully optimize their interactions with other parts of the device to maximize performance and efficiency [5]. The interaction of nanoparticles with other materials is critical in electrochemical devices since it directly affects stability, charge transport, and overall performance. The full potential of nanomaterials in energy conversion and storage technologies cannot be achieved without proper integration.

Hurdles in Integration and Large-Scale Manufacturing of nanomaterials for practical use is one of the major hurdles to their widespread adoption. Scalability, consistency, and cost-effectiveness issues still exist despite a large amount of research being conducted to improve manufacturing technologies such as solution-based, template-based, and microwave synthesis [6]. In addition, their incorporation should be improved for electrochemical devices since the performance is highly dependent on the interactions between nanomaterials and other device components [2].

Many microscopic approaches are used for characterization to better understand the properties of nanomaterials. The size and shape of silver nanoparticles are examined by electrostatic force microscopy (EFM) and scanning electron microscopy (SEM). SEM images have shown that most of the silver nanoparticles appeared spherical, as shown in Fig. 7.3b. In addition, the efficient production of silver nanoparticles is confirmed by energy-dispersive X-ray spectroscopy (EDS) analysis (Fig. 7.3a).

The silver nanoparticles coated with conductive PCPDTBT (Poly[2,6-(4,4-bis(2-ethylhexyl)-4H-cyclopenta[2,1-b;3,4-b′]dithiophene)-alt-4,7-(2,1,3-benzothiadiazole)]) is an organic semiconductor polymer mainly used in organic solar cells (OPVs) and organic electronics due to its excellent photovoltaic properties are subjected to SEM examination to better understand their structural and electrical characteristics. The Kelvin scanning probe microscopy (SKPM) topography of silver nanoparticles/PCPDTBT is simultaneously obtained with SEM and EFM images, as shown in Fig.7.3c, d. The results have illustrated that the silver nanoparticles remain

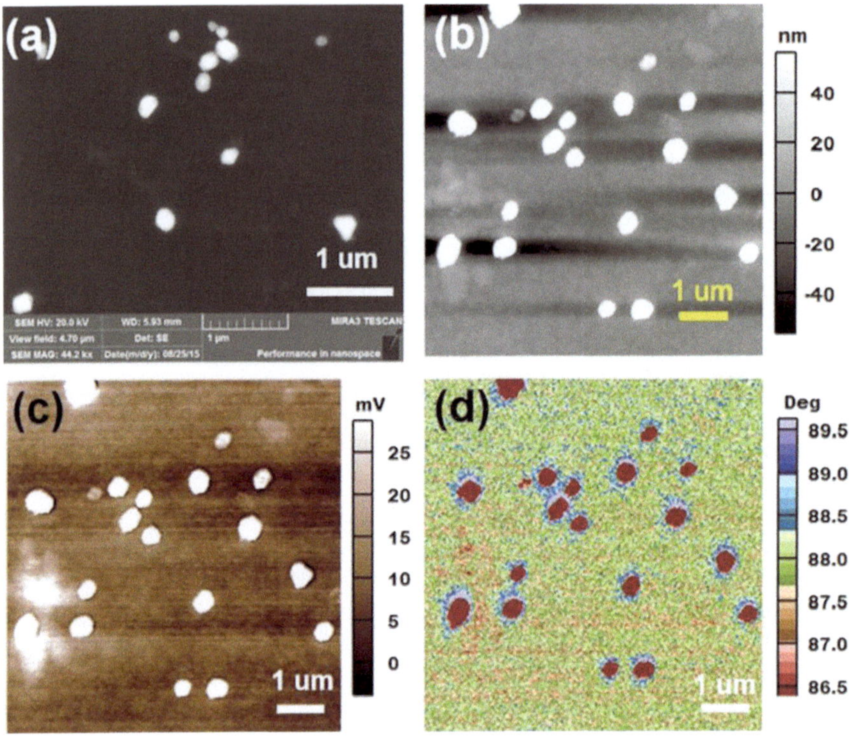

Fig. 7.3 **a** silver NPs imaged by SEM, **b** AFM, **c** surface potential, and **d** EFM image of silver nanoparticles/poly[2,1-b;3,4-b′]dithiophene)-alt-4,7-(2,1,3-benzothiadiazole)][2,6-(4,4-bis-(2-ethylhexyl)-4H-cyclopenta] (PCPDTBT) sample at 488 nm. VEFM is equivalent to 6 V. The Scale bar is 1 μm [7] [Open access]

the same size in both SKPM and EFM images, suggesting that AFM resolution matches SEM topography well. The robustness and reliability of data are highlighted by this consistency [7].

Nanomaterials hold great potential for improving electrochemical devices, especially those used for energy conversion and storage. However, challenges such as efficient integration and large-scale manufacturing need to be addressed to fully realize their potential. Improvements in characterization and synthesis methods are needed to overcome these obstacles. Nanotechnology will continue to transform materials science as research advances, opening the door to more sustainable and efficient electrochemical technologies.

Optimizing the integration of nanomaterials is crucial to maximizing their potential in electrochemical systems. Due to their special properties, nanomaterials can significantly improve the performance and efficiency of these devices [2]. Based on their dimensionality, nanomaterials can be classified into different types. Carbon nanotubes (CNTs) and other one-dimensional nanomaterials can be rods, wires, tubes, or fibers. Graphene and graphene oxide (GO) are examples of two-dimensional nanomaterials that are commonly found in sheets or films. In addition, the structure of

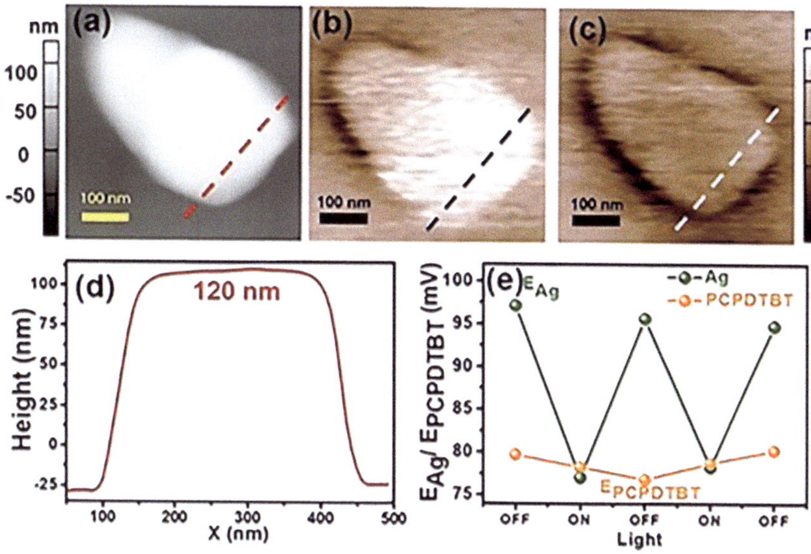

Fig. 7.4 The AFM images of surface potential of silver nanoparticles on PCPDTBT films obtained without light excitation and with light excitation at a wavelength of 488 nm are shown in (**a**) AFM topography of silver nanoparticles on PCPDTBT films, **b** 120 nm table; **e** Photomodulated surface potential of silver nanoparticles on PCPDTBT films [7] [Open access]

three-dimensional (3D) nanomaterials, such as diamond and graphite, is completely three-dimensional. As can be seen in Fig. 7.4, these nanomaterials can have different sizes and be either crystalline or amorphous.

The main topic of this section is carbon-based nanomaterials, which have attracted great interest due to their adaptive use in various sectors. The silver nanoparticles on PCPDTBT films are examined by SEM as shown in Fig. 7.4a. Figure 7.4d shows the nano size of silver nanoparticles, which is about 120 nm.

SKPM is used to measure the surface potential of Ag-PCPDTBT composite. The surface potential of the composite without light excitation is shown in Fig. 7.4b, and the surface potential under 488 nm irradiation is shown in Fig. 7.4c. The results indicate that the conductive polymer is continuously coated with silver nanoparticles, where the interface of the Ag-PCPDTBT composite cannot be protected from charge separation by the elevated surface potential of the film alone. This means that the charge separation and recombination induced by light at the observed surface potential are directly reflected at the interface.

The brighter areas on the silver nanoparticles in SKPM images (Fig. 7.4b) without light demonstrate that the surface potential of particles is greater than that of the surrounding medium. In particular, the PCPDTBT polymer has a slightly lower surface potential of 79 ± 2 mV than the silver nanoparticles, which have a surface potential of 96 ± 3 mV. This illustrates how the surface potential plays a critical role in how nanomaterials behave in electrochemical devices and highlights how important it is to maximize their interfacial interactions for better device performance [7].

7.2.1 Fullerene (0D)

Fullerenes are a special type of zero-dimensional (0D) nanomaterial consisting only of carbon atoms grouped in a hollow, closed configuration. Because it resembles a soccer ball, the best-known fullerene, C_{60}, is often called a buckyball. With 20 hexagons and 12 pentagons, this molecule has a perfectly symmetrical spherical shape.

Since their discovery in 1985 by Richard Smalley, Robert Curl, and Harold Kroto [8], fullerenes have attracted much interest due to their exceptional mechanical, optical, and electrical properties. Their unique structure allows them to exhibit high stability, electrical conductivity, and the ability to interact with a variety of chemical compounds, making them valuable in many scientific and industrial applications.

Despite their fascinating properties, fullerenes face certain challenges, particularly their low solubility in polar solvents, which limits their immediate use in some situations. However, fullerene structures have been modified by advances in nanotechnology, increasing their potential in areas such as energy storage, medicine, and nanotechnology. Fullerenes are currently being studied for their potential use in superconductors, drug delivery systems, and even nanomachines.

Along with carbon nanotubes (CNTs) and graphene, fullerenes continue to be key components in the development of carbon-based nanomaterials and modern nanotechnology. The buckyball is the most famous of many interesting carbon-based nanostructures known as fullerenes. Buckyballs are fullerenes that take on a spherical shape. Richard Smalley [9] discovered the fullerene molecule, which has attracted much attention due to its special properties. These molecules are large, triangular in shape, reminiscent of spheroids, and can form C–C bonds. The fullerene appears symmetrical due to its structure, which consists of twelve pentagons and twenty hexagons.

Fullerene is an interesting molecule to study due to its unique symmetry and reactivity with many substances. For example, the spherical shape of C_{60} fullerene is often used as a "wheel" in the construction of nano cars and nanovesicles due to the peculiar geometry of this structure. However, the insolubility of fullerene in polar solvents limits its wider application in various industries, which in turn limits its practical usefulness.

Fullerene derivatives have received less attention in recent years as research has shifted toward carbon nanotubes (CNT) and graphene-based materials [10–12]. However, fullerene research is still ongoing. Phenyl-C70-butyric acid methyl ester ($PC_{70}BM$)-rich domains undergo phase separation when fullerene derivatives are dissolved in a poorer solvent, as shown in Fig. 7.5. Combined with the decrease in PC70BM solubility, this effect is particularly noticeable when using a 1:3 mixture of ortho-dichlorobenzene and chloroform (oDCB and CF), which shows higher separation than a 1:1 mixture.

Further analysis of the emission spectra of fullerene mixtures, as shown in Fig. 7.5, reveals some intriguing results. A primary TQ1 is a hole-donating polymer (donor polymer) widely sought in organic solar cells (OPVs) and organic electronics due to

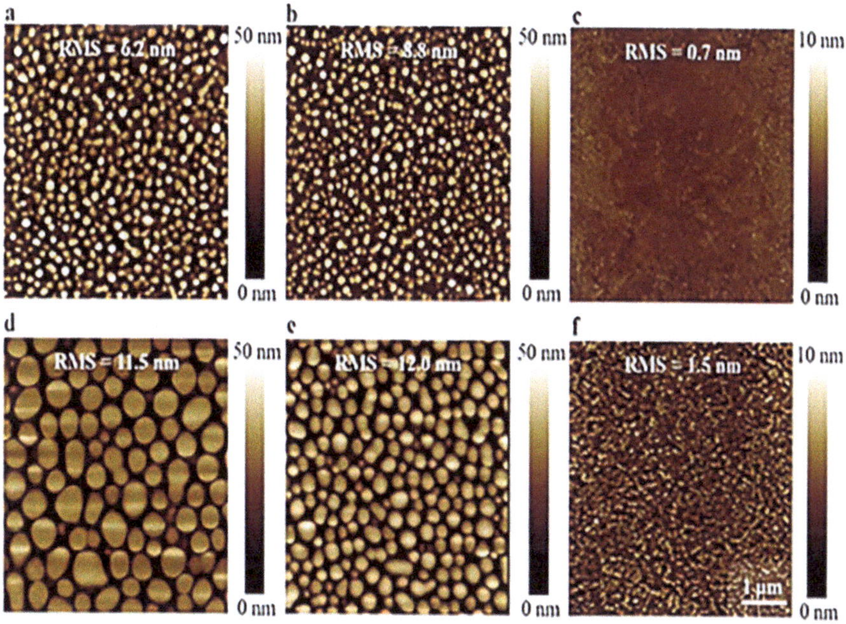

Fig. 7.5 TQ1:PC70BM (**a**–**c**) and 1:3 TQ1:PC70BM (**d**–**f**) thin films spin-coated with chloroform (CF) (**a**, **d**), chlorobenzene (CB) (**b**, **e**), and ortho-dichlorobenzene (oDCB) (**c**, **f**). The micrographs are obtained by AFM (5 × 5 μm). The scale bar represents 1 μm. The surface roughness (RMS) has been added to each image along with the corresponding height scale [3] [Open access]

its photovoltaic canceling properties. The emission peak at 700 nm is visible when the mixture films are excited at 380 nm. A broad emission band peaking between 520 and 550 nm also appears. Interestingly, this emission band is absent when TQ1 is in its pure film form, which shows no spectrum, or when it is dissolved in solvents such as CF, CB, or an analog-to-digital converter (ADC). This demonstrates the different interactions that fullerene derivatives exhibit under different solvent conditions [3].

By changing the spin-coating rate (Fig. 7.6). As can be seen from AFM images (Fig. 7.6a), thicker films with slower spin coating exhibit larger PC70BM-rich domains. The obtained emission spectra (Fig. 7.6b) show that the emission maximum changes from 520 nm to about 540 nm with increasing film thickness from 90 to 130 nm. The emission range does not change significantly with decreasing film thickness from 90 to 50 nm. This indicates that the domains must grow to a certain size for co-emission to become noticeable. The relationship between the morphology of TQ1:PC70BM blends and their in-depth emission spectrum analysis is investigated. Greater phase separation occurs when solvents with lower PC70BM solubility and higher PC70BM concentrations are used. PC70BM aggregates are expected to form in these larger domains, resulting in a fairly broad and soft emission at 380 nm excitation between 520 and 550 nm. The fullerene derivative remains well mixed with the polymer in superior PC70BM solvents and at low PC70BM concentrations, and no emission is observed between 500 and 550 nm [3].

Depending on the size of different shapes. C70 is ovoid rather than a sphere. Fullerene contains up to 2200 carbon atoms. In 1970, Osawa [13] was the first to mention fullerene in the literature after discovering that corannulene is a substructure of C60. As a result, two important isomers are formed during pyrolysis: C60 and C70. Various synthetic parameters can be modified to change their ratio. The flame process has been used to produce fullerene on a large scale. Fullerenes of different sizes are purified by another method called HPLC. It contains triply degenerate LUMO orbitals that accommodate six electrons upon electrochemical reduction due to its structure, exceptional electronic and electrochemical capabilities, and specific gravity. This characteristic makes C60 a specific electron acceptor for various purposes, especially in electronics. Due to the doubly degenerate LUMO orbitals in C70, which are capable of holding up to four electrons, the non-degenerate LUMO + 1 level is

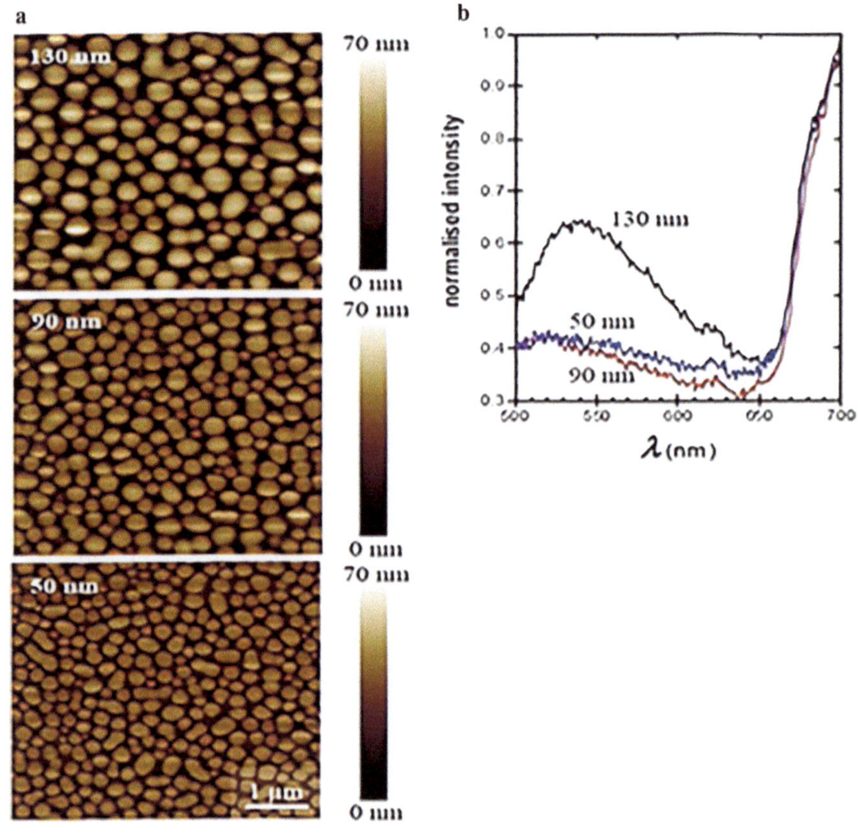

Fig. 7.6 **a** 5 μm AFM micrographs of 1:3 spin-coated TQ1:PC70BM CB thin films with the indicated film thickness. A height scale is added to each image. **b** Fluorescence emission spectra corresponding to the films with the corresponding thickness [3] [Open access]

almost ready to accept two more electrons. C60 is used as a building block of organic electronics.

7.2.2 Nanotubes of Carbon (1D)

Sheets of graphene are rolled to create hollow cylindrical structures that constitute a special class of one-dimensional (1D) nanomaterials known as carbon nanotubes (CNTs). These nanoparticles are among the most widely studied materials in nanotechnology due to their remarkable mechanical, electrical, and thermal capabilities.

Single-walled carbon nanotubes (SWCNTs), which have a single layer of graphene, and multi-walled carbon nanotubes (MWCNTs), which have multiple concentric layers of graphene, are the two main types of CNTs that were first identified in 1991 by Sumio Iijima [14]. CNTs have many applications in electronics, medicine, energy storage, and composite materials due to their high aspect ratio, low weight, and exceptional strength.

Depending on their structural configuration, the electrical conductivity of CNTs can be either metallic or semiconducting, making it one of their most amazing characteristics. This makes them very promising for conductive coatings, transistors, and nanoelectronics. In addition, their exceptional mechanical strength—much higher than that of steel—allows them to be used in flexible electronics, aircraft technology, and reinforced materials. Despite their potential, problems remain, such as the difficulty of producing them on a large scale, purifying them, and integrating them into devices.

CNTs are among the most significant nanomaterials in modern science and technology, although further study will expand the range of their applications. Materials with one-dimensional (1D) structures exist at two-dimensional nanoscale. Nanotubes, nanofibers, nanowires, nanorods, and nanothreads are just a few examples. Since electrons are limited to two dimensions, they cannot move freely. Figure 7.7 (pin 2 in Fig. 7.7) and Figures S1 and S2 (pins 1 and 3 in Fig. 7.7) in the supplementary material illustrate the results of successful stabilization. Scanning electron microscope (SEM) images of CNT fiber probes are shown in each of these figures before and after the treatment (Fig. 7.7e), S1e, and S2e show AFM images of CNT-coated silicon surface (prepared on silicon surface using chemical vapor deposition as described in Sect. 4.4), which are obtained to demonstrate the stable imaging of CNT fiber probes [4].

Like 0D materials, these 1D nanomaterials can be metallic, ceramic, polymeric, amorphous, mono- or polycrystalline. On the other hand, 1D nanomaterials can have pure or impure chemical composition (e.g., doped semiconductors). These elements can be displayed separately or integrated into another medium; their length exceeds the width of 1D material [14–16].

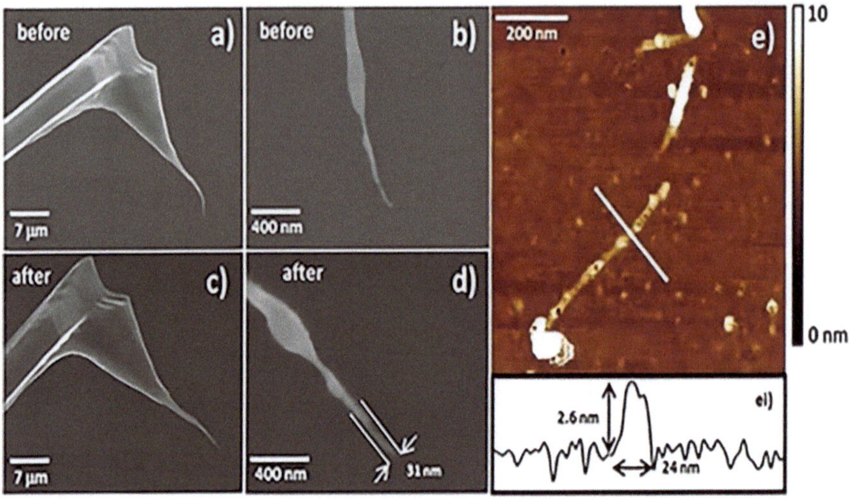

Fig. 7.7 SEM images of Tip 2 before (**a**, **b**) and after (**c**, **d**) processing. After processing, **e** cross section corresponding to the white line. Atomic number microscope image of a carbon nanotube on a carbon nanotube-coated silicon surface using Tip 2. The carbon nanotube fibers are significantly straightened, as can be seen from the comparison of SEM images before and after [4] [Open access]

7.2.3 Graphene and Its Derivatives (2D)

Since air permeates the entire world and is essential for all living things, it is one of the most important components of life. Compared to other types of pollution, air pollution has a more immediate and serious impact on human health, since breathing is a constant and necessary process [17]. By releasing hazardous chemicals into the atmosphere, such as carbon dioxide (CO_2), nitrogen oxides (NO, NO_2), hydrogen sulfide (H_2S), sulfur dioxide (SO_2), and particulate matter (PM), the increasing pace of industrialization has significantly increased air pollution. In addition to reducing air quality, these pollutants are extremely harmful to human health.

Numerous diseases, such as conjunctivitis, respiratory infections, allergies, tumors, dry eyes, vascular coagulation, and cardiovascular diseases, have been associated with air pollution [3]. While CO_2 is often considered a less hazardous gas, high concentrations of over 7.2% can cause respiratory discomfort, including rapid breathing. In addition, carbon monoxide (CO) is particularly harmful because it reduces oxygen delivery and causes toxicity by competing with oxygen (O_2) for binding to hemoglobin (Hb) in the blood.

In the initial morphological study, the molecular nucleation of submonolayer films ($t = 1$ nm) on a graphene substrate was studied to learn more about how air pollution affects materials and their interactions with the environment. The growth patterns and structure of organic thin films can be affected by industrial pollutants. The growth of organic thin films (Fig. 7.8a, b) typically occurs along substrate defects such as graphene wrinkles, which are revealed by morphological examination as shown in

Fig. 7.8c. Different organic materials had different nucleation densities; PDIF-CN_2 shows more sparse nucleation sites than PDI8-CN_2, indicating a stronger molecular effect. Moreover, the average height of characteristic line profiles of PDIF-CN_2 grains is 3.3–3.6 nm (Fig. 7.8e), which is consistent with previously documented organic molecular architectures [2].

We can learn more about how air pollution affects industrial materials, atmospheric chemistry, and human health by investigating it at the environmental and molecular levels. Creating effective air quality monitoring methods and advanced nanomaterials to detect and filter pollutants remains a vital scientific challenge as pollution levels continue to rise.

The main cause of climate change is CO_2, which is from flue gases, that must be captured and stored to reduce its concentration in the atmosphere. For this reason,

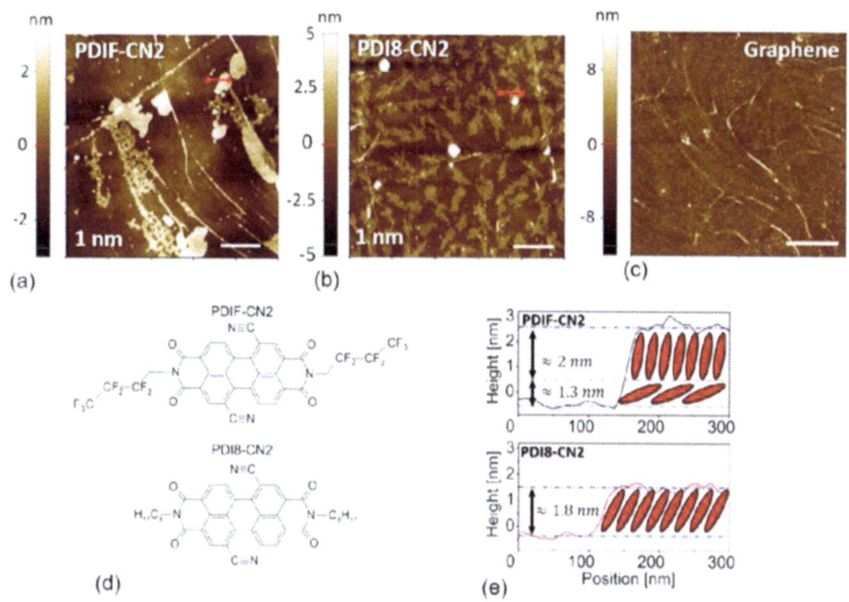

Fig. 7.8 **a**, **b** 5 nm × 5 nm NC-AFM topography of PDIF-CN_2 and PDI8-CN_2 submonolayers grown on graphene surfaces with a nominal thickness of 1 nm. The white lines of the panels serve as the 1 nm reference. **c** The reference surface is pristine CVD graphene (10 μm × 10 μm, where the white bar represents the 2.5 μm reference), highlighting the discontinuous occurrence of wrinkles. **d** Molecular structures of perylenediimide derivatives used in this study. **e** Line profiles of the two materials corresponding to the red lines in panels (**a**, **b**). In the PDIF-CN_2 example, the nearly downward-facing molecular wettability layer (≈1.3 nm) can be easily inferred [2] [Open access]

many techniques are being explored, including cryogenics, adsorption, absorption, and membranes [2, 4]. Adsorption is one such tactic that has received a lot of attention. The key factor here is how easy it is to regenerate the adsorbent. CO_2 adsorbents should have certain qualities such as a higher degree of CO2 adsorption capacity and selectivity, fast adsorption/desorption kinetics, moderate enthalpic heat, and good stability [5].

Many materials are reported to have the following properties: Activated carbon, carbon nanotubes, carbon spheres, metal–organic frameworks, graphene oxide, graphite surfaces, graphene-based porous adsorbents, zeolite-based molecular sieves, etc. are excellent options for CO_2 adsorption [6]. PM is a general term used to describe air pollutants that vary in composition and size as a result of a variety of anthropogenic activities [8]. PM consists of solid or liquid particles suspended in the air. PM2.5 and PM10 are two broad categories into which PM pollutants can be divided based on aerodynamic diameter. PM2.5 refers to pollutants with a diameter less than or equal to 2.5 μm, and PM10 refers to pollutants with a diameter less than or equal to 10 μm [15].

Given that PM2.5 damages the cardiovascular system immediately after exposure, it is more dangerous than PM10. In addition to its large surface area, PM2.5 contains a large number of pathogens, which increases the incidence of disease [3, 4, 7]. Their penetrating ability is great, and they are capable of transporting various harmful compounds that damage organs and cause cell death. In addition, it impairs renal function and causes renal failure [2]. In the presence of particulate matter, the rate of oxygen transfer by hemoglobin is reduced [14].

7.2.4 Graphite (3D)

The three-dimensional (3D) nanomaterial called graphite is made up of stacked layers of graphene, each made up of carbon atoms arranged in a hexagonal lattice. Its unique mechanical, electrical, and thermal properties make it one of the most popular and stable carbon-based materials.

In its natural mineral form, graphite is widely used in a variety of industries such as manufacturing, electronics, energy storage, and lubrication. It is a critical component in the production of pencils, batteries, and electrodes due to its layered structure, which facilitates easy interlayer sliding and makes it a good lubricant. The creation of energy storage devices such as lithium-ion batteries (LIBs) and supercapacitors is one of the most important applications of graphite. Due to its exceptional chemical stability and electrical conductivity, graphite is a preferred material for battery anodes, which are needed to increase storage capacity and energy efficiency. Its thermal stability also makes it a critical component in high-temperature industrial applications. Graphite research is still evolving despite its widespread use, especially in nanotechnology.

In the fields of sensors, energy storage, biomedicine, electronics, solar cells, optics, etc., they have found a wide range of applications [18]. By changing materials at the nanoscale, science can improve the performance and efficiency of energy storage systems. Figure 7.9a depicts an overview of vertically stacked graphene (VG) development mechanism using rf-PECVD on soda-lime glass. Initially, radio frequency (RF) discharge separates carbon source CH_4 into highly reactive carbon species (CHx, C_2Hy, C, and H atoms).

Developments in graphite exfoliation techniques have produced graphene, a single sheet of graphite with exceptional mechanical strength, flexibility, and electrical conductivity. Consequently, research is still being conducted to improve graphite's properties and expand its use in advanced technologies, keeping it at the forefront of materials science. The manipulation and control of matter at the nanoscale (size range, 1 to 100 nm), is the focus of the scientific and technological field known as nanotechnology [16]. Reactive species are first adsorbed onto the glass surface at a temperature of 600 °C, slightly below the softening point of soda-lime glass. Irregular

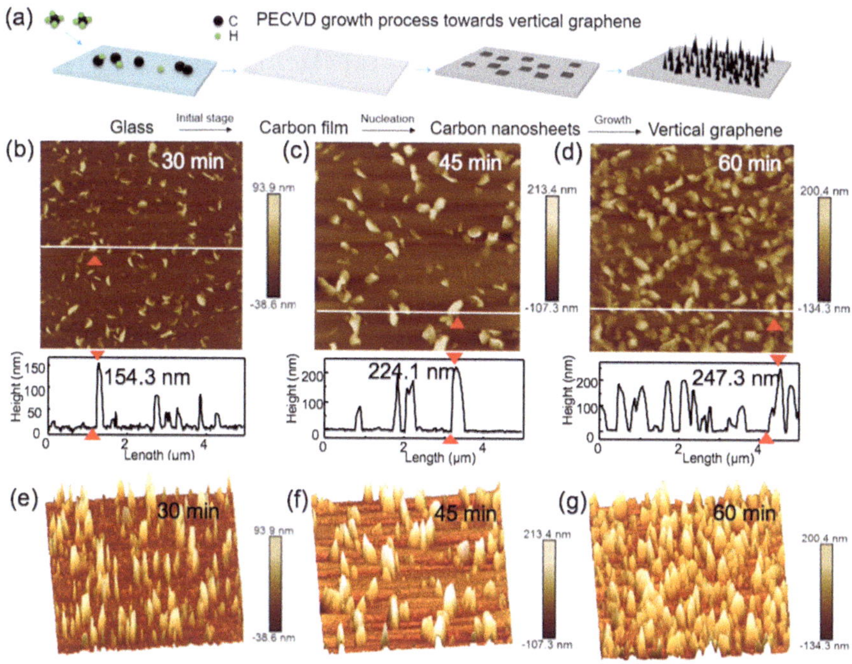

Fig. 7.9 Growth of VG using plasma-enhanced chemical vapor deposition (PECVD). **a** Schematic representation of VG graphene growth using rf-PECVD. **b–d** AFM height images of 5 μm × 5 μm plant graphene films directly formed on soda-lime glass are acquired at three different growth times (30, 45 and 60 min), all under the same growth conditions (10 standard cubic centimeters per minute (sccm) of CH_4, 600 °C, 300 W). Portions of images in insets are shown in direction of dashed white line. **e–g** corresponding 3D AFM height images [5] [Open access]

cracks, impurities, and dangling bonds serve as crystallization sites and cause the formation of a horizontal carbon buffer layer on the substrate surface [5].

Improving battery performance is one of the breakthroughs made possible by nanotechnology. Strong ion transport, high electron conductivity, fast ion diffusion, improved electrochemical capabilities, large surface area, and high-current resistance are just some of the advantages that nanomaterials offer over conventional battery materials [5]. These characteristics improve safety, cycle life, and efficiency by increasing energy density, charge and discharge rates, and operational capabilities [6].

Renewable energy research is becoming increasingly important as the world's energy needs increase. The widespread use of fossil fuels for energy production, including coal, oil, and natural gas, has worsened climate change and global warming. The development of sustainable alternatives that use nanotechnology to improve energy efficiency has increased as a result of this growing concern.

Thermoelectric heaters are constructed using samples that are collected to evaluate their thermoelectric properties and further explore these developments. Copper tapes are attached to both sides of the glass as electrodes. Analysis of thermoelectric (TE) performance of PbS monolayer shows that it outperformed bulk PbS by a large margin, with a maximum power factor of 6.1×10^{10} W/mK^2 and zT value, 1.0. This improvement is attributed to the monolayer's strong electrical conductivity and large Seebeck coefficient, which resulted from its narrow density of states (DOS) near the band gap edges and the band gap itself.

Pandit et al. [18] have conducted further theoretical studies to investigate the thermoelectric performance of 2D PbS monolayers using semi-classical Boltzmann transport theory coupled with density functional theory (DFT). According to their findings, the electronic structure is heavily influenced by spin–orbit coupling (SOC), resulting in higher zT values and lower thermal conductivity. It is found that one of the main reasons for the higher zT values of these materials is their low thermal conductivity values, as shown in Fig. 7.10 [5].

To solve climate change and other environmental problems, advances in renewable energy sources such as solar, wind, hydropower, bioenergy, geothermal energy, and others are inevitable [13]. The development of renewable energy technologies represents a viable replacement for fossil fuels, serving as an endless source of energy. However, not having enough energy storage devices leads to energy wastage, so it is important to save. Energy storage technologies with high-power density and energy density will inevitably be developed to provide energy for future generations [15].

Redox reactions (reduction–oxidation) are used in electrochemical energy storage devices to convert chemical energy into electrical energy. Apart from this mechanism, the main electrochemical energy storage devices that are the focus of much research in sustainable energy storage technologies are batteries, supercapacitors, and fuel cells. Supercapacitors are one of them, which have emerged as an important energy storage technology with a wide range of potential uses in many sectors. Supercapacitors are energy storage devices that have significant capacity for both rapid discharge and

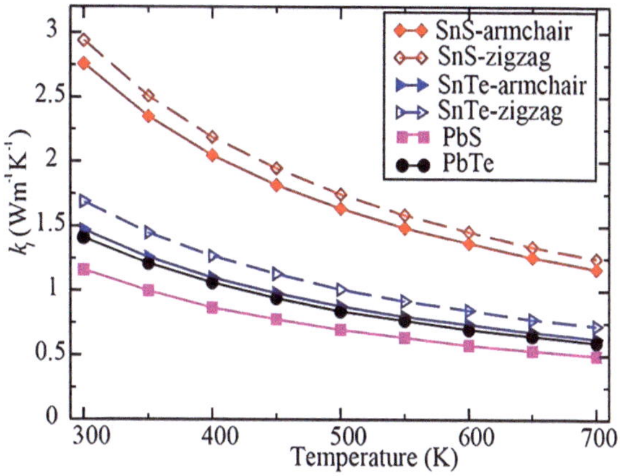

Fig. 7.10 The lattice thermal conductivity (κl) of SnS, SnTe, PbS, and PbTe MLs is dependent on temperature [5] [Open access]

energy storage. They differ from regular capacitors in that they have much higher capacitance values, which allows them to store more energy. They have better power density compared to batteries [6, 13–15].

7.3 Nanomaterials for Application in Engineering

The ability to alter the structure of materials on incredibly small scales to achieve specific properties at the heart of many benefits of nanotechnology, greatly expands the range of tools available to materials scientists. Among other qualities, materials can effectively be made stronger, lighter, more resilient, reactive, sieve-like, or better electrical conductors using nanotechnology. Currently, a large number of commercial products available on the market and used daily depend on materials and processes at the nanoscale:

- In personal body armor, nano-additives or surface treatments can help the materials resist wrinkles, stains, and bacterial growth, or they can provide easy deflection of ballistic energy.
- Transparent nanoscale coatings can be applied to windows, computer and camera screens, glasses, and other surfaces to make them anti-reflective, self-cleaning, water and residue-repellent, scratch-resistant, antibacterial, anti-fog, UV, or infrared-resistant. And electrically conductive.
- Nanoscale materials are making possible washable, durable "smart fabrics" with flexible nanoelectronics and sensors that can monitor health, harvest solar energy, and control motion.
- Fuel savings can be significant if vehicles such as cars, trucks, airplanes, ships, and spacecraft become lighter. Baseball bats, tennis rackets, bicycles, motorcycle helmets, car parts, luggage, and power tool bodies are becoming lighter, stiffer, more flexible, and more durable by adding nano-additives to polymer composite

materials. Sheets of carbon nanotubes are now being created for use in next-generation aircraft. For example, their light weight and conductivity make them ideal for heat control and electromagnetic shielding.

- High-quality NASA micrograph of a polymer-silicate nanocomposite.
- The high-resolution photograph shows a nanocomposite of polymers and silicates. This material can be used in food and beverage containers, automotive and aircraft fuel tanks, and aerospace components due to its excellent thermal, mechanical, and barrier properties.
- The goal of enzyme nanoengineering is to enable the conversion of cellulose from sources such as wood chips, corn stalks, unfertilized perennial grasses, etc., into ethanol for fuel. Numerous industrial fields, including electronics, construction, packaging, food, energy, healthcare, automotive, and defense, have demonstrated the potential use of cellulose nanoparticles. Among its many advantages, cellulose nanoparticles are expected to be less expensive than many other nanomaterials and have an exceptional strength-to-weight ratio.
- Automotive products that use nanotechnology materials include high-capacity batteries, thermoelectric materials for temperature control; tires with low rolling resistance; low cost, high-performance sensors and electronics; thin film smart solar panels; and fuel additives for longer, cleaner emissions.
- When it comes to machine parts, nanostructured ceramic coatings are much more durable than traditional wear-resistant coatings. Wear on moving parts in everything from power tools to industrial equipment can be significantly increased by using lubricants and motor oils enhanced with nanotechnology.
- Stabilizing, reducing, and capping agents. Various techniques, such as UV–vis spectroscopy, SEM, transmission electron microscopy (TEM), X-ray diffraction (XRD), and energy-dispersive spectroscopy (EDS), are used to characterize these synthesized trimetallic nanoparticles (Fig. 7.11) [19].
- Superior household products, including degreasers and stain removers, air purifiers, filters, and environmental sensors, are made from nanotechnology materials. Another example is antibacterial cleansers, and specialty paints and sealants, such as house paints, that clean themselves and won't stain or scratch.
- Nanomaterials are classified into several categories, such as organic, inorganic, and carbon nanomaterials, based on various parameters such as structure, shape, size, and characteristic physical and chemical properties (Fig. 7.12) [19].

7.3.1 Cosmetics and Individual Items

In addition to supporting healthy skin, makeup should offer a stunning rainbow of shades to suit all skin tones. The use of high-quality plant ingredients and finely ground minerals represents the latest achievement in classic cosmetics. These ingredients must be completely natural and 100% pure to be suitable for the most delicate skin. However, most dyes used in mass-produced cosmetics are synthetic. In fact, the Food and Drug Administration has outlawed all natural color additives in the US

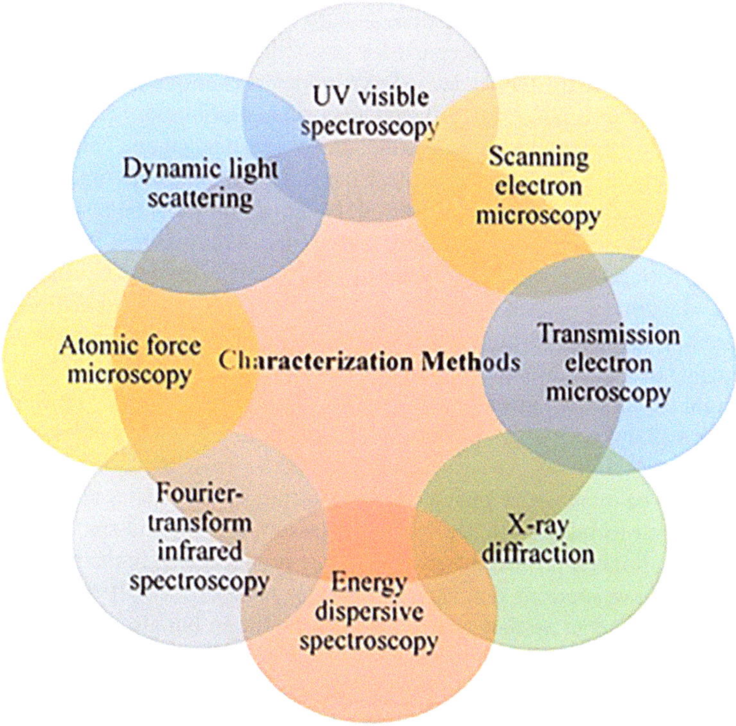

Fig. 7.11 Various techniques for characterizing nanoparticles [19] [Open access]

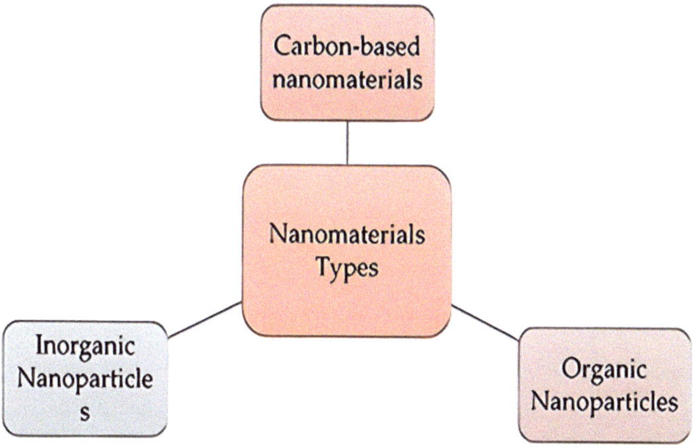

Fig. 7.12 Various kinds of nanomaterials [19] [Open access]

unless they are explicitly approved (like annatto seeds) to prevent allergic reactions and irritation. Color additives intended for cosmetic use, both inorganic (powders) and organic (dyes and pigments), are safe for human consumption and do not contain heavy metals such as cadmium, lead, and hexavalent chromium.

Inorganic pigments include ultramarine, iron oxide, manganese, chromium oxide, titanium dioxide, and ferric ferrocyanide. Cosmetic pigments, both organic and inorganic, do not dissolve in either water or oil. However, they can be dispersed microscopically in a liquid medium, giving the appearance of dissolution. Because soluble dyes can be harmful to your health, especially when used in excessive concentrations, very few of them are used in cosmetic products. For example, since different metal salts (sodium, calcium, barium, and aluminum ions) impart different shades, lakes are pigments produced by precipitating a soluble dye with these salts.

Bismuth oxychloride, guanine, or titanium dioxide can be used with color additives. When mixed with titanium dioxide-coated mica, it produces incredibly beautiful pearls in two colors. A soft substance used as frost in lipsticks is bismuth oxychloride. The main use of guanine obtained from herring scales is in nail paints. Apart from their technological applications, titanium dioxide and zinc oxide are the most common mineral powders used in cosmetic products sold in the market. These powders may have soothing and anti-inflammatory properties [20]. These nanomaterials do not create an aesthetic translucent composition but also effectively block ultraviolet (UV) rays. In consumer products, silver nanoparticles are used as antibacterial and preservative agents. Cosmetic products in the United States cannot make antibacterial claims because these claims involve physiological processes; as a result, they are used in medicines. Moreover, nanotechnology has had a major impact on renewable energy sources, namely energy storage and conversion. Table 7.1 lists the applications of nanomaterials in various fields, and some of these applications are briefly described in the next part [21].

Table 7.1 Current research on cosmetics that use nanoparticles

Nanomaterial	Description	pH	Penetration
Quantum dot-655/COOH	Shape: Ellipsoid Diameter: 6–12 nm Hydrodynamic diameter: 18 nm Skin: Pig	9	Negative
Quantum dot-655/PEG	Skin: Human stratum corneum	8.3	Negative
TiO2 nanoparticles	Particle size: 200 nm Skin: Pig	–	Negative
TiO2 nanoparticles in oil/water	Particle size: 10 × 50 nm Skin: Pig	–	Negative
ZnO nanoparticles	Skin: Human epidermal Size: 15–30 nm Duration: 24 h Skin: Human	–	0.03%—Negative

7.3.2 *Transportation and Automobiles*

Due to the special material, physical and chemical properties of products, nanotechnology is used, and these properties greatly benefit the global materials industry. The manipulation of matter between 0.1 and 100 nm in size is known as nanotechnology [1, 3, 7]. The distinctive physical and chemical characteristics of nanotechnologies could lead to significant improvements in (photo) catalysis, optical sensitivity and mechanical strength, enabling their use in energy storage, sensing and other applications. Thus, through controlled manipulation of shape and size measured in nanometers, nanotechnology helps in the development of tools, instruments, and structures [15]. In fact, due to properties of materials, particle size is very important and inevitably affects the nanoscale. This is because enhanced material properties are achieved at the nanoscale, where the electrostatic forces and quantum phenomena cancel out the effects of gravity. Many disciplines, including chemistry, biology, and engineering, have undergone revolutionary changes as a result of the nanotechnology revolution [16]. Due to its higher efficiency, nanotechnology is increasingly being applied in various sectors, including biomedicine, robotics, electronics, automobiles, and civil engineering (including transportation) [17, 20, 22, 23]. The following is a quick description of nanomaterials associated with cars:

7.3.2.1 Body Parts with Low Weight

Reducing your vehicle's weight offers amazing benefits that benefit your passengers as well as your business's bottom line. It is critical to adapt to new manufacturing regulations to make your vehicles lighter and overall, more efficient. So, how can you effectively shorten this method? Some useful tips you should know about. This is the information you will need to produce premium cars by learning about car weight reduction.

Realize the importance of car weight loss first of all. Why should losing weight in your car be a top priority for you? Lighter cars use less fuel and produce fewer emissions. Your customers will be grateful for the reduced burden on their pockets, as well as the fact that your vehicles will comply with the latest environmental regulations. By taking into account the weight reduction of the vehicle during construction, you can create a more economical and environmentally friendly vehicle. In addition, these modifications improve vehicle safety, which benefits all drivers [1, 3, 4, 7].

In addition, polymer nanocomposites, such as clay nanocomposites reinforced with polyamide, polycarbonates, and polypropylene, are widely used. These nanoparticles will become less flame-retardant and more heat-resistant when mixed with polymers [2, 5, 6]. Using nanocomposite and hybrid solutions, a car that normally weighs 900 kg of steel can weigh as little as 300 kg. On average, weight savings when using nanocomposite plastics are 25% compared to densely filled plastics and

80% compared to steel. Vanadium oxide (VW) uses long fiber reinforced thermoplastic (LFRT), which provides superior design flexibility, high strength-to-weight ratio, impact resistance, and less corrosion compared to metals [1, 7].

7.3.2.2 Paint-Coating

Manipulating matter at the nanoscale is the focus of nanotechnology. At this level, materials undergo significant physical and chemical changes due to a high surface-to-volume ratio. With its applications in science and technology to create new materials at the nanoscale useful for various fields such as economics and environment, nanotechnology has become increasingly popular in recent years. Recent studies [3] have identified potential uses of TiO_2 nanoparticles (NPs) in some industries, including sensor technologies, photocatalysis, solar energy conversion, pigments, and biological activities such as antifungal, antibacterial, and anticancer properties. Despite their widespread use, TiO_2 NPs have many other characteristics, including high chemical stability, high redox potential, and availability and environmental friendliness [3, 4]. In addition, the process has been used to create NPs that have a significant impact on these characteristics, including the specific surface area, crystal structure, particle shape, and crystallite size. There are three different polymorphic crystalline forms of TiO_2 NPs: rutile (tetragonal structure), brookite (orthorhombic structure), and anatase (tetragonal structure). Anatase and brookite can be heated to high temperatures to produce rutile, which is a stable phase [6]. Depending on the structure and particle size, each crystalline phase has unique physical characteristics that make it suitable for different applications [13]. For example, bandgap, 3.2 eV, means that exposure to ultraviolet light can activate the anatase phase, so it is often used for photocatalysis [17].

7.3.2.3 Scratch–Resistance

Nowadays, the automotive coating system—a multilayer polymer system—plays an important role in bringing the appearance of a car in line with the evolving aesthetics of people. In addition, the system protects key components from radiation and environmental degradation. Each layer in a multilayer system has a specific function and is used in a sequence of drawing commands. A previous study [23] has provided detailed descriptions of functional features of each layer. Overall, the resistance of coating systems to damage directly affects the long-term appeal of a vehicle and the durability of its beautiful body. Over the past few decades, maintaining the integrity and appearance of a vehicle has become a top priority for original equipment manufacturers (OEMs), especially for premium vehicles.

On another hand, used vehicles are easily scratched during daily use and resulting damage to the finish can seriously degrade the appearance and lifespan of the vehicle. Customers now consider scratch resistance one of the most important requirements for automotive coating systems. However, research on automotive coatings, which typically have more complex structures, is rather sparse.

Divided scratches into two extreme groups depending on their diameter. Just a few microns in width and depth constitute one category of scratch that is considered scratch damage [24]. Usually, the cause of deposits is wear and tear on the automatic car wash brushes when washing the car. However, due to its microscale, a single mark is difficult to detect with the naked eye except under special viewing conditions. Moreover, if dirt damage accumulates in large quantities, the gloss of the vehicle's paint surface will decrease, even if it cannot penetrate the coating [19].

Over 20 years ago, two fundamental methods were widely used to determine the abrasion resistance of coating surfaces: the meter test [5] and advanced sensors and measurement technologies at AMTEC and Kistlercar wash test (ISO/FDIS 20566) [15]. In addition, several studies [17, 20, 22, 23] have been published to date on the subject of nano scratch testing, which has been used to quantitatively measure the scratch resistance of automotive coatings, and give some idea of their mechanical properties [17].

7.3.2.4 Tires

Commercially produced nanoparticles, known as carbon black (CB), are produced by partial combustion and vapor-phase pyrolysis of hydrocarbons under carefully controlled conditions, resulting in a variety of target grades. It consists of pure elemental carbon (EC) in the form of a fine black powder. CB is mainly used as a black pigment in inks, a reinforcing agent in vehicle tires, and as a way to improve the electrical conductivity of plastics. 800,000 tons of CB are added to plastic every year, and the demand for plastic is growing worldwide.

Since automobile tires typically contain CB as a hardening agent, these tires may emit particulate matter (PM) during braking, and tire wear is an example of a non-emission mechanism. Most rubber particles that form non-exhaust PM resulting from tire wear on road surfaces include carbon black, zinc oxide, and various heavy metals. It has been shown that inhalation of non-exhaust particulate matter from worn tires and exhaust particulate matter from road-going vehicles can lead to respiratory diseases caused by oxidative stress and inflammatory pathways. The non-exhaust particulate matter emission factor associated with tire wear on gasoline-powered vehicles is twice the exhaust particulate emission factor of 6.1 mg/km. Although particulate matter risk management and toxicity testing have been extensively studied, nothing is known about the health effects of CB-containing particulate matter that is generated during tire wear.

The main particles that make up CB are typically nanosized and can aggregate to sizes ranging from a few microns to 100 nm. Due to their special characteristics and small size, these nanoparticles may have negative effects on the environment and human health. Therefore, an adequate risk assessment should be carried out regarding the clean and free particle-related wear of CB tires. Many studies [14] have been conducted in laboratory settings or on CB production workers to investigate the possible health effects of CB. According to others [15], CB may be directly genotoxic and harmful to reproduction. However, some toxicological data mistakenly identify carbon black (CB) with black carbon (BC). Characteristics are comparable to CB in terms of component content, aggregate shape, and black color. Given their similar names, it is not surprising that CB and BC are often misused [15].

7.3.2.5 Other Uses in Automobiles

Due to their unique chemical and physical properties, nanomaterials have attracted much attention. The result of nanotechnology and nanoscience is nanomaterials. While nanomaterials are a vast field in themselves, they have the potential to completely change the way materials and products are produced. The purpose is to report on the latest advances in the use of nanotechnology in the automotive industry. The results have demonstrated that using more nanomaterials in automobiles improves vehicle performance and makes transportation more environmentally friendly. Nanomaterials, including graphene, carbon nanotubes, aluminum–silicon carbide composites, and aluminum-carbon nanotube composites, are often used in automobiles. This essay looked at how these resources are used for future applications as well as in many automotive industries [14]. Fluid viscosity increases as a result of the orientation of magnetic nanoparticles in magnetorheological fluids, which are dispersed in a solvent and respond to small changes in pressure and magnetic fields.

7.3.3 Agriculture

The use of agrochemicals has increased agricultural production, but at the expense of the health of farmers and the general population who consume chemically produced food, as well as the soil and water systems with their associated flora and fauna. Applications of nanoparticles in the automobile sector are listed in Table 7.2 [21].

The risk assessment of this technology occurs much later than its use. Therefore, this chapter will cover hazards, uncertainties, and nanoecotoxicology. The use of nanotechnology in agriculture raises ethical issues. We will talk about how the public views and understands nanotechnology at the moment [16]. Thus, the development of nanoagrochemicals such as nano pesticides, nano fertilizers, and nanosensors to improve solubility, bioavailability, and targeted delivery, as well as their controlled release, will have many potential benefits, such as precise dosage of fertilizers, enhanced control of disease vectors and pests, reducing chemical pollution, and ultimately increasing agricultural productivity. Regarding the toxicity and safety of these systems for human health and the environment, many unresolved problems remain. Graphene oxide (GO), titanium dioxide (TiO_2), silver (Ag), and zinc (Zn) [15]. Therefore, the approaches of this promising method include protection against viruses and diseases, as well as increasing the effectiveness of pesticides, fungicides, and herbicides, which improves plant development and maintains

Table 7.2 Applications of nanomaterials in the automotive industry [21] [Open access]

Nanomaterials	Merits	Demerits	Applications
Long fiber reinforced thermoplastic	Extensive impact resistance, high stability	Production is challenging	Body parts
Silicon dioxide (SiO_2)	Lower paint toxicity, supports self-healing	Uniform dispersion is challenging	Paint coating
Copper dioxide (CuO_2), Molybdenum disulfide (MoS2), Titanium dioxide (TiO2)	The coefficient of friction can be lowered up to 50%	Life span is shorter	Engines
Layered double hydroxides (LDH)	Absorbs heat, enhances char growth	Stability concerns	Paint coating
Carbon nanotubes (CNTs)	High mechanical stability and lightweight	Expensive	Body parts
Soot, carbon black, silica	Extended durability, higher fuel efficiency	Uniform dispersion is challenging	Vehicle tires
Copper oxide (nano-fluids)	Upgrades the heat transfer coefficient	Agglomeration issues	Engines
Carbon nanotubes (CNTs), copper (Cu), zinc oxide (ZnO), Gold (Au)	Zero-toxicity, good efficiency	Relatively expensive	Vehicle interior

discharges [22]. Since nanomaterials apply to almost all aspects of the agricultural industry, including transportation, processing, production, and storage, they improve the efficiency, productivity, and safety of agriculture [16].

From a practical point of view, there are several options for using nanoparticles in agriculture. The main goal is to improve the delivery of fertilizers, which, in turn, leads to an increase in the absorption of elements by plant cells and a decrease in nutrient losses. They provide a synchronous introduction of macro- and microelements [20]. Because nanostructured fertilizers have a large surface area, a precise delivery mechanism, and gradual, continuous release in response to environmental signals and biological needs, they improve nutrient use efficiency. Nanophosphorus iron (Fe), nanomagnesium (Mg), and nanozinc (Zn) are a few examples of typical nanofertilizers.

In addition, since CNTs have competitive thermal, chemical, electrical and mechanical capabilities, they can be used to produce micro- and macronutrients while reducing inputs, which can yield good results in agriculture [18]. The second goal is to control insect pests. Some nanoparticles, such as zinc oxide, titanium dioxide, and gold, are effective in pest control due to slow and active release of active chemicals as an affordable and reliable alternative to industrial drugs that have adverse side effects. Nano-pesticides reduce the movement of unwanted pesticides and the release of organic solvents. In addition, nanoparticles can detect, identify, and report crop diseases using rapid analytical methods and smart field systems, prompting producers

Table 7.3 A few instances of nanoparticles used in the field of agriculture [21] [Open access]

Nanomaterial	Function of Nanomaterial	Industrial Name	Application
Zinc (Zn)-nanoparticles, Copper (Cu)-nanoparticles, Boron (B)-nanoparticles, Iron (Fe)-nanoparticles, Biohumus	Fertilizer	Motawazen, Saula, Solocross, Green Earth NanoPlant	Nano-fertilizers
Copper-doped montmorillonite	Detection of propineb fungicide	–	Nano-sensors
Copper (Cu)-nanoparticles, Polymeric-nanoparticles, Zinc (Zn)-nanoparticles	Bactericide, controlled release, and fungicide	Bio-nano technology	Nano-pesticides
Poly-epsilon caprolactone	Controlled release	–	Nano-herbicides

to apply the necessary ingredients before symptoms appear [16]. Nanofungicides perform another function. Chemical fungicides damage plants, which are sensitive to the environment and health. Considering the size and shape of nanoparticles, nanofungicides provide a valuable answer to the above-mentioned problems. The antibacterial properties of Ag nanoparticles make them widely known nanofungicides. While TiO_2 nanoparticles have antibacterial and photocatalytic activities that provide plant protection, ZnO nanoparticles also exhibit antifungal, antibacterial, and antimicrobial properties [18, 23, 24]. The latest application is nanosensors, which are highly sought after in the agricultural industry for their ability to monitor environmental, aquifer, and soil contamination, easily and accurately [24]. Ag nanoparticles are a well-known nanomaterial used as a sensor. Typical nanomaterials used in agriculture are listed in Tables 7.3 and 7.4 [16].

7.4 Risks of Nanotechnology

Nanotechnology is said to be the next big thing in chemistry, medicine, materials science, engineering, and other fields. It is necessary to talk about the various risks associated with using nanotechnology. Breakthrough developments can come from molecular manufacturing, a complex form of nanotechnology. As the production of nanomaterials increases worldwide, toxicologists and regulatory experts are increasingly focusing on nontoxicity and nanodisks. Despite the novelty of using nanostructures, concerns have already been raised about the possible toxicity of nanomaterials used in various aspects of human existence. To evaluate the possible harmful and toxic effects of nanomaterials on biological ecosystems and human health, more research is needed to understand the physicochemical characteristics of nanomaterials and their effectiveness. The effect of nanoparticles on the human body and associated biosystems varies depending on their type, size, shape, substituents, and coatings.

Table 7.4 A few other uses of nanomaterials in a variety of domains [21] [Open access]

Applications	Examples
Biomedical and health care	– Gold nanoparticles for targeted delivery of drugs directly to cancer tumors – Liposome nanoparticles for controlled and targeted drug delivery – Quantum dots used for the detection and diagnosis of cancer – Graphene and carbon-based nanomaterials as wearable multifunctional sensors for monitoring skin temperature, wrist pulse, breathing, and pressure mapping
Construction materials	– CNT, GO, and SiO2 nanomaterials used as reinforced cement composites – CNT as reinforced metal matrix composite to produce stronger, ultralight, and corrosion-resistant nanocomposite materials – ZnO nanoparticles used as coatings to reduce ultraviolet radiation exposure – Silver nanoparticles embedded in paint to provide antimicrobial properties – TiO2 nanoparticles used as glaze coating for anti-fouling properties
Electronics and electrical applications	– Carbon-based nanomaterials used as electrodes in supercapacitors, offering large storage capacities and fast charge/discharge rates
Environmental remediation	– CNT, graphene oxide, and other nanomaterials are used as adsorbents to remove heavy metals and pollutants from water bodies – Functional nanomaterials such as TiO2, Al2O3, CNT, and zeolites incorporated into filtration membranes to improve mechanical and thermal stability, permeability, fouling resistance, and pollutant adsorption capabilities
Food packaging	– Nanoclay is used for food packaging due to its strong mechanical and thermal properties – Nanoemulsion used as an edible coating for preserving and storing food – Gold nanoparticles are used as food sensors to monitor food quality and safety
Food processing	– SiO2 nanomaterials are used as anti-caking agents to maintain flow properties in powdered products – Titanium dioxide (TiO2) is used as a food color additive and flavor enhancer – SiO2-gallic acid nanoparticles used as antioxidants – Silver (Ag) and zinc oxide (ZnO) nanoparticles used as antimicrobial agents in the food packaging industry
Agriculture	– Fertilizers coated with nanocapsules or nanoparticles for improved delivery of agri-chemicals and nutrients – Nanomaterials used as pest control agents for enhanced effectiveness – Zinc oxide (ZnO) quantum dots used for pesticide detection – Magnetic nanoparticles used for smart agrochemical and herbicide delivery systems – MWCNT nanomaterials used to increase plant production rates – TiO2 nanomaterials used as sensors for monitoring soil conditions, pesticides, and crop growth
Renewable energy	– Carbon nanomaterials used for direct conversion of sunlight into electricity – Graphene used as a metal-free catalyst in fuel cells – Metal–organic frameworks used for hydrogen storage
Sensors	– Metal oxide nanomaterials used for various sensor applications, such as gas sensors, chemical oxygen demand (COD) sensors, and biosensors

The nanoparticles are generally considered harmless in specific contexts. With its ability to manipulate matter at the atomic and molecular level, nanotechnology has enormous potential in some disciplines, including electronics, environmental protection, and medicine. However, it also comes with its risks and challenges. Here are some of the main challenges:

Ingestion, skin absorption, or inhalation are three ways nanoparticles can enter the body. Because of their small size, they can penetrate cells and tissues, which can lead to unknown health effects. Long-term exposure can cause problems such as cancer, oxidative damage, and inflammation. Plants and animals can be harmed by the accumulation of nanoparticles in the environment and ecosystems. It is not yet clear how they affect air, water, and soil quality. They can have unexpected interactions with natural processes [8, 19, 25].

Unintended consequences: Because of their special properties, nanoparticles can cause unintended effects or interactions. For example, nanoparticles that deliver drugs may unintentionally interfere with other biological functions. The potential of nanotechnology for ethical dilemmas associated with its use. Issues include concerns about potential misuse as a weapon, privacy issues with nano-monitoring devices, and broader implications of human enhancement. Due to rapid development and significant commercial potential, nanomaterials have today become one of the highest priority areas of research. Recent developments in engineered nanomaterials have opened new possibilities for a variety of industrial applications, from electronics and energy storage to biotechnology and healthcare. Full implementation of nanotechnology in industrial applications requires addressing and successfully regulating the toxicity and other associated hazards of nanomaterials despite their superior capabilities.

The improved properties and unique characteristics of nanomaterials compared to bulk materials have made them increasingly important in the development of new technologies. Their exceptional properties are driving innovation in a wide range of industrial applications. This chapter on carbon nanomaterials provides a detailed overview of the fundamental engineering principles involved in nanomaterials. It investigates various technical applications of nanomaterials in several important sectors such as biotechnology, food processing, biomedicine, environmental remediation, construction, electronics, renewable energy, and energy storage. The chapter briefly discusses the potential impact of nanoparticles on human health, highlighting the need for continued research to ensure efficacy and safety. In short, nanomaterials have enormous potential to advance the field of nanotechnology, providing revolutionary opportunities in various industries and driving advances in advanced technological applications.

Problems

1. Describe the importance of nanomaterials in engineering applications.

2. List the materials of flexible electrodes in case a triple error occurs when comparing the sizes of nanomaterials.
3. What are carbon nanotubes' characteristics and characterization methods (1D)?
4. What role do nanomaterials play in engineering applications?
5. Describe the different uses in preparing and classifying various materials.

References

1. Giannazzo, F., Schilirò, E., Greco, G., & Roccaforte, F. (2020). Conductive atomic force microscopy of semiconducting transition metal dichalcogenides and heterostructures. *Nanomaterials, 10*(4), 803.
2. Chianese, F., Aversa, L., Verucchi, R., & Cassinese, A. (2022). Molecular doping of CVD-graphene surfaces by perfluoroalkyl-substituted perylene diimides derivatives. *Nanomaterials, 12*(23), 4239.
3. Lindqvist, C., Moons, E., & Van Stam, J. (2018). Fullerene aggregation in thin films of polymer blends for solar cell applications. *Materials, 11*(11), 2068.
4. Slattery, A. D., Shearer, C. J., Shapter, J. G., Quinton, J. S., & Gibson, C. T. (2017). Solution based methods for the fabrication of carbon nanotube modified atomic force microscopy probes. *Nanomaterials, 7*(11), 346.
5. Cui, L., Cui, K., Ci, H., Zheng, K., Xie, H., Gao, X., ... & Liu, Z. (2019). Transparent electrothermal heaters based on vertically-oriented graphene glass hybrid materials. *Nanomaterials, 9*(4), 558.
6. Wang, Z., Hu, T., Liang, R., & Wei, M. (2020). Application of zero-dimensional nanomaterials in biosensing. *Frontiers in Chemistry, 8*, 320.
7. Wu, Y., Wang, D., Liu, J., Cai, H., & Zhang, Y. (2020). Atomic force microscope study of ag-conduct polymer hybrid films: Evidence for light-induced charge separation. *Nanomaterials, 10*(9), 1819.
8. Fang, Z., Xing, Q., Fernandez, D., Zhang, X., & Yu, G. (2020). A mini review on two-dimensional nanomaterial assembly. *Nano Research, 13*, 1179–1190.
9. Paramasivam, G., Palem, V. V., Sundaram, T., Sundaram, V., Kishore, S. C., & Bellucci, S. (2021). Nanomaterials: Synthesis and applications in theranostics. *Nanomaterials, 11*, 3228.
10. Wang, K., Ma, Q., Qu, C. X., Zhou, H. T., Cao, M., & Wang, S. D. (2022). Review on 3D fabrication at nanoscale. *Autex Research Journal.*
11. Gentselev, A. N., & Baev, S. G. (2022). Production of planar elements of terahertz optics by means of deep X-ray lithography. *Optoelectronics and Instrumentation Data Processing, 58*, 198–205.
12. Lohse, M., Heinrich, M., Grützner, S., Haase, A., Ramos, I., Salado, C., Thesen, M. W., & Grützner, G. (2021). Versatile fabrication method for multiscale hierarchical structured polymer masters using a combination of photo- and nanoimprint lithography. *Micro Nano Engineering, 10*, Article 100079.
13. Pandit, S., Behera, P., Sahoo, J., & De, M. (2019). In situ synthesis of amino acid functionalized carbon dots with tunable properties and their biological applications. *ACS Applied Bio Materials, 2*, 3393–3403.
14. Yan, Y., Gong, J., Chen, J., Zeng, Z., Huang, W., Pu, K., Liu, J., & Chen, P. (2019). Recent advances on graphene quantum dots: From chemistry and physics to applications. *Advanced Materials, 31*, 1808283.
15. Khan, Z. H. ed. (2017). *Nanomaterials and their applications* (Vol. 84). Springer.
16. Dao, T. C., & Luong, T. Q. N. (2020). Fabrication of uniform arrays of silver nanoparticles on silicon by electrodeposition in ethanol solution and their use in SERS detection of difenoconazole pesticide. *RSC Advances, 10*, 40940–40947.

17. Kang, S. M., Kim, H., Choi, J. S., & An, J. H. (2021). Bioinspired omniphobic microchamber structure. *Advanced Materials Interfaces, 8*, 2100027.
18. Haghanifar, S., Tomasovic, L. M., Galante, A. J., Pekker, D., & Leu, P. W. (2019). Stain-resistant, superomniphobic flexible optical plastics based on nano-enoki mushroom-like structures. *Journal of Materials Chemistry A, 7*, 15698–15706.
19. Roy, A., Kunwar, S., Bhusal, U., Alghamdi, S., Almehmadi, M., Alhuthali, H. M., ... & Azlina, M. F. N. (2023). Bio-fabrication of trimetallic nanoparticles and their applications. *Catalysts, 13*(2), 321.
20. Baig, N., Kammakakam, I., Falath, W., & Kammakakam, I. (2021). Nanomaterials: A review of synthesis methods, properties, recent progress, and challenges. *Materials Advances, 2*, 1821–1871.
21. Radnik, J., Kersting, R., Hagenhoff, B., Bennet, F., Ciornii, D., Nymark, P., ... & Hodoroaba, V. D. (2021). Reliable surface analysis data of nanomaterials in support of risk assessment based on minimum information requirements. *Nanomaterials, 11*(3), 639.
22. Kolahalam, L. A., Kasi Viswanath, I. V., Diwakar, B. S., Govindh, B., Reddy, V., & Murthy, Y. L. N. (2019). Review on nanomaterials: Synthesis and applications. *Materials Today: Proceedings, 18*, 2182–2190.
23. Vollath, D. (2013). *Nanomaterials: An introduction to synthesis, properties and applications* (2nd ed.). Weinheim, Germany.
24. David, N., Djilali, N., & Wild, P. (2012). Fiber Bragg grating sensor for two-phase flow in microchannels. *Microfluidics and Nanofluidics, 13*, 99–106.
25. Hang, D. R., Huang, C. F., & Cheng, K. A. (2009). Probing semiclassical magneto-oscillations in the low-field quantum Hall effect. *Physical Review B, 80*, Article 085312.

Chapter 8
Thin Film Synthesis and Its Applications

Because of its intriguing surface features and numerous applications in the fields of microelectronics to optics, space science to aircraft, photovoltaic, solar cells to superconductivity, thin film technology is a key area of modern scientific research. Synthesis kinetics and thermodynamics regulate the atomic structure, composition, microstructure, defects, and interfaces, all of which influence the characteristics of films. The ability to create flexible and transparent films, as well as simple and low production temperatures, are the main advantages of thin film production. Deposition processes that are sustainable, affordable, and highly efficient can be used to produce crystalline thin films that are compact, high quality, and multi-color. However, creating desirable films for emerging industrial and technological applications is fraught with challenges. This chapter will review some thin film technologies, obstacles, and promising results.

8.1 Introduction

One of the possible technologies for obtaining energy from fuels and reducing pollutant emissions into the atmosphere in industrial processes is catalytic combustion. Methane, carbon monoxide, and volatile organic compounds (VOCs) emissions from industrial gases can be effectively controlled by catalytic total oxidation [1, 2]. The ecology and living beings are seriously affected by these constant emissions, which mainly consist of hazardous gases such as carbon monoxide, methane, and other volatile organic compounds (VOCs) such as formaldehyde, benzene, toluene, and chlorinated compounds. These pollutants contribute to global warming and harm the environment.

To get rid of these hazardous materials by oxidation processes, various catalysts have been used, especially those containing transition metal oxides or noble metals such as Pd or Pt [3–5]. However, the long-term exposure of catalysts to combustion conditions may lead to minor deactivation caused by high operating temperatures

R. A. Al-Samarai and Y. Al-Douri, *Engineering of Nanomaterials*,
https://doi.org/10.1007/978-981-95-5499-7_8

during sintering [6]. For methane oxidation, low-temperature catalysts are particularly important because they reduce the deactivation process. Methane is used as a benchmark for evaluating combustion activity parameters due to its high stability and notable polluting properties [7]. Several methods for material preparation are used to create composite materials with enhanced catalytic activity. These methods emphasize:

- Increasing the accessible surface area.
- Improving the redox properties.
- Incorporating more species to improve performance.

The goal is to create oxidation reaction catalysts that are extremely active. Depending on the choice of composite components, the development of thin films for active materials has shown impressive results in improving catalyst stability and combustion efficiency. One of them in particular, thin film deposition, is an attractive method for catalyst development because it provides precise control over the thickness and size of material. The catalytic activity depends heavily on this control, which also facilitates the interaction of composite layers and improves the redox capabilities.

This aims to overcome the difficulties in producing active catalysts for methane and carbon monoxide combustion using thin film technology. Organic thin films are easier to produce and require lower temperatures for their preparation than ceramic or mineral materials. There are many popular thin film fabrication methods, such as:

- Sputtering.
- Electrodeposition process.
- Spin coating.
- Coating and electron deposition.
- Physical vapor deposition (PVD).
- Chemical vapor deposition (CVD).
- Sol–gel method.

A specially designed heater to hold the sample [3]. The sample temperature is monitored using a K-type chromium-aluminum thermocouple, which has a temperature coefficient of 25 °C/mV. Figure 8.1 shows a schematic illustration of the Ag/CdTe/Al structure [1].

These methods typically produce thin films based on material characteristics. The synthesis and production of powders and thin films at atmospheric pressure is made possible by the sol–gel process. Nowadays, modern CVD and PVD processes operating at maximum efficiency are used to create high quality, compact, and multi-color crystalline thin films. Figure 8.1 depicts the general procedure for cardiovascular disease.

Thin films of various materials have been prepared using a variety of traditional procedures. In CVD, a precursor is heated to produce vapor, converted to gas phase by ultrasound, and transported to a heating chamber using a carrier gas to deposit the gas on the surface of the substrate at a predetermined temperature. Therefore, high temperatures in vacuum chambers are necessary for the production of thin films by

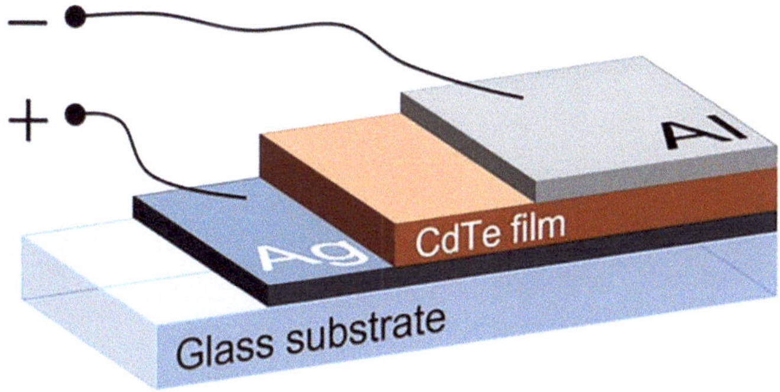

Fig. 8.1 Schematic representation of Ag/CdTe/Al sandwich configuration [1] [Open access]

CVD. It is believed that large surface thin films can be produced on an industrial scale using CVD. Atomic force microscopy (AFM) is used to investigate the surface topography of both undoped and Li-doped films. A 1 μm × 1 μm area is scanned in tapping mode to evaluate the surface roughness. Figure 8.2 shows 2D AFM images. As Li concentration increases, the average size of the dense, rounded grains in each sample varies from about 150 nm to about 90 nm [1].

Similarly, PVD involves heating a precursor material to the gas phase, which is deposited onto a substrate to form a thin film using an electron beam or resistive heating. These methods are generally divided into several groups, including thermal evaporation, magnetron sputtering, and pulsed laser deposition. These methods for controlling the artificial grain size on the thin film surface require improved operating protocols. For other reasons, the formation of grains of a certain orientation and structure on the surface of thin films requires using an expensive single-crystalline substrate. Given the high cost of a single-crystal substrate, growing this type of thin film on large surface areas such as wafers is expensive and requires sophisticated equipment [4–6]. To investigate the distribution of silver particles on the surface of ceramic substrates further, scanning resistance diffusion microscopy is used, which creates a potential difference between the material under study and the probe to measure the resulting current. A scanning electron microscopy (SEM) image of NiO: Li (10%) film deposited on glass is shown in Fig. 8.3a. The film shows well-packed, spherical, and cubic agglomerates with an average size of 100 nm, which is consistent with what is seen in AFM images of other samples. The preferred orientation shown in the XRD pattern for this film is consistent with a regular and highly oriented structure. A cross section of the same film is shown in Fig. 8.3b; it has a thickness of 450 nm and a flat, compact surface with good adhesion [1].

In heterogeneous catalysis, thin film catalysts play a crucial role as support materials, stabilizing the smallest metal particles as reactive species. The extreme adaptability of thin film catalysts makes them useful for a variety of processes, including CO preferential oxidation (Prox) and gas displacement with water (water–gas shift), oxygen storage, hydrocarbon oxidation, three-pass catalysts, and hydrogen production (biomass and bio-oil) [8–11].

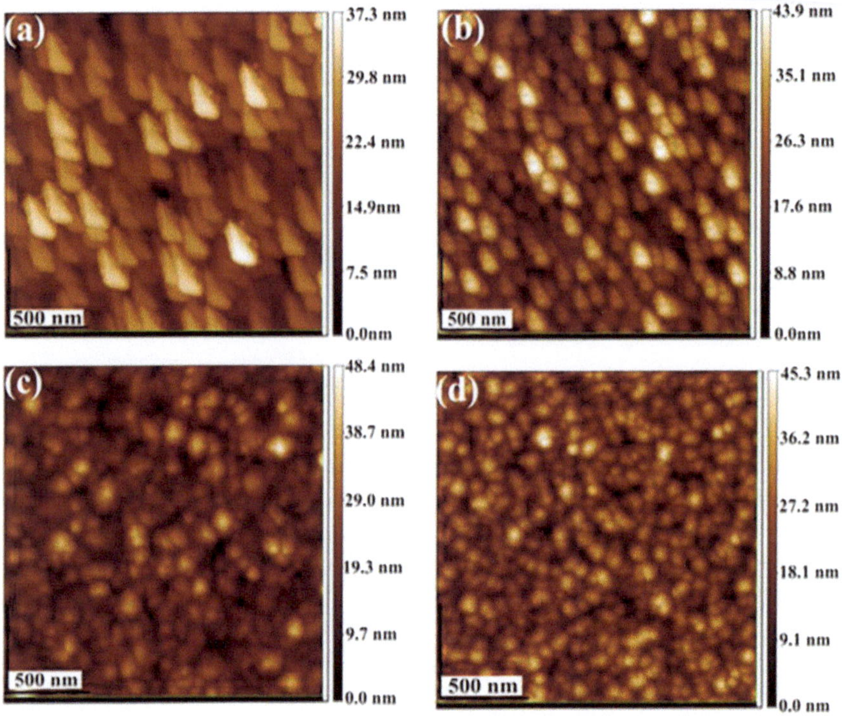

Fig. 8.2 AFM images of undoped and lithium-doped NiO thin films. RMS values of **a** NiO: Li (0%); 5.47 nm, **b** NiO: Li (5%); 5.97 nm, **c** NiO: Li (15%); 6.29 nm, and **d** NiO: Li (20%); 6.40 nm [1] [Open access]

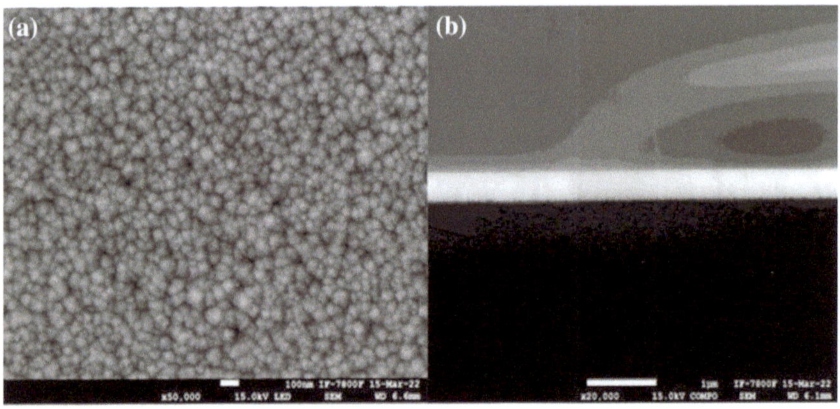

Fig. 8.3 **a** SEM micrograph of NiO: Li (10%) deposited on glass. **b** SEM image of the cross section of the same film [1] [Open access]

8.2 Nanostructured Thin Films

With roots in the ancient metal age, thin film technology is one of the newest sciences and oldest arts. Thin films are made up of incredibly thin layers of material ranging in thickness from one micron (10^{-6} μm) to the nanoscale. The thickness of deposited layers is the primary difference between the thick film and thin film depositions. Individual atoms or molecules are deposited onto a surface to form thin films, while particles are deposited to create thick coatings. The composition of thin films can be either uniform or non-uniform, depending on the desired properties. There are several important steps involved in the production of thin films:

- Deciding what material to unload.
- Transporting the material from the substrate source.
- Depositing the material as a thin layer onto the substrate.
- Post-deposition procedures to enhance the film, including annealing or heat treatment.
- Nanostructured thin films of metallic or composite conductors are of great interest from a technical and technological point of view.

The optical properties of thin films make them extremely important for various applications such as spectrally selective solar absorbers, solar control windows, angle-selective filters, optical biosensors, and decorative coatings. Nanograined metallic layers, thin metallic coatings on nanostructures, tilted nanometallic columns inside oxides, and polymer-based composites with metallic, oxide, or boride nanoparticle conductors are examples of thin film morphologies.

Under certain circumstances, surface plasmon polaritons and other generated currents may pose a threat to quasi-static hypotheses about the optical response of the effective medium. Surface plasmon interactions in nanoparticle arrays can initiate a wide range of processes leading to new polarization regimes. By varying the thickness of nanostructured conducting layers on top of dielectric nanoparticles, it is shown that broadband tuning can be achieved. Since metallic nanoshells are a common component of nanostructured film systems and often form hierarchical nanostructures that collectively control the final optical behavior, they offer additional advantages over metallic particles.

Today, thin films are used in almost every field to improve the chemical and physical properties of materials. Thin solid films have been modified to perform a wide range of applications and used in many technical systems. For example, significant and noteworthy advances have been made in thin film technology, allowing rapid miniaturization of electrical devices. The retention of electrical charge in these devices [2].

The films have a granular structure with grains that are nearly spherical and have an average radius between 18 and 63 nm as determined by AFM images of the studied samples, which are shown in Fig. 8.4. In the (I)-693 K series, the average grain size is 63.1 ± 8.6 nm, while the roughness and porosity are 7.4 ± 1.9 nm and 19.3 ± 2.4%, respectively.

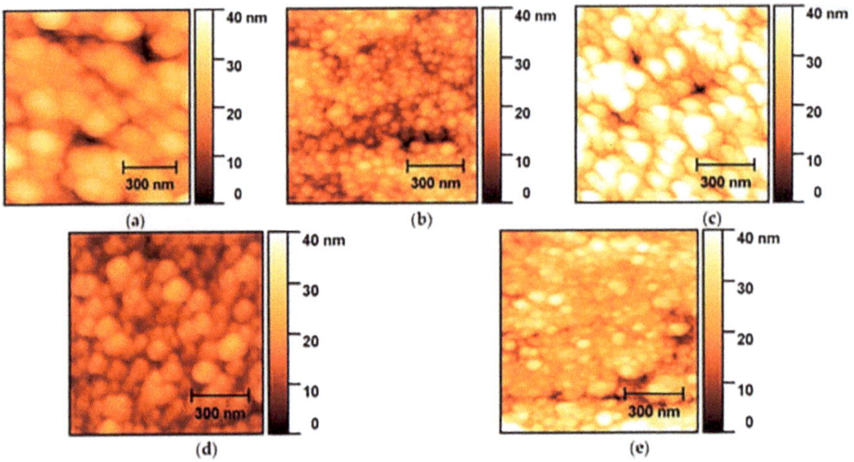

Fig. 8.4 Samples include series (I)-693 K (**a**), series (II)-693 K (**b**), series (III)-723 K (**c**), series (IV)-693 K (**d**), and series (V)-723 K (**e**) in AFM pictures [2] [Open access]

The (II)-693 K series film (Fig. 8.4b) has illustrated a slight decrease in roughness and an increase in porosity by 1.6 times compared to the (I)-693 K series. The average grain size is probably reduced by 1.8 times as a result of this. The addition of rare earth elements has been shown to reduce the grain size in metal oxide semiconductors according to the literature.

8.3 Importance of Nanostructured Thin Films

This can be a bit confusing as the term thin films is often used to refer to coating layers that range in thickness from a single atomic layer to a significant fraction of a millimeter. Nanostructured thin films are often classified according to their crystallinity, shape, and function. In this book, thin films are divided into types based on their composition and shape.

There are many methods and approaches for creating nanostructured thin films, all aimed at optimizing their various qualities and scalability. In general, there are two methods for producing thin films: top-down and bottom-up.

This chapter provides an overview of the evolution of thin film technology for optoelectronic applications. AFM and SEM images show nanogold thin films created at different times of deposition from a solution of $HAuCl_4$ with $Pb(CH_3COO)_2$ acting as a structure-directing agent. The Ag/AgCl electrode (3 M KCl) is subjected to a constant voltage of 0.05 V to fabricate the films. When the deposition period is increased from 360 to 540 s, the peak height increases sharply to 302 ± 57 nm. In addition, the size and shape of nanostructures can be modified by changing the concentration of the structure-forming chemical.

After applying −1.6 V for 30 s and −1.2 V for 60 s to a 50 mM potassium dicyanide solution, SEM images of nanogold films (NGF) are obtained, illustrating how the deposition parameters affect the fabrication of thin films.

Additionally, a reflectance-based biosensor device is investigated, which collects the reflected light using an optical probe and uses an incident light to activate plasmons on the surface. This highlights the significance of metallic layer arrangement in bimetallic films. The surface-enhanced Raman scattering (SERS) response is increased when a gold layer is placed on top of a silver layer, but it is decreased when a silver layer is placed on top of a gold layer.

Top-down methods are based on Exfoliation methods, physical or chemical, which are often used to create laminar nanostructures such as nanoflakes and nanosheets. This field offers a wide range of chemical exfoliation methods for the synthesis of hexagonal. For many years, researchers have been exploring nanomaterials such as nanostructured thin films. Observing, measuring, and working with materials at nanoscale is one of the most challenging tasks for scientists and engineers working in the field of nanotechnology and nanoscience.

Traditionally, the focus of much research on nanostructured materials has been collective behavior over large areas rather than at the nanoscale. In such situations, the characteristics and actions noted and quantified are primarily. Moreover, surface engineering of nanostructured films is considered a cost-effective strategy for important sectors and clarifies the mechanism of many activities throughout the product layer development process. More research is being conducted to fine-tune the surface technology for future developments. In modern countries, the market for surface treatment technologies occupies a decent share. According to a recent study [2], UK has spent more than £50 billion on surface development, of which more than half is used to improve products' surface resistance and protect them from surface corrosion. As shown in Fig. 8.5, the improvement of the structure of bimetallic films may involve the formation of a hybrid structure that mixes gold nanoparticles (AuNPs) with a silver film on top of the nanospheres (AgFONs). AgFON films made from 505 nm polystyrene beads, AuNP-AgFON hybrid structure, and the cross-sectional view are shown in SEM images. Three distinct Raman bands for AgFON-505, AuNP-AgFON-505, and AgFON-Ag films are visible in the SERS spectra of benzenethiol. The SERS enhancement between AuNP-AgFON-505 and AgFON-505 is about 30 times higher, suggesting a marked improvement in the response.

An excitation laser with a wavelength of 785 nm is used to obtain Raman images of AgFON-505 and AuNP-AgFON-505. AuNP dimers and slit gap in the AuNP-AgFON-505 system are visible in high-magnification images. In addition, the electric field distribution around AuNP clusters on AgFON-505 is imaged using finite-difference time-domain (FDTD) simulations, which shed light on the improved optical properties of the hybrid structure.

The goal of producing nanostructured films is to meet the needs of automotive, aerospace, agricultural, consumer electronics, and electronics industries [2, 3].

First, AFM images of nanostructured ZnO thin films, which are deposited under different circumstances, i.e., for Ar/O_2 flow ratios ranging from 3:0.3 to 30:3 (in sccm), are obtained. However, based on the corresponding AFM images shown in

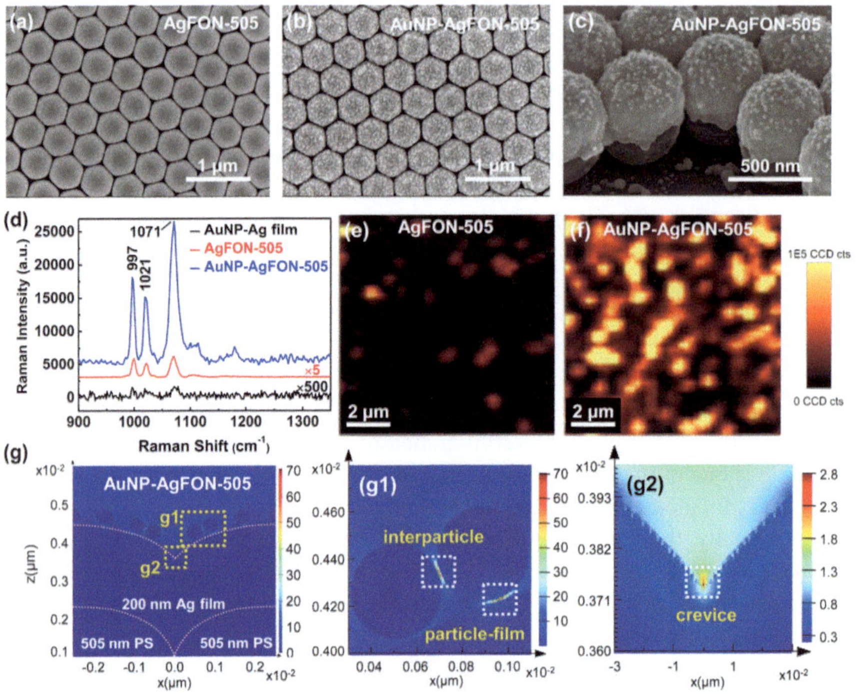

Fig. 8.5 SEM images of **a** AgFON-505, **b** AuNP-AgFON-505, **c** AuNP-AgFON-505 cross sections, and **d** both Raman spectra of AgFON-505, AuNP-AgFON-505, and AuNP-AgFON-505 film substrates. **e**, **f** Raman images of AuNP-AgFON-505 compared to AgFON-505 under 785 nm laser excitation. **g** FDTD simulations at 785 nm laser wavelength of the E fields on AuNP clusters on AgFON-505 substrate. Slightly magnified photograph of the AuNP-AgFON-505 system of doubly bonded (g1) and gap-gap (g2) AuNP particles [3] [Open access]

Fig. 8.6, a detailed analysis of surface morphology for these two selected nanostructured ZnO thin films is proposed since the most noticeable and obvious difference in AFM images is noticed for ZnO samples deposited under completely different conditions, i.e., for the Ar/O_2 flow ratios of 3:0.3 and 30:3 (in sccm), respectively [3].

8.4 Methods of Thin Film Production

Thin film research is essential in many advanced technologies, such as thermoelectric (TE) applications that directly convert thermal energy into electrical energy using the Peltier and Seebeck effects. One of the important elements that influences the efficiency and performance of these films is the flow of charge carriers within them. Solvent evaporation, which is often temperature dependent, is one of the important factors affecting thin film production. In deposition techniques requiring precise

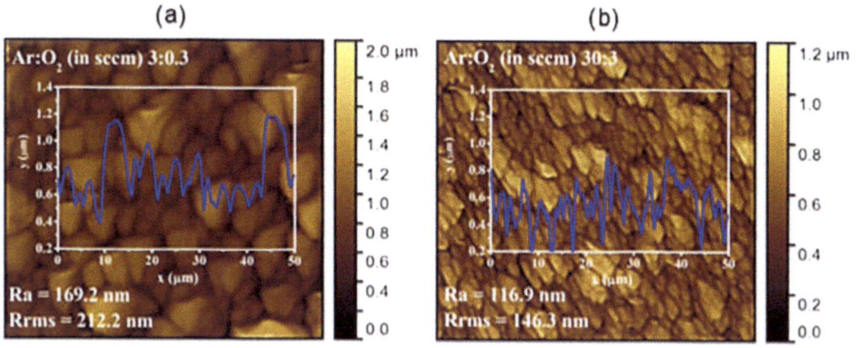

Fig. 8.6 Arithmetic mean deviation of the estimated profile is presented as Ra, while the root mean square roughness parameter is Rrms. AFM images of nanostructured ZnO thin films deposited under Ar/O_2 gas flow of 3:0.3 (**a**) and 30:3 (**b**), respectively (in sccm) [3] [Open access]

control to obtain the desired electrical and optical properties, the solvent evaporation rate plays a critical role in controlling the thickness and uniformity of thin films.

Experiments are conducted on various sputtered, sputtered plus carbon, and non-sputtered plus carbon substrates to maximize the thin film formation temperatures during pentacene deposition, 200, 300, and 350 K. These factors have a direct impact on the molecular configuration and subsequently, performance of the high-mobility pentacene transistor.

The gold contact, SiO_2 channel region, and key gold/SiO_2 transition region are analyzed using characterization techniques such as AFM. The mechanisms of base layer development that determine the electrical performance of the final device are elucidated through these analyses. The impact of different deposition conditions on the thin film morphology and ultimately, the device performance is depicted in Figs. 8.7, 8.8, and 8.9.

The researchers can improve the functional performance of devices, whether for power conversion, optoelectronics, or high-performance transistors, by combining thermoelectric principles with thin film engineering. The development is made possible by combining sophisticated characterization techniques and precise deposition procedures, developing a scalable and more efficient thin film technology for a variety of applications.

We begin this chapter by covering the basics of TE effect and TE devices. Next, we present TE thin films, which have received much interest due to their special properties including flexibility, high crystallinity, and preferential orientation. A number of ion coating methods for creating thin TE films have been collected, and corresponding TE films for various purposes. These methods include molecular beam epitaxy, magnetron sputtering, and pulsed laser deposition. Connecting TE devices also requires ion plating technology to fabricate the electrode and transition layer. Unlike the electroplating method, the advantages of magnetron sputtering are low contact resistance, good adhesion and smooth surface of the electrode and transition layer [4].

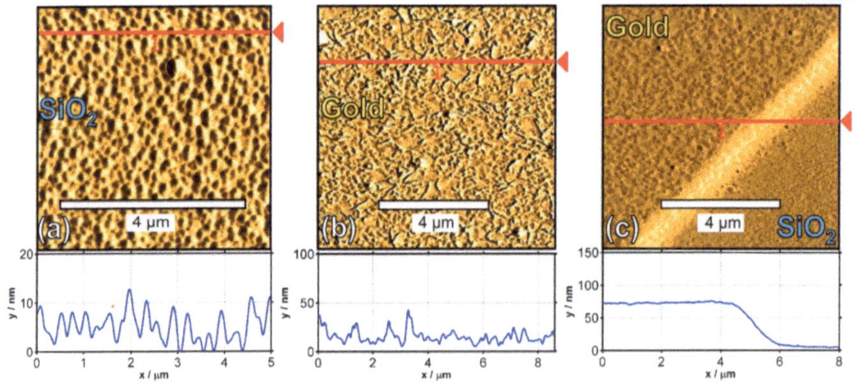

Fig. 8.7 Pentacene deposition at 200 K: 7.5 monolayers on gold contact–channel transition region (**c**) and 1.5 monolayers on SiO_2 (**a**) and gold (**b**). In AFM scans, the line denotes corresponding cross sections, which are then displayed below [4] [Open access]

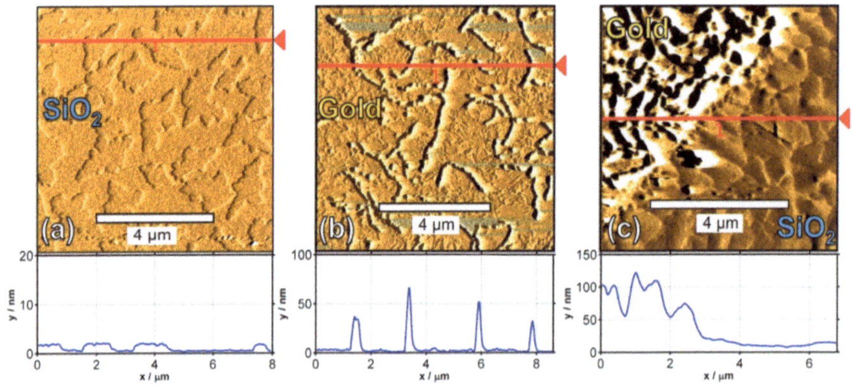

Fig. 8.8 Pentacene deposition at 300 K: 7.5 monolayers on the gold contact–channel transition region (**c**) and 1.5 monolayers on SiO_2 (**a**) and gold (**b**). In AFM scans, the line denotes the corresponding cross sections, which are shown below [4] [Open access]

8.4.1 Deposition of Physical Vapor

PVD, which transfers material at the atomic level in a vacuum, is a popular coating deposition technique. This technique, chemical vapor deposition (CVD), is somewhat similar. The main difference is that PVD starts with the starting materials in solid form, while CVD introduces them into the reaction chamber in a gaseous state.

TiO_2-NRs-based coatings that radically alter surface morphology are an example of the beneficial use of thin film deposition. The surface morphology of an untreated stone sample (PLA) and a treated sample (PLD) is compared in Fig. 8.10. Subtle changes in PL morphology are caused by TiO_2-NRs coating. In particular, in PLD,

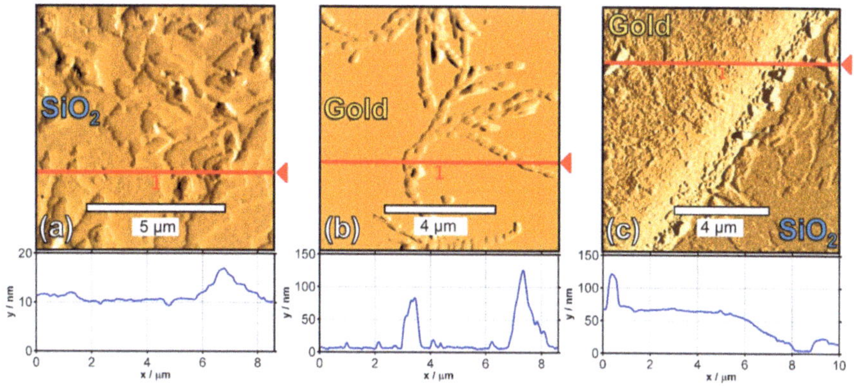

Fig. 8.9 Pentacene deposition at 350 K: 7.5 monolayers on the gold contact–channel transition region (**c**), SiO_2 (**a**), and gold (**b**). In AFM scans, the corresponding cross sections are indicated by the line and shown below it [4] [Open access]

the coating created granular aggregates that are tightly packed together, covering the stone surface with a uniform, crack-free layer. Sputtering a sufficient amount of TiO_2 NRs for two minutes and thirty seconds allowed the coating to adhere to the base stone morphology and create this continuous coating.

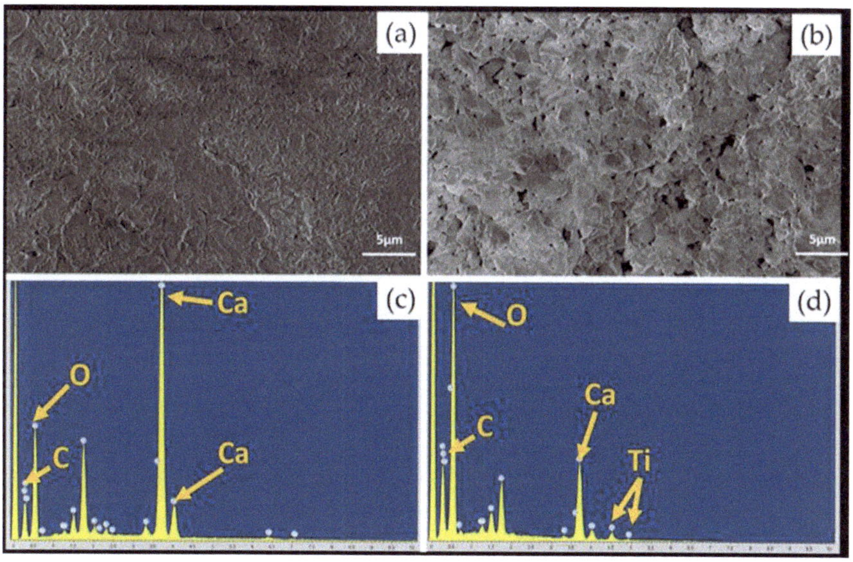

Fig. 8.10 SEM micrographs of untreated PL stone (**a**) and PLD stone treated by sputter coating (**b**), together with the corresponding EDS spectra (**c, d**). A working distance of 3.7 mm, an aperture size of 20 µm, and an accelerating voltage of 1 kV are used for SEM measurements [5] [Open access]

Figure 8.10c, d clearly shows the differences in surface shape and chemical composition between the treated and untreated samples. In line with the diffuse reflectance spectroscopy (DRS) analysis, the corresponding energy-dispersive spectroscopy (EDS) spectra confirm the presence of TiO_2-NRs coating. In addition to showing more Ti peaks ($K\alpha = 4.508$ keV), the spectra demonstrate that the treated sample has a higher O/Ca ratio (Fig. 8.10d) than the untreated stone (Fig. 8.10c).

By combining advanced deposition techniques such as PVD with nanostructured coatings, scientists can improve surface performance for a range of applications, including enhanced optical performance, self-cleaning, and protection.

This process consists of four steps; (i) transporting steam to the substrate to be coated; (ii) reaction between the metal atoms and suitable reactive gas such as oxygen, nitrogen or methane during the transport stage; (iii) evaporation of deposited material using high energy source such as an electron beam or ions—this evaporates atoms from the surface; (iv) coating the surface of substrate.

PVD offers various advantages, such as: (i) ability to form coatings with properties that may be better than those of substrate material; (ii) environmental friendliness of the process compared to many other processes, such as electroplating, and (iii) ability to use all forms of inorganic and some forms of organic materials. However, PVD has certain disadvantages, such as (i) problems in coating complex structures; (ii) high process costs and low productivity; and (iii) complexity of the process [4–6, 12].

8.4.2 Evaporation-Based Physical Vapor Deposition (PVD)

A popular thin film deposition technique called evaporative PVD uses the phase transition of solid materials into vapor and condensation on a substrate to create a uniform coating. To reduce contamination and ensure the creation of high-purity films, this process occurs in a vacuum atmosphere.

In a variety of applications where precise control of film thickness, composition, and shape is required, such as electronics, optics, and surface engineering, this approach is indispensable. Temperature-sensitive materials can benefit from evaporative PVD's lower processing temperatures and higher deposition rates compared to alternative deposition methods such as CVD.

Researchers and engineers can tailor thin film properties by changing process variables such as evaporation temperature, deposition rate, and vacuum pressure to meet specific application requirements. Evaporation-based PVD continues to be a vital method for creating high-performance coatings with improved mechanical, electrical, and optical properties as nanotechnology advances.

The patent for plasma PVD was granted about 80 years ago, while physical vapor deposition (PVD) was discovered more than a century ago. The pretreatment smoothes the substrate surface, but adds tiny topographical irregularities and micrometer-scale defects such as pits, scratches, and ridges.

Organic substances such as PCBM (phenyl-C61-butyric acid methyl ester), NPD (N, N′-Di(1-naphthyl)-N, N′-diphenyl benzidine), and Alq_3 (tris(8-hydroxyquinoline)aluminum) are used to study the applicability of filtered cathodic vacuum arc PVD (FCPVD). According to the results, films produced by FCPVD had extremely smooth surfaces free of particles and obvious defects, with root mean square (RMS) roughness less than 0.4 nm. On the other hand, the films produced by standard cathodic arc PVD (SCPVD) show highly concentrated, extremely rough surfaces, RMS roughness >10 nm, and high concentration of aggregated particles (Fig. 8.11). In addition, various substrates such as Si, quartz, mica, and ITO substrates are subjected to FCPVD technology. These results indicate that by suppressing nucleation, FCPVD technology can be used to create ultrasmooth films on various organic compounds and substrates [6].

Given the rapid development of computer numerical control (CNC) machining processes and the emergence of new processing methods, the development of PVD deposition technologies in recent decades has mainly focused on tool coating [4]. The wide range of defect forms and causes, as well as the different application areas in which growth defects are studied, are sources of diverse classifications. The development of well-known technologies, including sputtering, vacuum, plasma technology, magnetic fields, gas chemistry, thermal evaporation, arcs, and power

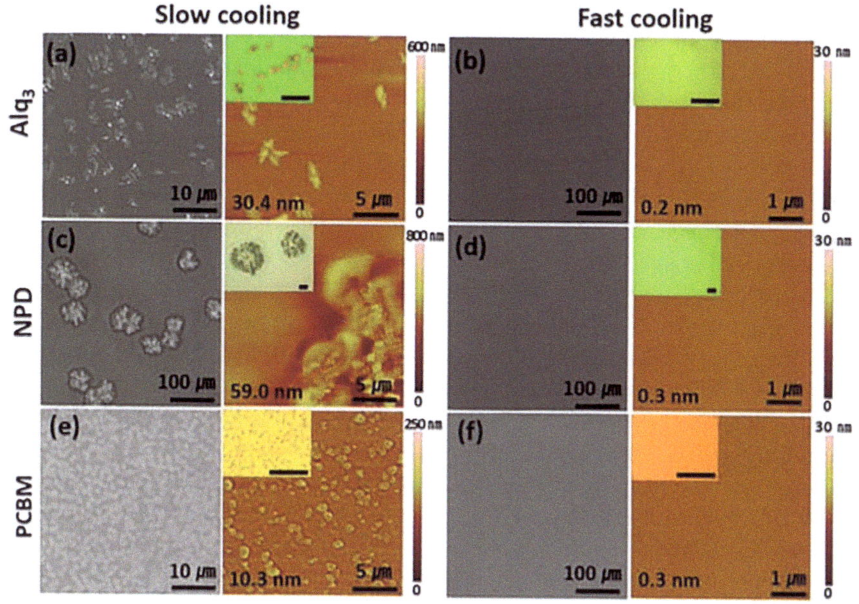

Fig. 8.11 Various organic films as a function of substrate cooling rate. SEM and AFM images of Alq3 films (**a**, **b**), NPD films (**c**, **d**), and PCBM films (**e**, **f**) are shown as a function of cooling rate. SCPVD and FCPVD are used to create the movies on the left (**a**, **c**, and **e**) and on the right (**b**, **d**, and **f**), respectively. Optical microscope images of each film (scale bar: 10 μm) are shown as insets [6] [Open access]

supply control, was necessary at that time for the development of vacuum coating processes, as stated in Powell's book [2].

It is not until the 1960s that the phrase "physical vapor deposition" came into use. However, many new surface treatments have emerged to meet commercial and industrial requirements, improving the functionality of a huge number of objects. The ability of the FCPVD method to successfully suppress nucleation is confirmed by a grazing incidence wide-angle X-ray scattering (GIWAXS) study. According to these results, the smooth films are amorphous, and the rough films are crystalline (Fig. 8.12).

The in-plane and out-of-plane scattering vectors are denoted as qxy and qz, respectively. The colors, which represent the scattering intensity, range from high (red) to low (blue). Assigned to a mixture of orthorhombic and triclinic phases, the rubrene film obtained by the SCPVD approach exhibits strong and anisotropic scattering peaks, and large aggregated particles form thermodynamically favorable crystalline forms (Fig. 8.12a) [6]. On the other hand, the rubrene film prepared by FCPVD confirms the kinetically favorable amorphous phase by exhibiting large peaks at q ~ 0.68 Å − 1 with isotropy (Fig. 8.12b) [4].

While the films prepared using the SCPVD approach show strong and anisotropic scattering peaks indicating crystalline phases (Fig. 8.12c–g), Alq3, NPD, and PCBM films, like the rubrene films, show nucleation suppression using the FCPVD method (Fig. 8.12d–h). The as-deposited Alq3 and NPD films are described as thermodynamically favorable forms consisting of the triclinic phase and α-phase, respectively [4, 6]. Moreover, the crystalline PCBM powder and PCBM film are in good agreement with each other .

In contrast, the kinetically favorable amorphous phase is confirmed by large scattering peaks in Alq3, NPD, and PCBM films fabricated by the FCPVD approach (Fig. 8.12d–h) [6]. These results indicate that rapid substrate cooling effectively suppresses the nucleation of adsorbed molecules, which contributes to the fabrication of ultrasmooth organic thin films [6].

The past thirty years have seen significant developments in the field of plasma photovoltaic power (PAPVD). These include direct current (DC) diodes, triodes, radio frequency (RF) coatings, pulsed plasma, and ion beam coatings. Initially, the procedure is difficult to understand at a basic level, and necessary modifications are made to provide benefits such as superior coating-to-substrate adhesion, structure control, and low-temperature material deposition [3].

Using the PVD technique, a thin film deposition technique, the coating is applied to the substrate atom by atom. PVD is the process of sputtering or vaporizing a solid starting material, commonly called a target. The layer thicknesses in thin films typically range from a few microns to several atomic layers. As a result, the surface characteristics change, as well as the transition zone between the sprayed material and substrate. Conversely, substrate characteristics can influence the quality of films. The atomic deposition method can be used in electrolytic, gas, vacuum, or plasma environments. In addition, the vacuum atmosphere of the deposition chamber will significantly reduce the number of gaseous impurities during operation [5].

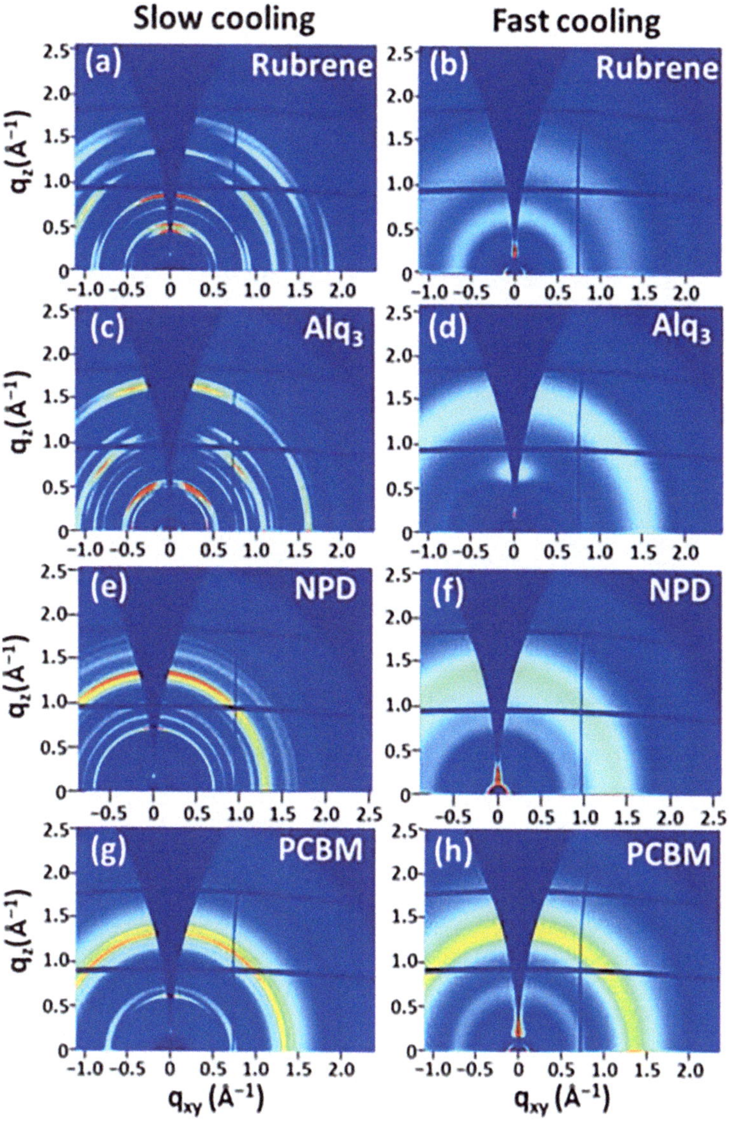

Fig. 8.12 Structure of organic thin films. Rubrene films (**a**, **b**), Alq3 films (**c**, **d**), NPD films (**e**, **f**), and PCBM films (**g**, **h**) grown by SCPVD (left column) and FCPVD (right column) methods are shown in their GIWAXS patterns [6] [Open access]

8.4.3 Physical Vapor Deposition (PVD) by Sputtering

Most of this research focuses on the production of nanomaterials and nanodevices, their utility as nanocarriers for the controlled release of active ingredients, and their interactions, both beneficial and harmful, with microbes, plants, and animals.

The number of research works on the use of nanomaterials in environmental, agricultural, and medical fields has recently been growing rapidly. The area of research that researchers are primarily focusing on is the production of nanomaterials through green chemistry approaches using microorganisms, plants, or their metabolites. XRD patterns of Cu-Se thin films with different sputtering powers are shown in Fig. 8.13. So, XRD patterns have shown a broad halo without strong diffraction peaks. The lack of long-range order in the amorphous structure caused by the disordered arrangement of atoms is the reason why the XRD pattern shows no identifiable peaks. In addition, compared with the ordered arrangement of atoms in the crystalline material, the large halo in the XRD pattern indicated that the atoms in amorphous Cu-Se thin films are distributed in somewhat disordered form at the short-range scale [12].

What makes this development particularly exciting is that it reduces the negative impacts that integrated nanoparticles caused by using a more environmentally friendly method. Green synthesis of nanomaterials and their use in several industries are the main topics of this chapter. Application methods can be critical to achieving the desired results from nanocarrier production, so advantages and disadvantages must be carefully considered. As the need to expand our horizons increases, we must start looking for genuine, workable solutions, and the gaps will be filled [6]. Figure 8.14 shows field emission scanning electron microscopy (FESEM) images of the surface and cross section of Cu–Se films deposited at different sputtering powers. Each thin film has an identical amorphous structure with no visible crystals or grain boundaries. Regardless of the sputtering power used, FESEM images of the

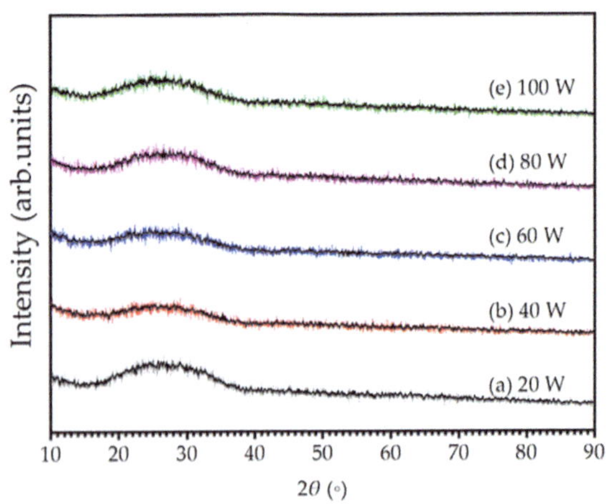

Fig. 8.13 XRD patterns of Cu-Se thin film coatings after deposition with different sputtering powers: **a** 20, **b** 40, **c** 60, **d** 80, and **e** 100 W [12] [Open access]

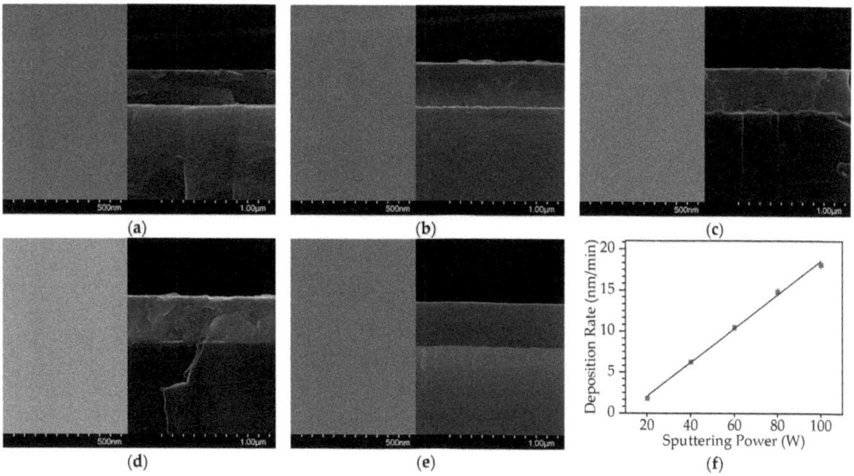

Fig. 8.14 Cu-Se thin films with different sputtering powers **a** 20, **b** 40, **c** 60, **d** 80, and **e** 100 W are shown in Fig. 1 in cross-sectional and surface images obtained by FESEM. **f** The linear trend line plots deposition rate of Cu-Se thin films as a function of sputtering power [12] [Open access]

surface of each sample depict a smooth and featureless morphology with no noticeable change in the film morphology. The amorphous structure of thin films remains unchanged with increasing sputtering power, even at high sputtering powers up to 100 W, despite the significant increase in deposition rate. As a result, the thin films have no grain boundaries [12].

8.4.4 Chemical Vapor Deposition (CVD)

CVD is a powerful technology that produces durable thin films and coatings of the highest quality. Despite its widespread use in modern industries, it is constantly evolving as new materials require adaptation. The precise synthesis of high-purity polymer thin films that can be conformally deposited on a variety of substrates, as well as inorganic thin films from 2D materials, is now taking CVD synthesis to new heights. Initially, 500 μm thick pure metal substrates (Cu and Ni) are used for CVD growth. Raman spectra of the resulting films are obtained (Fig. 8.15). With an I2D/IG ratio of about 1.8 and a 2D FWHM of 30, graphene formed on Cu shows good agreement with the Raman fingerprint obtained for CVD-grown monolayer graphene [7].

This implies that future advances may focus on improving the process scalability, efficiency, and suitability for temperature-sensitive materials. The researchers can improve throughput, reduce failures, and enhance applicability in areas such as electronics and nanotechnology by assessing current challenges and improving methods.

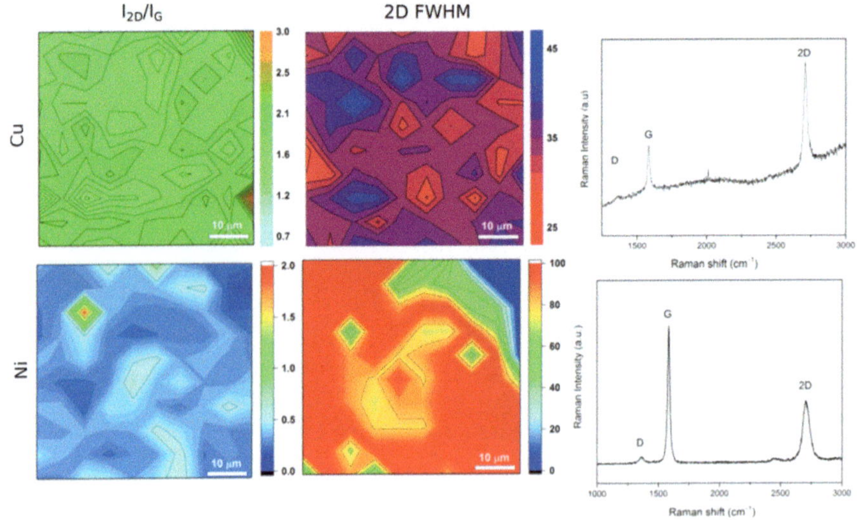

Fig. 8.15 Raman characteristics of graphene obtained by CVD on 500 μm thick copper and nickel foil at 1000 °C for 30 min and a growth pressure of 0.1 mbar [7] [Open access]

Then, Cu-Ni alloys with a weight ratio of Cu70-Ni30, Cu55, Ni45, and Cu33-Ni67 compositions are subjected to vapor deposition under the same conditions. The Raman spectra and maps of the obtained graphene sheets are shown in Fig. 8.16. It is obvious that with increasing nickel content in the alloys, Raman spectra change from those similar to those of pure copper to those of pure nickel [7].

Best practices for substrate pre-treatment, high-temperature growth, and post-growth techniques for experiments with graphene, two-dimensional transition metal dichalcogenides (TMDs), and polymer thin films are described as representative examples. Additional attention is given to current events and challenges in scaling up operations [6, 7, 12]. Future development strategies should focus on optimization and addressing current limitations. In particular, it is recommended to improve the reactor architecture to facilitate low-temperature thin film growth (lower temperature thin film deposition) and high throughput (faster production).

8.4.5 Process of Conventional CVD

The earliest known cases of cardiovascular disease date back to the Syrian and Egyptian cultures around 300 BC. CuS and CuAs are precipitated by alchemists and priests, respectively, by gas–solid–phase interaction of metallic copper and vapors containing sulfur or arsine. The silvery-shiny CuAs films on copper are misinterpreted as copper turning to silver, while the black CuS film is supposedly created for artistic purposes. The earliest known industrial use of cardiovascular disease dates

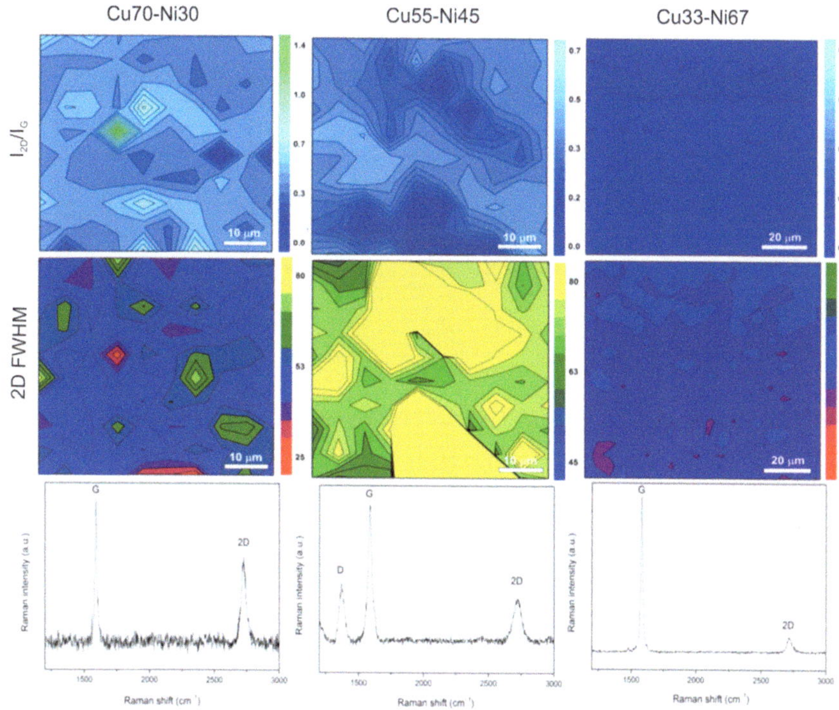

Fig. 8.16 Raman I2D/IG and 2D FWHM bands of Cu-Ni alloys. The growth conditions are 1000 °C growth temperature, 0.1 mbar growth pressure, and a high cooling rate [7] [Open access]

back to the 1880s. Two examples of such processes are pyrolytic carbon deposition and carbon metal coating. Lamp filaments and the Mond process, which uses the chemical transport of nickel tetracarbonyl to produce high-purity nickel:

$$Ni(CO)_4(g) \rightarrow Ni(s) + 4CO(g) \tag{8.1}$$

Chemical vapor deposition (CVD) has become a popular industrial thin film deposition technique due to two important developments in electronics technology over the past decade. Figure 8.17 shows the chemical vapor deposition of hydrides. CVD is one of the oldest and most widely used chemical vapor deposition techniques. It uses dilute gases including nitrogen (N_2), silane (SiH_4), diborane (B_2H_6), phosphorus (PH_3), and oxygen (O_2), and operates at temperatures ranging from 240 to 500 °C.

The morphology of a monolayer of pure graphene (Gr) deposited on an alumina-silicon oxide (AlO_3-Si) substrate is compared in atomic force microscopy (AFM) images. The thickness of AlO_3 obtained by atomic layer deposition at 250 °C is about 22 nm (Fig. 8.17b). High-quality thin film growth is achieved due to a flat surface and uniform, pore-free graphene coating obtained by optimized seed layer deposition.

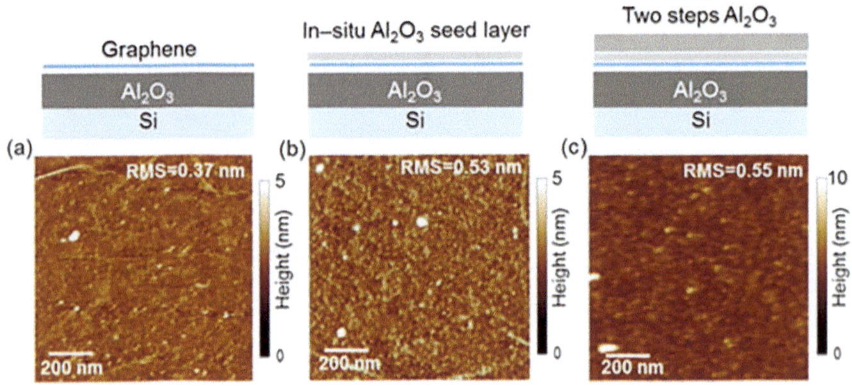

Fig. 8.17 Morphological images of **a** CVD graphene deposited on an Al_2O_3−Si substrate, **b** after deposition of optimal seed layer at 100 °C, and **c** after two-step ALD growth at 250 °C [13] [Open access]

The overall film and seed layer have low surface roughness and a conformal Gr coating devoid of pinholes [13].

First, silicon has replaced germanium as the fundamental semiconductor. Second, the production of integrated circuits is made possible by the invention of planar technology. During this period, heteroepitaxial films on oxide substrates, heteroepitaxial films of silicon, III-IV semiconductor compounds, and heterojunctions are developed.

The deposited Al_2O_3 film with a thickness of 12 nm is shown in Fig. 8.18a as a high-resolution cross-sectional SEM image. It has an amorphous structure and constant contrast throughout the thickness, indicating a uniform density. ACM image of the deposited Al_2O_3 film on a large scale, 20 μm × 20 μm, is depicted in Fig. 8.18b. It shows a uniform and consistent layer with small depressions corresponding to bilayer Gr regions. The boundary region between the 2 L and 1 L regions is shown in Fig. 8.18c, which provides further clarification of these morphological aspects of Al_2O_3. Figure 8.18d shows that Raman spectroscopy studies of EG before and after Al_2O_3 deposition indicate that the ALD process does not significantly affect the deformation and addition of EG [13].

8.4.6 Design of CVD Reactor

It presents a new computational fluid dynamics (CFD) model to study the formation of SiC film over a combustion nozzle in a commercial hot-wall CVD reactor. To account for gas-phase chemistry and surface reactions, a comprehensive three-dimensional model is constructed, including the reaction mechanism and compression transport models. The optical microscope images of CVD graphene obtained by Ar heating followed by H_2/Ar (Fig. 8.19a, d), H_2/Ar heating and annealing (Fig. 8.19b, e), and

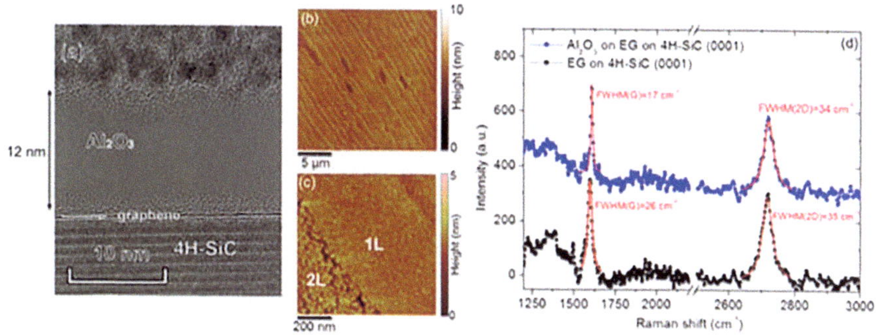

Fig. 8.18 **a** Cross-sectional TEM image of a 12 nm thick Al_2O_3 film grown on single-layer SiC graphene (EG). **b** AFM morphology over a 20 μm × 20 μm scan area showing small pits on 2 L Gr and uniform and conformal Al_2O_3 coating on 1 L Gr. **c** High-resolution AFM morphology showing larger grains separated by small pits (up to 2 nm) on 2 L EG and compact Al_2O_3 film with small grains on top of 1 L EG. **d** Normalized Raman spectra of 1 L EG before and after Al_2O_3 deposition [13] [Open access]

Ar heating and annealing (Fig. 8.19c, f) at high and low magnification are shown in Fig. 8.19a, f. The largest Cu domain and lowest graphene density among all methods are obtained by Ar heating followed by H_2/Ar (Fig. 8.19a), demonstrating that this approach is an ideal combination of heating and annealing. The reduction of Cu domain boundaries during the current heating and annealing process is the main factor in reducing the graphene density because the surface oxidation state (Ar annealing for 20 min) is the same for all samples in Fig. 8.19, which acts as active nucleation sites for graphene [8].

This results in flow recirculation, which ultimately leads to the destruction of film uniformity. A sensitivity analysis is performed to determine. As a result, the binary alloys such as Cu/Ni, which have variable carbon solubility, would be ideal. More precisely, Cu/Ni alloy can be created by depositing metal films (e.g., Ni and Cu) on SiO_2/Si substrates to create a Cu/Ni/Si/Si multilayer structure. This process involves the separation of graphene films from dissolved carbon atoms (Fig. 8.20a). An overview of possible kinetic processes is given below. Part of the dissolved carbon atoms comes to the metal surface to initiate surface-mediated diffusion after dynamic diffusion throughout the volume. Then, in the development stage, graphene domains crystallize on catalyst defects or step regions and grow (Fig. 8.20b). This is the initial step when homogeneous graphene sheets are obtained using Cu/Ni alloy as a catalyst (Fig. 8.20c). Since the atomic ratio of nickel can be precisely determined by manipulating the thickness of nickel films during the deposition process, the possibility of controlling the number of layers of graphene films is investigated by the separation method . Increasing the nickel concentration from 5.5% to 10.4%, respectively, resulted in a uniform monolayer graphene film with 95% coverage and a bilayer graphene (BLG) film with 89% coverage, as shown in Fig. 8.20d–k [9].

Numerical predictions are made regarding the temperature, velocity, and concentration profiles inside the reactor. It is known that multimodal transport functions

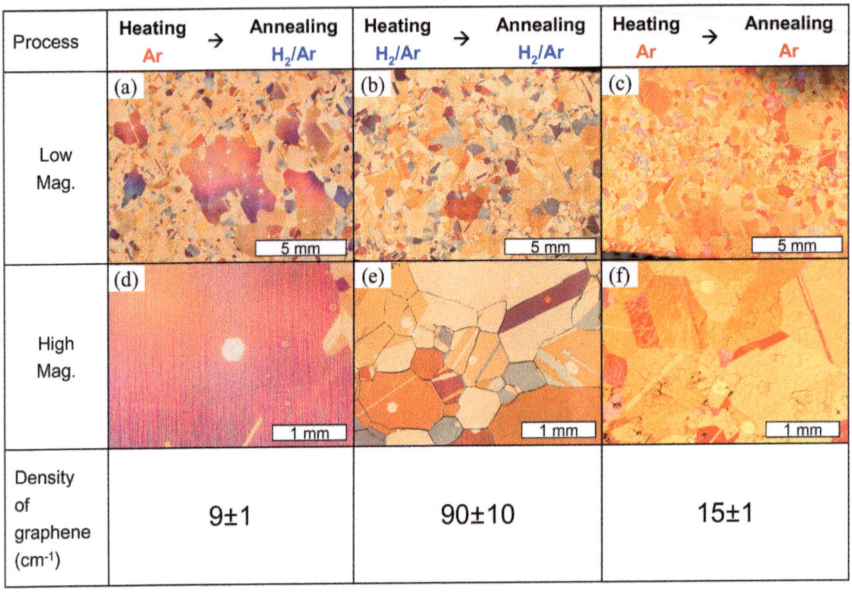

Fig. 8.19 Low-magnification (**a–c**) and high-magnification (**d–f**) optical microscope images of chemical vapor-treated graphene on copper are obtained by heating in argon followed by hydrogen/argon (**a**, **d**); heating and annealing in hydrogen/argon (**b**, **e**); and heating and annealing in argon (**c**, **f**) [8] [Open access]

correctly in the reactor when interacting with gas and surface processes. Several dimensionless groups (Re, Pr. Pe and Gr/Re2) are used to study natural convection and hydrodynamic stability of flow. It is noted that at high Reynolds numbers (Re) and Gr/Re2 ratios, buoyancy-driven flow predominates, and how important parameters influence the deposition process. Additionally, the integration of the Nelder–Mead algorithm, flexible supervised machine learning, and support vector machines helps optimize the CVD reactor by improving film quality. The current approach successfully achieves ideal operating conditions, significantly improving film performance. An example of a typical CVD reactor design that can be used for a variety of CVD applications. Reactors, both horizontal and vertical, are used to increase atmospheric and low pressure. Hot- and cold-wall reactors are two types of horizontal reactor systems. Most CVD work in industry is done in single-plate cold-wall reactors, while hot-wall reactors are mainly used for small-scale academic research [4–6].

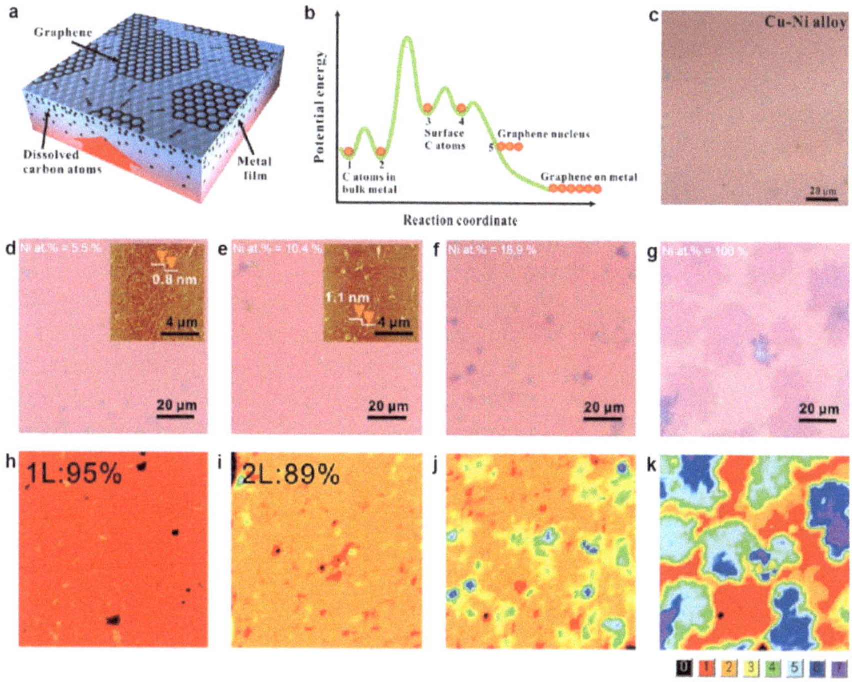

Fig. 8.20 Graphene growth on Cu/Ni alloy catalysts. **a** Schematic illustration of the graphene film preparation process using segregation. b)Energy profiles of carbon atoms during the segregation process. **c** Optical image of a graphene layer formed on SiO_2/Si with Cu/Ni alloy catalyst. (a–c). Graphene films prepared on Cu/Ni alloy substrates with nickel atomic ratios of **d** 5.5%, **e** 10.4%, **f** 18.9%, and **g** 100% are shown in optical images (**d**–**g**). AFM images are shown in the appendices in (**d**, **e**). Layer number distributions (h–k) are obtained from corresponding optical contrasts (**d**–**g**) in the order given (d–k) [9] [Open access]

8.5 Different Approaches to Chemical Vapor Deposition

Thermal CVD uses higher temperatures to supply the energy required for a chemical reaction that produces a thin layer on the substrate. An inert gas such as nitrogen carries gaseous reactants into the reactor chamber during thermal CVD procedures at atmospheric pressure. High temperatures are commonly used for CVD to speed up processes that would otherwise be impractical, increase density, improve crystallinity, and increase deposition rates. It is noteworthy that when the film is deposited at low temperatures and annealed at high temperatures, achieving certain film properties becomes difficult. Stress generation, increased rate of transfer of elements from substrate to film, and potential destruction of substrate are the main disadvantages of high-temperature deposition. Gold, aluminum, and other metallization processes are often achieved using low-temperature CVD under air pressure [12].

Atmospheric thermal CVD is based on a chemical reaction between precursors or volatile chemicals at atmospheric pressure. The carrier gas carries these substances into the reactor, where the reaction forms thin layers on the substrate due to thermal energy. Homogeneous and adhesive films at low temperatures can be obtained by using all the starting components as gases in a basic reactor with hot walls and one temperature zone. This technology is quite cost-effective due to its high productivity. The low temperature prevents self-doping and prevents impurities from entering the reactor walls. The main disadvantage of thermal atmospheric CVD is the lack of highly non-pyrophoric, harmless, and volatile substances. The main factors determining the rate and characteristics of film deposition are the rate of precursor delivery and the temperature of the substrate and evaporator [10]. Thermal low-pressure chemical vapor deposition (LPCVD) is based on a chemical reaction between volatile substances at low pressures (0.001 to 100 Torr). Unlike CVD at atmospheric pressure, many precursors can evaporate at lower temperatures due to the reduced pressure. These volatiles enter the reactor either by diffusion or by a carrier gas, which provides the thermal energy necessary to carry out the deposition process. In addition, since the precursor and reaction gases rapidly diffuse to the deposition surface, the CVD process tends to occur in a kinetically controlled region at lower pressures, and the rate of surface reaction controls the reaction. SEM images of sandwich structures, double-layer graphene (DLG), and dingle-layer graphene (SLG), and calculated the area percentage of cracks and holes under different tensile loads to investigate the degradation process of conductivity and stability (Fig. 8.21). When SLG is subjected to 10% stress, more short cracks are formed than in the original SLG. At 20% stress, the cracks expanded and connected with their neighbors to form longer cracks and even holes. At 30% stress, the holes filled a larger area, and cracks became longer and wider [10].

Laser flare manufacturing (LFM) images of sandwich structures, DLG and SLG are shown in Fig. 8.22a, while Fig. 8.22b shows the graph of normalized friction force. Compared with SLG, the normalized friction force is lower. DLG, which is consistent with the results of previous studies [14], is caused by the random misorientation of artificially stacked chemically deposited graphene between its layers, which reduces their interaction. Therefore, the wrinkling effect can increase the friction of DLG. During expansion, the decrease in contact between layers can lead to a sliding effect between layers [10].

Compared to CVD at atmospheric pressure, film quality under such circumstances is often improved. There are several different methods available now for CVD thin film deposition, including

- Chemical vapor deposition aided by aerosol
- Chemical vapor deposition assisted by electric field-directed chemical vapor deposition at atmospheric pressure
- Epitaxy using chemical beams
- Infrared Vapor Analysis
- Vapor deposition facilitated by electrostatic means chemical vapor deposition using lasers
- Chemical vapor deposition at low-pressure vapor deposition of metal-organic compounds

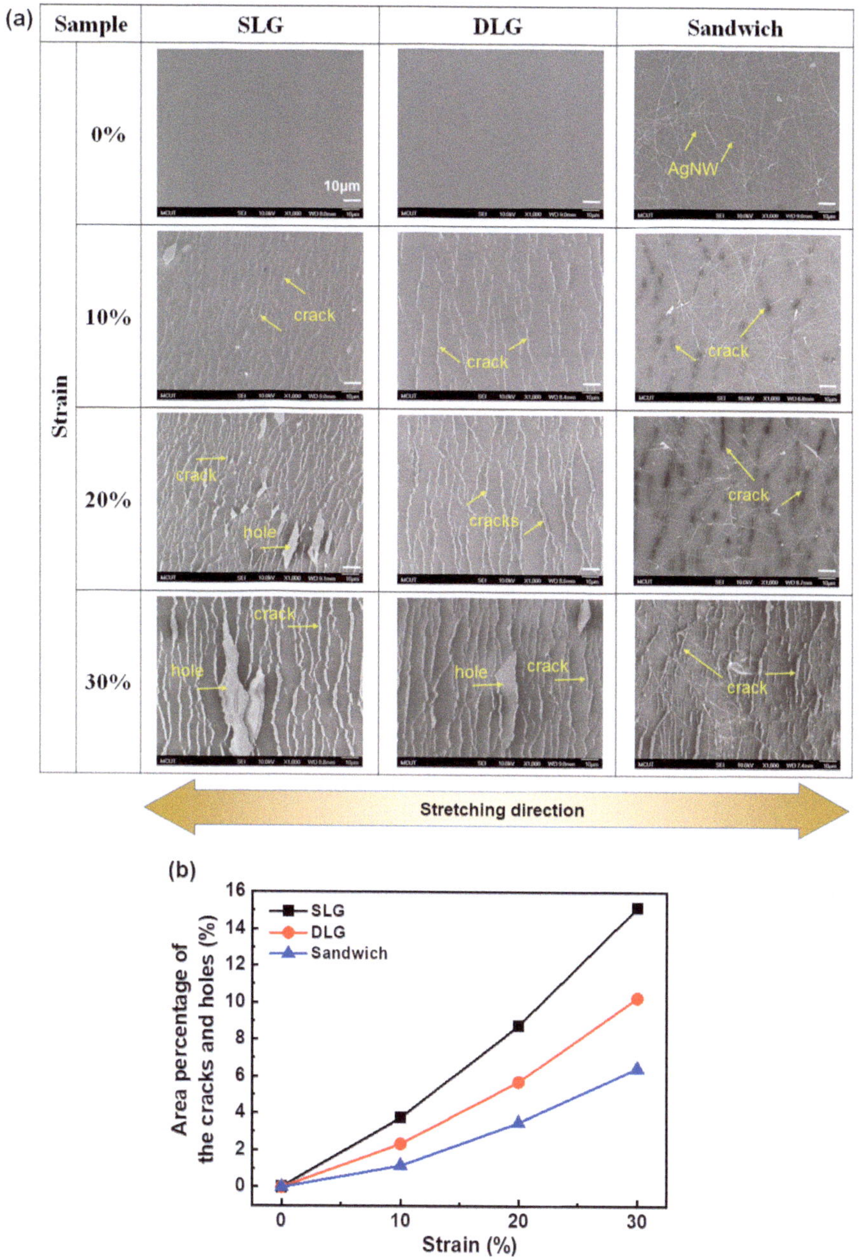

Fig. 8.21 Area percentages of sandwich structure, DLG and SLG at different tensile stresses, as well as (**a**) SEM images [10] [Open access]

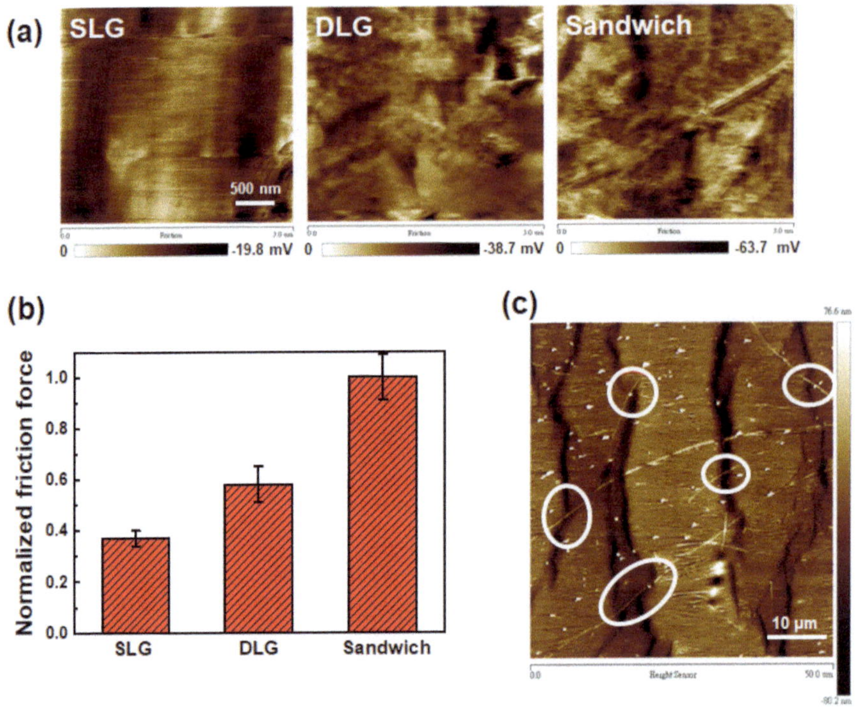

Fig. 8.22 **a** AFM image of sandwich structure at 30% strain, **b** normalized friction forces of SLG, DLG, and sandwich structure, and (c) LSM images [10] [Open access]

- Chemical vapor deposition with plasma assistance
- Chemical vapor deposition with thermal activation.

8.5.1 Aerosol-Assisted Chemical Vapor Deposition's (AACVD) Fundamentals

By spraying a precursor solution into tiny submicron-sized aerosol droplets, aerosol-assisted (AA) CVD creates the desired products through evaporation, decomposition, and homogeneous or heterogeneous chemical reactions in a heated reaction zone. AACVD is a variation of traditional CVD methods that deals with the delivery and availability of chemical precursors. Since instability is no longer an issue, a wide range of precursors can be used, opening up more opportunities to create affordable, high-quality CVD products. Surface imaging under a certain constant force between the tip and the self-assembled monolayer (SAM) surface is the basis of the AFM technique. Using AFM with tapping mode prevents damage to SAM. Figures 8.23 and 8.24 show the corresponding AFM topography images of Os-tripod SAM and Ru-tripod SAM, respectively. Two images are presented in each figure. After preparing

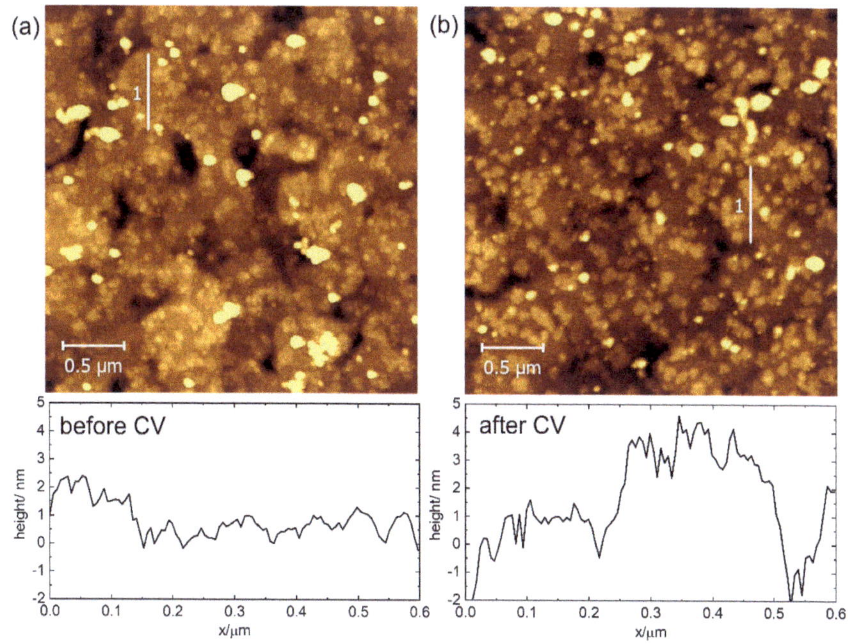

Fig. 8.23 AFM topography image of 3 × 3 µm square plot and height plot of oxygen-supported SAM on Au(111) along line 1 (top): before and after cyclic voltammetry experiment in acetonitrile with 0.1 mol TBAPF6 at a scan rate of 0.1 V/s [11] [Open access]

the SAM surface, the left image (a) is taken. The surface area that is subjected to the CV experiment (red curves) is imaged in Fig. 8.23b for Os-tripod SAM and Fig. 8.23b for Ru-tripod SAM, respectively, and this is where the right image (b) is taken [11].

The resistance plots of the Au|Os-rack and Au|Ru-rack SAM systems are shown in Fig. 8.25. These plots are obtained by analyzing 500 I-V curves at each given current. AFM image of the analyzed sample is shown in Fig. 8.25. In general, the wider the resistance range and the larger the Gaussian peak width, the lower the given current (and the farther the gold tip is from the SAM surface). In addition, the shorter the distance between the Au|SAM surface and the gold tip, the more accurately the I-V curves should reflect the SAM conductivity without significant contribution from the wind tunnel [11].

Several other variants of AACVD have been created, including electrostatic atomization vapor deposition (ESAVD), electrostatic aerosol jet deposition (EAAJD), and AA(C) combustion CVD. These changes make AACVD-based procedures more adaptable and effective. There is growing interest in AACVD-based procedures. They are widely used in most CVD-related fields for the synthesis of various films, coatings, powders, composites, nanotubes, and nanowires, among other materials. This focuses on the fundamentals, practical uses, and recent advances of AACVD-based procedures [15]. The schematic AACVD is displayed. Single source molecular

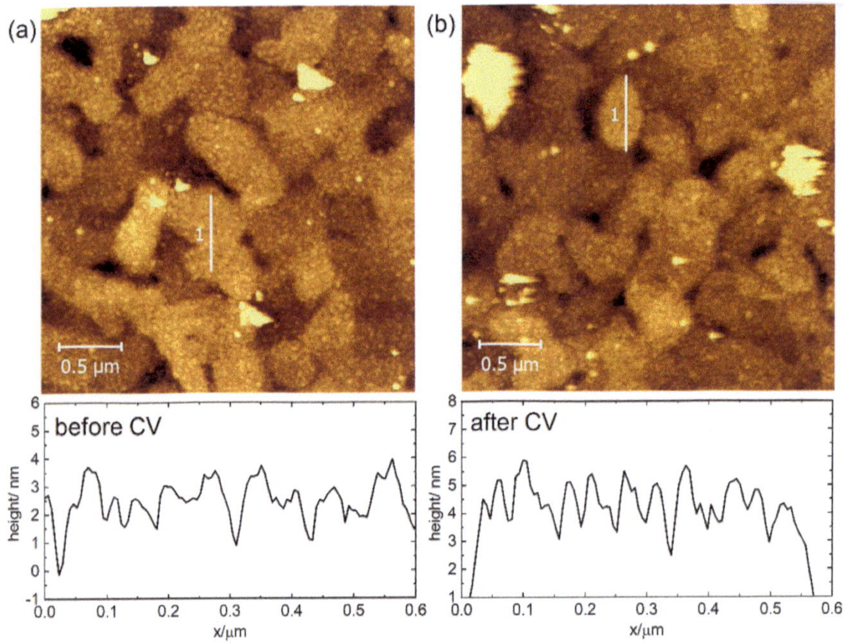

Fig. 8.24 AFM image of 3 × 3 µm square and height plot of ruthenium-based AFM on Au(111) along line 1 (top): before and after the cyclic voltammetry experiment in acetonitrile with 0.1 M TBAPF6 at a scan rate of 0.1 V/s [11] [Open access]

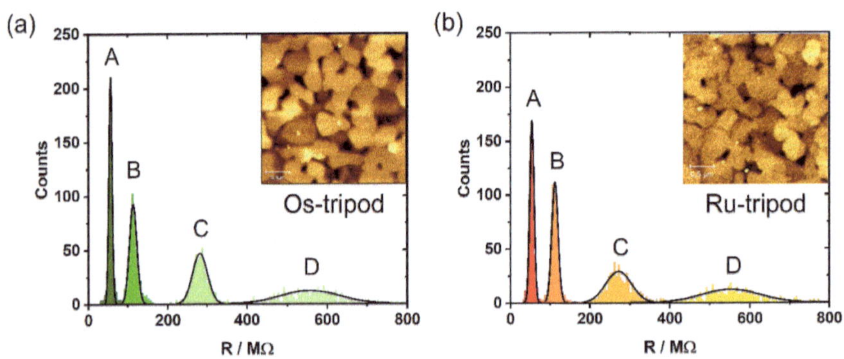

Fig. 8.25 One-dimensional resistance plots of Os-tripod SAM and Ru-tripod SAM, respectively. The information is collected from 500 I-V curves with a container size of 5 MΩ and measured at different tuning currents of 1 nA (**a**), 0.5 nA (**b**), 0.2 nA (**c**), and 0.1 nA (**d**) and a constant bias voltage of 0.05 V. The best Gaussian fit is indicated by black lines. AFM images of corresponding samples are shown in Fig. 8.25 [11] [Open access]

precursor in AACVD process dissolved in an organic solvent, the precursor solution is sprayed at room temperature, resulting in submicron-sized aerosol droplets that are dispersed using a carrier gas, which transports the resulting aerosols to thermal chamber, where the solvent quickly evaporates to form the finished product applied to the surface of the substrate [4, 7].

8.5.2 Aspects Technical to the AACVD Process

The key technological elements of the AACVD process for thin film deposition are listed below.

8.5.2.1 Pressure of Operational

Although the AACVD process can be carried out at atmospheric or low pressure, the deposition of thin films of oxides or sulfides must be carried out externally, without the use of a reaction chamber [7, 12, 13].

8.5.2.2 Precursor

Because main-group hydrides are extremely toxic and metal alkyls are often pyrophoric, they are mainly used as precursors. As a result, special tools are required to protect these precursors from environmental and safety risks [8, 9, 13].

8.5.2.3 Carrier Gas and Atomisation

Using an ultrasonic humidifier, the AACVD process produces an aerosol from a solution of precursors dissolved in a solvent. The nitrogen carrier gas carries the precursor mist into the CVD reactor (cold-wall reactor), which contains the heated substrate. However, if required by the AACVD process, the reactive gases such as hydrogen can be used in addition to the carrier gas [9–11, 16].

8.5.2.4 Vaporisation of Precursor and Evaporation of Solvent

The chemical and physical characteristics of liquids play an important role in solid–liquid interactions. Thus, untapped analytical opportunities exist to study surface wetting and evaporation behavior. Multilayer double oxide (LDO) microparticles are sprayed onto a fluoropolymer film to create highly hydrophobic thin films with a rough surface. The evaporation of ethanol/water mixtures from low-energy (8.4 ± 2.6 mJ/m^2) composite surfaces is investigated using weight loss monitoring, infrared imaging, and high-speed visible imaging simultaneously. Depending on the solvent composition, the wetting behavior changed from Cassie (pure water) to Wenzel (pure ethanol). Three stages of the evaporation process can be distinguished by consistent evaporation rates, and CT = $1768/(m^2K)$ is the heat transfer coefficient between water and the layer under study. A strong correlation is observed between the specific physical properties of the solvent and the parameters of measured evaporation profiles. The liquid's surface tension, boiling point, and viscosity can be calculated from the total time it takes for the liquid to evaporate. With this innovative solvent identification method, such low-energy hybrid surfaces could find new applications [13].

8.5.2.5 Heterogeneous Reaction

Real-world field observations, computer models and simulations, and basic laboratory research work closely together to study the heterogeneous reactions of atmospheric aerosols. While each element offers valuable information, complete understanding can only be achieved through a collaborative effort. Although much of this chapter is devoted to modeling and laboratory studies of heterogeneous oxidation reactions, it is important to advance in understanding real-world chemical data. However, there are still several unanswered questions. For example, there is growing evidence that the energetics of heterogeneous reactions can be influenced by the surface characteristics of liquid droplets and particles, which makes transformations possible that are not predicted by either gas-phase or condensation-phase processes.

Changes in Solvent Environment and Local Increases in Concentration Surface partitioning have been demonstrated to cause a significant increase in reaction rate and a change in chemical equilibrium. Although early studies [7] have examined these effects from a basic perspective, new subtleties are beginning to emerge. Research shows that ozonolysis and Fenton-type reactions at water interfaces, for example, can proceed orders of magnitude faster than those predicted for bulk solution [10, 13]. Moreover, organic compounds in molecular films at interface exhibit photochemical reactivity, allowing suggests that the interface plays another role in controlling chemical reactions [11]. These findings raise intriguing concerns for future research because the mechanisms underlying them are not yet fully understood [6, 7, 12].

8.5.2.6 Reaction of Homogeneous

The stretched flow of a viscoelastic fluid in two dimensions is considered using magnetohydrodynamics and a curved sheet. To study the heat and mass transfer, thermal radiation analysis, Joule heating, and homogeneous–heterogeneous reactions are used. A nonlinear ordinary differential system is created by applying the proper transformations, and numerical solutions are obtained using the quasi-linearization and implicit finite-difference approaches.

The findings indicate that while the temperature rises as radiation parameters rise, fluid concentration falls as Schmidt number rises. Compared to a viscoelastic fluid, a viscous fluid loses velocity and concentration more quickly. Furthermore, the increased surface resistance force determines the second-order fluid parameter [6, 7, 12].

8.6 The Fundamental Idea Behind EFDAACVD, or Electric Field-Directed Aerosol-Assisted Chemical Vapor Deposition

The aerosol is sprayed onto the surface of either ordinary glass or a fluorine-doped titanium dioxide glass substrate to achieve the required temperature under the influence of an electric field. The complete configuration of the in-house generated electric field-assisted chemical vapor deposition (EFDAACVD) setup. The main goal of the synthesis method is to create porous tin(IV) oxide nanospheres with a diameter range of 195–632 nm and a well-defined, repeatable structure with nanosized pores in the range of 33–35 nm. The films had excellent adhesion properties to the substrate and were stable in the presence of air and moisture after deposition at 350 °C and annealing at 400 °C.

8.7 Thin Films Applications

Modern silicon photovoltaic (PV) technology is well complemented by thin film PV technology. Thin film photovoltaic (PV) cells are flexible and lightweight, allowing them to be mounted on a variety of surfaces and produce electricity from ambient or sunlight. The bulk of thin film photovoltaic cells is previously produced using the vacuum method. Solution-processed thin film photovoltaic cells are very popular because they reduce the cost of energy production. In most thin film photovoltaic systems, metal oxides serve as important functional layers in the photovoltaic cell architecture. When thin film photovoltaic technology entered the solution production stage, metal oxide solution deposition became mandatory. The two most studied thin film photovoltaic technologies, organic and perovskite, are used as examples in this chapter to illustrate solution processing. In these photovoltaic cells, metal oxides serve as layers to extract and transport charge (electrons and holes). Several materials based on metal oxides have been described. Metal oxides are introduced not only to transport charges but also as electrodes and other functional layers [17–19,].

8.7.1 Thin Films in Optics

Numerous nanophotonic and plasmonic applications including high-sensitivity sensors [3], plasmonic circuits [4], nanolasers [5], optical metamaterials and metasurfaces [6, 7], and photonic hypercrystals [7] depend on nanoscale thin metal films [8]. Therefore, to deeply understand the relationship between the structure of metal films and intrinsic properties of plasmonic/nanophotonic devices, it is necessary to study the optical absorption of metal due to interband and intraband electronic transitions. It has been shown that in the case of polycrystalline metal films, losses affecting the complex dielectric function [7] are significantly influenced by electron scattering at grain boundaries and surfaces [8, 13]. Thorough investigation of structural and optical characteristics of a wide range of materials, including alternative plasmonic materials and traditional plasmonic metals (Au, Ag, Cu, and Al) [2–4].

Thin films thinner than 50 nm, for which changes in their optical characteristics depending on thickness under constant evaporation modes [6, 7, 12] (TiN, TCO) have been reported with great interest [5]. In addition, recent studies have examined the temperature-dependent optical characteristics of mono- and polycrystalline gold thin films [7, 13]. This plasmonic application can use metal sheets ranging in thickness from several tens to hundreds of nanometers. Therefore, to improve theoretical studies, numerical simulations and overall performance optimization of plasmonic devices, it is a critical to understand how the thickness of metal films affects their optical properties. In this regard, it is noteworthy that some guidelines and formulas have already been published [2] to help scientists create thick metal films with optical and structural characteristics tailored for plasmonic applications. It is important to remember that even though several materials are potentially good for plasmonic applications due to its low loss in the visible and near-infrared spectra, comparatively high-temperature stability and oxidation resistance, gold is still the most commonly used metal.

Thus, the purpose of this study is to resolve some unresolved issues related to the optical and structural characteristics of thin gold films. The standard electron beam evaporation (EBE) is used to obtain high-quality gold films on silicon substrates in a thickness range of 20–200 nm. It is critical to remember that in the range of film thicknesses, the islanded or extremely rough film surfaces can occur if the film thickness drops below 20 nm, and quantum confinement effects must be taken into account [15, 20–22]. In addition, the layers thicker than 200 nm are guaranteed to behave like bulk metals. The structural morphology of films is studied by X-ray diffractometry (XRD) and AFM; electrical properties are checked using a four-point probe; optical constants in the spectral range, 300–2000 nm, are determined by spectroscopic ellipsometry. To analyze observations of optical constants, the Mayadas-Shatzkes (MS) theory, which predicts optical losses and DC conductivity values based on the structural characteristics of gold films, and the Drude free electron model are used [8, 14].

8.7.2 Thin Film Photovoltaic Cells

Thin film solar cell technology represents a second-generation improvement over c-Si modules. These cells are created by coating various surfaces with one or more layers of thin photovoltaic materials such as glass, metal, plastic, or a mixture of these materials. Compared to thin layers of first-generation c-Si technology, more flexible and lightweight applications are possible. Photovoltaic (PV) energy is still needed to operate thin film solar panels, although fewer semiconductor materials are used in their production. The semiconductor of the thin film photovoltaic module changes the arrangement of its electrons, creating a useful electrical current that can be used by an external circuit to produce energy [17, 18].

Compared to c-Si versions, thin film panels have a lower cost but lower efficiency. Evolving technologies and semiconductors, among other materials, have a direct impact on the efficiency of thin film solar energy. In operation, despite their reputation for low output efficiency, thin film solar modules have shown promise due to recent technological advances. This chapter provides a detailed overview of the surface technology and chemistry perspectives of nanostructured thin film production methods, highlighting their potential uses. An in-depth discussion is organized to highlight the surface chemistry of thin films for efficient and controlled growth using multiple techniques. Particular attention is paid to various processes of formation of nanostructured films and the influence of growth factors on the surface of the substrate. Optimization of processing costs is related to the function played by the mechanical properties and chemical composition of thin film methods. Surface interactions and chemical modification of the surface of thin films for wider applications are being explored in industrial processes. New advances and potential applications of nanostructured films are opening up many new areas of possible use [19].

Due to its unique surface properties and wide range of applications in various sectors, thin film technology remains an important area of research. Thin films are vital to modern technologies ranging from microelectronics and optics to space exploration, aviation, solar cells, photovoltaics, and superconductivity. The properties of thin films, including the atomic structure, composition, microstructure, defects, and interfaces, are controlled by the kinetics and thermodynamics of synthesis. One of the most important advantages of thin film technology is that it is capable of producing transparent and flexible films and can do so at relatively low temperatures. In addition, improvements in deposition techniques allow the production of compact, high quality, multi-color, crystalline thin films in an efficient, low cost, and sustainable manner.

Despite these advantages, the challenges remain in creating thin films that meet the requirements of new technological and industrial applications. This chapter reviews a range of thin film technologies, focusing on the challenges encountered as well as encouraging results achieved. As research progresses, thin film technology is expected to continue to improve to address these issues and expand its applications and impact.

Problems

1. Describe thin film synthesis and its applications.
2. What is the solution if thin film production methods suffer a triple failure?
3. How important are characterization techniques and nanostructured thin films?
4. Why and how does physical vapor deposition occur?
5. Explain different nano-thin film techniques from the point of view of mechanical strength.

References

1. López-Lugo, V. H., García-Hipólito, M., Rodríguez-Gómez, A., & Alonso-Huitrón, J. C. (2023). Fabrication of Li-doped NiO thin films by ultrasonic spray pyrolysis and its application in light-emitting diodes. *Nanomaterials, 13*(1), 197.
2. Maksimova, N. K., Malinovskaya, T. D., Zhek, V. V., Sergeychenko, N. V., Chernikov, E. V., Sokolov, D. V., ... & Korusenko, P. M. (2024). Advanced SnO_2 thin films: Stability and sensitivity in CO detection. *International Journal of Molecular Sciences, 25*(23), 12818.
3. Bhattarai, J. K., Maruf, M. H. U., & Stine, K. J. (2020). Plasmonic-active nanostructured thin films. *Processes, 8*(1), 115.
4. Yunus, Y., Mahadzir, N. A., Mohamed Ansari, M. N., Tg Abd Aziz, T. H., Mohd Afdzaluddin, A., Anwar, H., ... & Ismail, A. G. (2022). Review of the common deposition methods of thin-film pentacene, its derivatives, and their performance. *Polymers, 14*(6), 1112.
5. Petronella, F., Pagliarulo, A., Truppi, A., Lettieri, M., Masieri, M., Calia, A., & Comparelli, R. (2018). TiO_2 nanocrystal based coatings for the protection of architectural stone: The effect of solvents in the spray-coating application for a self-cleaning surfaces. *Coatings, 8*(10), 356.
6. Yoon, Y., Lee, J., Lee, S., Kim, S., & Choi, H. C. (2021). Ultrasmooth organic films via efficient aggregation suppression by a low-vacuum physical vapor deposition. *Materials, 14*(23), 7247.
7. Al-Hilfi, S. H., Kinloch, I. A., & Derby, B. (2021). Chemical vapor deposition of graphene on Cu-Ni alloys: The impact of carbon solubility. *Coatings, 11*(8), 892.
8. Suzuki, S., Terada, Y., & Yoshimura, M. (2017). Suppression of graphene nucleation by turning off hydrogen supply just before atmospheric pressure chemical vapor deposition growth. *Coatings, 7*(11), 206.
9. Li, Y., Sun, L., Liu, H., Wang, Y., & Liu, Z. (2020). Rational design of binary alloys for catalytic growth of graphene via chemical vapor deposition. *Catalysts, 10*(11), 1305.
10. Huang, C. H., Wu, H. C., Chen, B. F., & Li, Y. C. (2021). Graphene/silver nanowires/graphene sandwich composite for stretchable transparent electrodes and its fracture mechanism. *Micromachines, 12*(5), 512.
11. Kocábová, J., Vavrek, F., Nováková Lachmanová, Š, Šebera, J., Valášek, M., & Hromadová, M. (2022). Self-assembled monolayers of molecular conductors with terpyridine-metal redox switching elements: A combined AFM. *STM and Electrochemical Study. Molecules, 27*(23), 8320.
12. Kim, S., Lee, Y. S., & Kim, N. H. (2023). Homogeneity-and stoichiometry-induced electrical and optical properties of Cu–Se thin films by RF sputtering power. *Materials, 16*(18), 6087.
13. Schilirò, E., Lo Nigro, R., Roccaforte, F., & Giannazzo, F. (2019). Recent advances in seeded and seed-layer-free atomic layer deposition of high-K dielectrics on graphene for electronics. *C, 5*(3), 53.
14. Valášek, M., & Mayor, M. (2017). Spatial and lateral control of functionality by rigid molecular platforms. *Chemistry—A European Journal, 23*, 13538–13548.

15. Šebera, J., Kolivoška, V., Valášek, M., Gasior, J., Sokolová, R., Mészáros, G., Hong, W., Mayor, M., & Hromadová, M. (2017). Tuning charge transport properties of asymmetric molecular junctions. *Journal of Physical Chemistry C, 121*, 12885–12894.
16. Durante, O., Di Giorgio, C., Granata, V., Neilson, J., Fittipaldi, R., Vecchione, A., ... & Bobba, F. (2021). Emergence and evolution of crystallization in TiO_2 thin films: A structural and morphological study. *Nanomaterials, 11*(6), 1409.
17. Kolivoška, V., Šebera, J., Sebechlebská, T., Lindner, M., Gasior, J., Mészáros, G., Mayor, M., Valášek, M., & Hromadová, M. (2019). Probabilistic mapping of single molecule junction configurations as a tool to achieve desired geometry of asymmetric tripodal molecules. *Chemical Communications, 55*, 3351–3354.
18. Šebera, J., Lindner, M., Gasior, J., Meszáros, G., Fuhr, O., Mayor, M., Valášek, M., Kolivoška, V., & Hromadová, M. (2019). Tuning contact conductance of anchoring groups in single molecule junctions by molecular design. *Nanoscale, 11*, 12959–12964.
19. Nováková Lachmanová, Š, Vavrek, F., Sebechlebská, T., Kolivoška, V., Valášek, M., & Hromadová, M. (2021). Charge transfer in self-assembled monolayers of molecular conductors containing tripodal anchor and terpyridine-metal redox switching element. *Electrochimica Acta, 384*, Article 138302.
20. Tu, M., Xia, B., Kravchenko, D. E., Tietze, M. L., Cruz, A. J., Stassen, I., Hauffman, T., Teyssandier, J., De Feyter, S., Wang, Z., et al. (2021). Direct X-ray and electron-beam lithography of halogenated Zeolitic Imidazolate frameworks. *Nature Materials, 20*, 93–99.
21. Bokov, D., Turki Jalil, A., Chupradit, S., Suksatan, W., Javed Ansari, M., Shewael, I. H., Valiev, G. H., & Kianfar, E. (2021). Nanomaterial by sol-gel method: Synthesis and application. *Advanced Materials Science and Engineering, 2021*, 5102014.
22. Valášek, M., Lindner, M., & Mayor, M. (2016). Rigid multipodal platforms for metal surface. *Beilstein Journal of Nanotechnology, 7*, 374–405.

Chapter 9
Nanofiber Applications in Tissue Engineering

Because they improve tissue and organ creation, nanofibers, which have diameters between 1 and 1000 nm are essential for tissue engineering. They are ideal for stimulating cell proliferation and creating artificial tissues due to their large surface area and texture that resembles biological tissue. Modeled after natural tissues, nanofibers enhance biological interactions, supporting cell differentiation and proliferation while offering a stable environment. The viability of transplanted tissues is enhanced by improved oxygen and nutrient delivery. The development of muscle, bone and skin tissues using nanofibers is already a breakthrough in tissue engineering, with great promise for improved healing outcomes and tissue regeneration.

9.1 Introduction

One of the most promising methods of tissue regeneration is autologous cell/tissue transplantation for tissue repair. However, autografts have several disadvantages including donor site morbidity and insufficient supply. Autografts can be replaced by allografts, which involve using tissue from another member of the same species. Although there is no shortage of allografts, there is a possibility that they could cause an immunological reaction and spread disease. Tissue engineering has become a highly effective method for repairing and regenerating damaged tissues, allowing one to overcome all the disadvantages associated with restoring allogeneic and autologous tissues [1–4]. The goal of tissue regeneration is achieved through using biological, chemical and engineering concepts in the emerging interdisciplinary field of tissue engineering. Tissue engineering techniques use factors, cells and biomaterials, individually or in combination, to improve maintain, or restore tissue function. The standard tissue engineering approach removes healthy cells from a patient and grows them in culture. Once expanded, these cells are placed into a three-dimensional (3D) biodegradable scaffold that serves as a reservoir for bioactive substances such

R. A. Al-Samarai and Y. Al-Douri, *Engineering of Nanomaterials*,
https://doi.org/10.1007/978-981-95-5499-7_9

as growth factors and provides structural support. Over time, the scaffold gradually breaks down and is replaced by a newly developed tissue composed of seeded cells. Differentiation of stem cells derived from bone marrow, adipose tissue, hair follicles and umbilical cord into other designated cell types such as myocytes, chondrocytes, osteocytes and many others can be achieved using various methods. Using the principles of tissue engineering, it is possible to restore the following tissues; muscles, liver, blood arteries, bones and even nerve conductors. The surface morphology of gelatin nanostructures before and after air-pressurized plasma treatment is evaluated using scanning electron microscopy (SEM) and atomic particle microscopy (APM) techniques. The SEM results of untreated and plasma-treated gelatin nanofibers are shown in Fig. 9.1a and b. After plasma treatment, the nanofibers' average diameter, shape and surface area did not change significantly. It is found that the formation of bead-like patterns on electrospun nanofibers is suppressed by the surface tension, viscoelastic properties and electrostatic repulsion because of using acetic acid solvent. Therefore, to obtain smooth gelatin nanofibers, it used optimized electrospinning settings based on our previous studies [1].

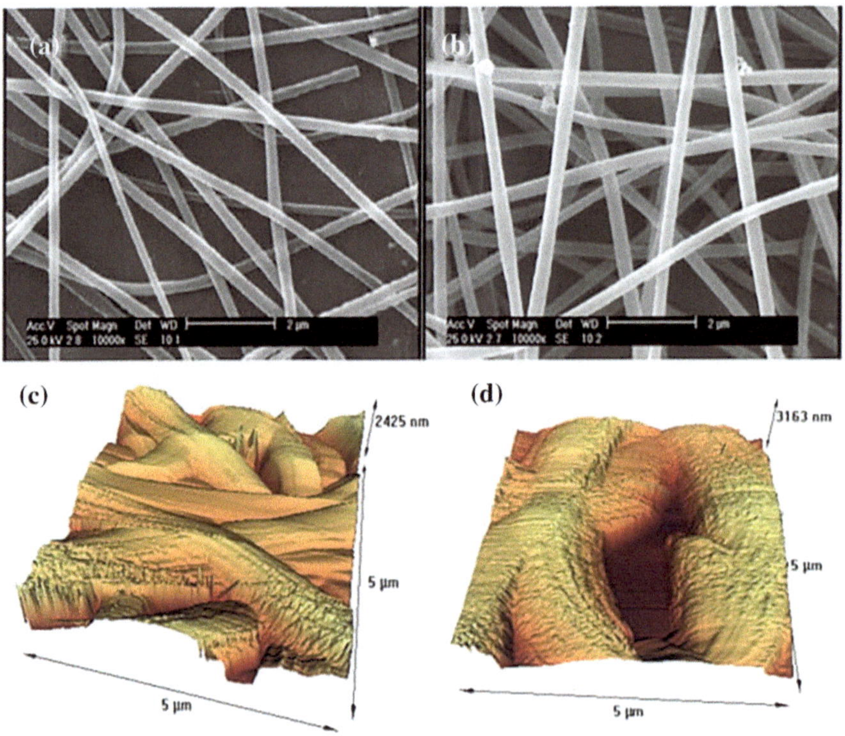

Fig. 9.1 Untreated gelatin nanofibers (**a**), air plasma-treated gelatin nanofibers (**b**), untreated gelatin nanofibers and (**c**) air plasma-treated gelatin nanofibers (**d**) SEM and AFM images, respectively [1] [Open access]

9.2 Historical Perspective

Providing health is essential to maintaining quality of life and global need. Demands in the fields of drug applications, tissue engineering, medical device manufacturing and permanent or temporary implants have led to the ongoing and growing interest in the study and expansion of development of biomaterials for medical or healthcare applications. The research is aimed at improving biomaterials in traditional applications and finding new, effective bioactive materials for various advanced applications such as regenerative medicine or other health disorders. One such application is theranostic biomaterials, which provide combination therapy, monitoring and diagnostic capabilities [3]. The demand for precision and customized treatments is expected to increase. From the beginning of tests, it is observed that the self-assembly process of elastin progressed from nanofibers to fibers and finally to fibrils. It took at least two weeks for the first nanofibers to form after the preparation of the initial suspension of elastin in 1% w/v ultrapure water at 20 °C (Fig. 9.2d). These nanofibers are self-assembled to form fibers after about two weeks (Fig. 9.2c). Eventually, the self-assembled fibers formed elastin fibers about six weeks after the preparation of the initial aqueous suspension (Fig. 9.2b). Thus, Fig. 9.2a shows 10×10 μm atomic force microscopy (AFM) topography indicating the presence of four fibrils and one elastin fiber (three on the right and three on upper left of fiber). The height difference between the nanofibers (4.5 ± 0.7 nm) and fibers (88.8 ± 5.7 nm) made it difficult to observe the weak nanofibers despite the presence of many of them. A high-magnification AFM topography image of same 5×5 μm elastin fiber is shown in Fig. 9.2b [2].

Biomaterials play an important role in modern medicine, both in promoting healing and restoring the original biological or beneficial effect. Biomaterials can be artificial or natural. Since they are biocompatible, they are used for various medical purposes. Accordingly, biomaterials are defined as "any materials designed to interact with the biological systems of the human body to assess, support, treat, enhance, augment, repair or replace any damaged tissue, organ or bodily function" (definition put forward by the European Society for Biomaterials Consensus Conference II) [4]. Biocompatibility is defined as "the study and knowledge of interactions between living and nonliving materials" [5]. Because of biomaterials are in close contact with body fluids and tissues, they must meet several requirements including stability, biocompatibility and safety.

9.3 Current Situation

Today, many biomedical and bioengineering companies are interested in repairing injured or damaged tissue and unlike large pharmaceutical companies, these companies are very interested in collaborating. One of these is Life Cell, a large business known for creating tissue grafts for transplantation and preserving blood products

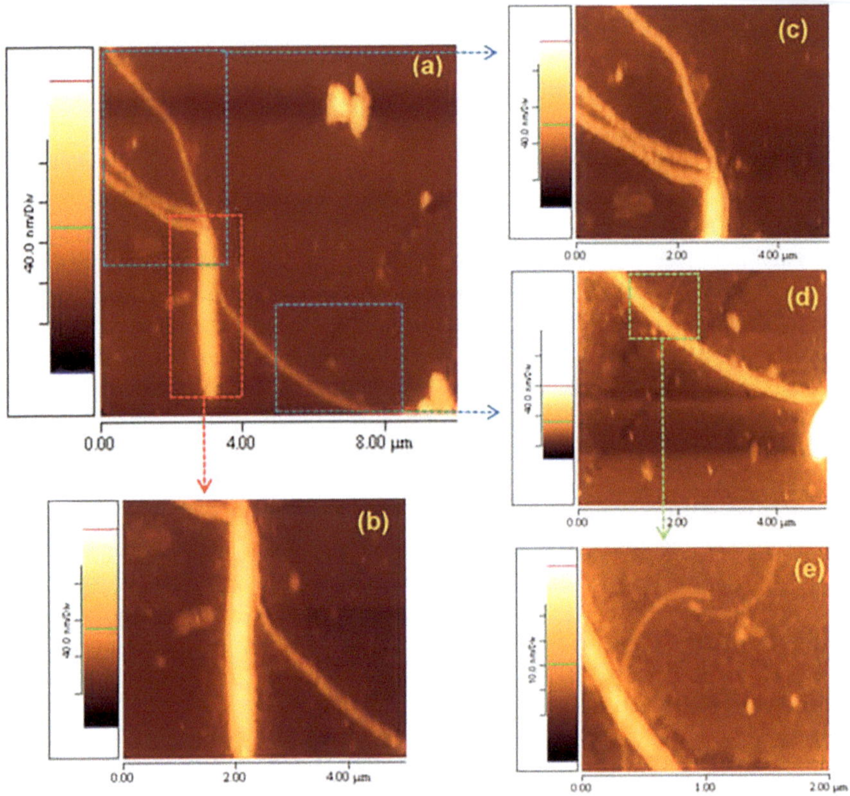

Fig. 9.2 Topographic images of elastin self-assembly obtained using AFM: (**a**) Fiber (**b**) fibrils (**c**, **d**) nanofibrils (**d**) [2] [Open access]

for future use. Specially designed heart valves for pigs have also been created by Life Cell in collaboration with Medtronic, a renowned firm for human heart valve replacement. Medtronic produces heart-related surgical devices including tissue and technological artificial heart valves. Life Cell and Medtronic have worked together to develop important and readily available tissue engineering products.

Figure 9.6 shows the topography of Nafion 115 at different relative humidities. The scanned area of each image is 5 μm × 5 μm. The color intensity of each image corresponds to surface height, which is displayed as a color bar to the left of image. Figure 9.3a–f dipects the morphology and relative line profile at 15, 25, 45, 55, 65 and 75% relative humidity, respectively. Random cluster formations with surface sizes ranging from a few tens to a few hundred nanometers are visible in each image. However, the images do not show any significant morphological changes at high relative humidity. The numerical data is provided by the line profile, which displays the change in the reported line height. The surface height is from −10 nm to 20 nm at 15% relative humidity (RH), which is a fairly large height dispersion compared

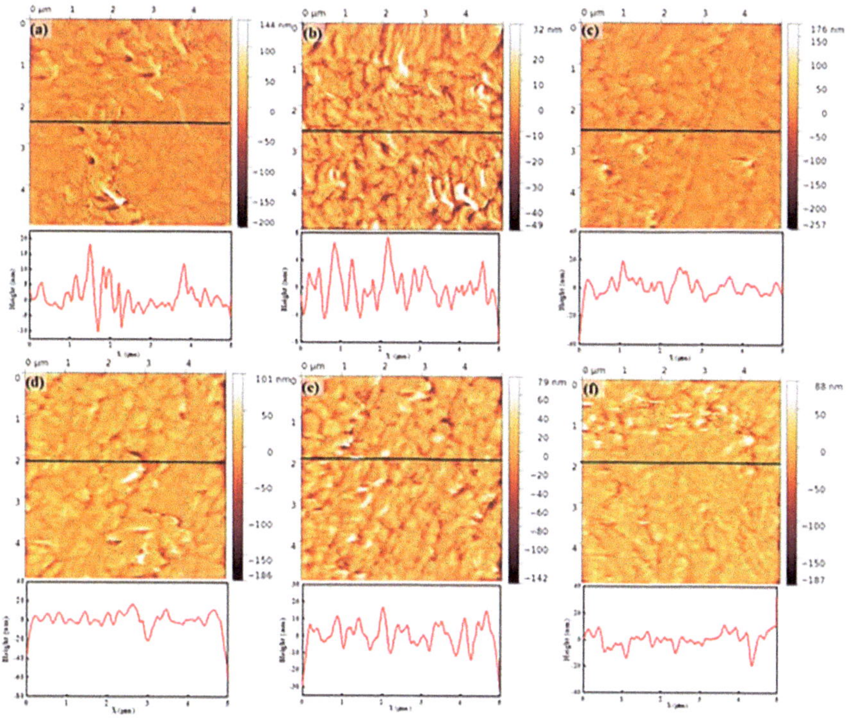

Fig. 9.3 Topography of Nafion 115 and corresponding linear profiles at 15%, 25%, 45%, 55%, 65% and 75% relative humidity [3] [Open access]

to low RH. The current map of Nafion 115 at different relative humidity levels is shown in Fig. 9.4. The brightness of current map indicates the magnitude and color bar next to shows the numerical value of current at each brightness [3].

Partnerships such as these point to different path to the commercialization of tissue engineering products than the pharmaceutical sector. Military medicine for armed forces during conflict is the primary application for tissue-engineered products. For example, the US Army Chemical Defense Institute has funded clinical trials of Dermagraft-TC, artificial human skin tissue mixed with a synthetic dermal layer to treat chemical burns. Although patients have significant concerns such as protection from burns and reduction of microbial infections, this graft mainly protects the skin from acute immunological rejection, limited supply of artificial grafts and possibility of pathogen transmission [6]. The current maps and linear profiles associated with the membrane at 15%, 25%, 45%, 55%, 65% and 75% relative humidity are shown in Fig. 9.4a–f. The current map shows a very low distribution of current value on the membrane surface at 15% relative humidity. It can be seen that current differences are in the range of a few tens of picoamperes. The measured current can be very close to zero if we take into account the presence of instrument noise, which is usually a

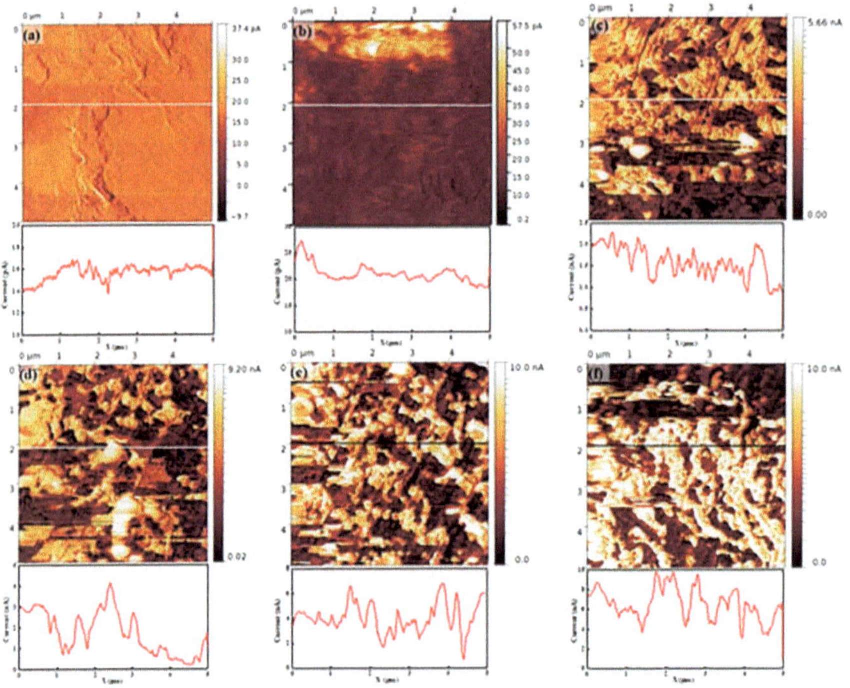

Fig. 9.4 Current Nafion 115 map and associated linear profile at 15%, 25%, 45%, 55%, 65% and 75% relative humidity [3] [Open access]

few tens of picoamperes. In addition, the organization of current map resembles a terrain in terms of prominent areas [3].

Creating three-dimensional (3D) synthetic scaffolds as called scaffolds or structures for cell adhesion, proliferation and differentiation that guide the development of new tissues or organs is a critical function of biomaterials in tissue engineering. Several innovative techniques have been developed for creating 3D structures based on biomaterials. Therefore, 3D nanofiber-based scaffolds have been developed for tissue engineering applications.

9.4 Tissue Engineering Triad

Tissue engineering substrates can now have greater mechanical strength and better cellular differentiation and proliferation thanks to nanomaterials [3]. Soft tissue engineering has shown a great interest in carbon-based nanomaterials, including graphene oxide (GO) [6], fullerenes, carbon nanotubes (CNTs) , carbon dots (CDs) [7] and graphene [8], composite scaffolds and nanoparticles [9]. In another study, the researchers [4] have printed 3D images of nerve canals using graphene and poly(lactide-co-glycolide) (PLGA) nanoinks to improve neurorenewal [4] before use, AFM probe tip profiles are first visualized by scanning electron microscopy (SEM) analysis (Fig. 9.5a, c and e). In contrast, the modified and introduced probes are used to generate SEM scans of 2% agar hydrogel surface (Fig. 9.5b, d and f). The high-resolution image of the agar hydrogel surface obtained with this tip (Fig. 9.5b) and SEM image of standard ScanAsyst Fluid High-Resolution probe (Fig. 9.5a) are shown in Fig. 9.5a and b. Here, the smallest pores of the gel, down to 50 nm in diameter and 200 nm in depth could be easily resolved [4].

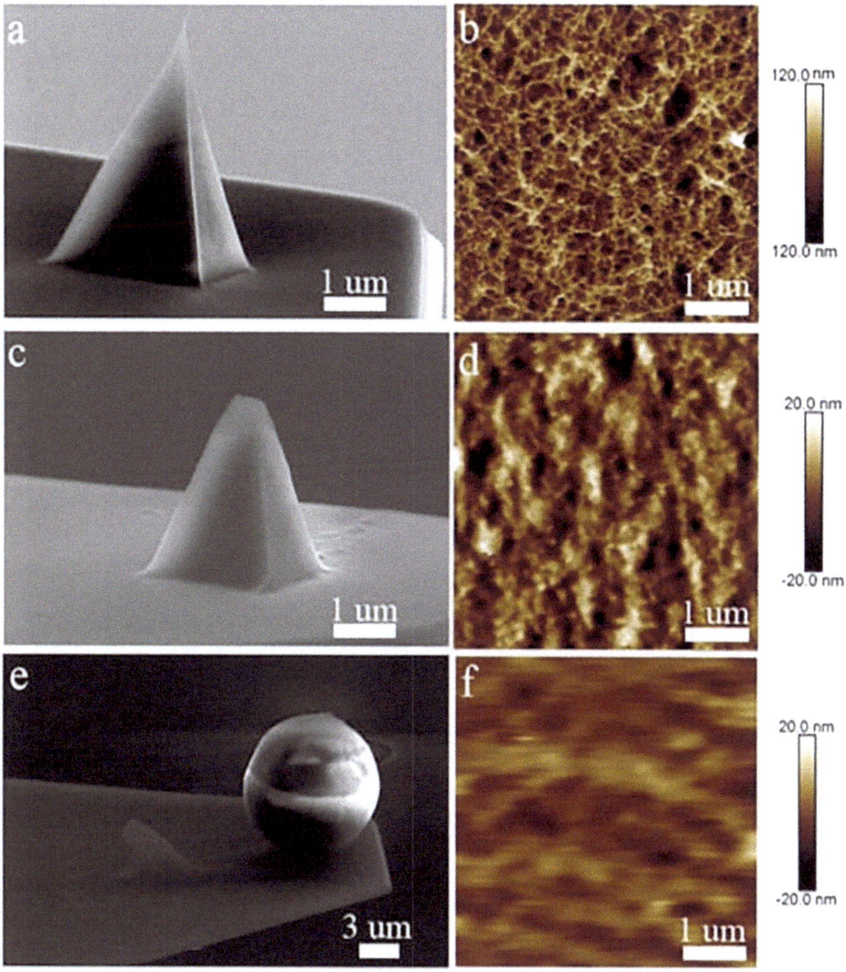

Fig. 9.5 SEM showing comparative properties of AFM probes: **a** original ScanAsyst Fluid High-Resolution probe; **b** laser-modified ScanAsyst Fluid High-Resolution probes; and **e** NP-10 probe with microsphere attached. **b**, **d**, **f** AFM morphology maps are obtained using above probes on surface of 2% agar gel [4] [Open access]

These materials offer new opportunities for improving soft tissue engineering due to their distinctive mechanical, geometric, optical and conductive properties. Various functionalization approaches have been developed by several groups for carbon nanomaterials to improve their biocompatibility and promote cellular attachment and differentiation. For soft tissue regeneration and healing, artificial cellular substrates must have mechanical, chemical and physical characteristics similar to those of natural extracellular matrix structures. In addition, the conductivity properties of electrical signals can be applied to tissues using carbon-based nanocomposite scaffolds [4]. Some conductive nanomaterials have been used in biomanufacturing such as nanocomposite bioinks. For example, conductive bioinks called single-walled CNTs coated with bioactive DNA are used in a two-step process on various supported platforms to print conductive flexible electronics in 3D [2]. Another example of using nano inks is the 3D printed model of "heart organs on a chip" (thermoplastic polyurethane filled with carbon black nanoparticles used as conductive ink in combination with other inks) [3].

Histocompatibility, mechanical compatibility and biocompatibility are believed to be critical elements for brain regeneration both in vitro and in vivo. Polymers with acceptable biocompatibility and low protein adsorption capacity can be created by treating them with oxygen or plasma. In this regard, it has been demonstrated that the combination of chitosan and PLGA, together with the highly biocompatible laminin, enhances the adhesive and proliferative properties of cells [22]. In addition, the in vivo effectiveness of growth factors can be enhanced through the use of nanomaterials, which can also prolong their local release and provide protection from enzymatic degradation. Numerous nano delivery systems, most notably lipid nanoparticles, polymer nanoparticles and nanofibrous structures have been investigated as ways to improve wound healing. Growth factors and other biological chemicals have been investigated to optimize their controlled release using biocompatible polymer systems [6, 7].

This printing method allows the creation of structures with high conductivity, excellent biocompatibility and piezoresistance. The high cell viability observed in the various sections as demonstrated by LSM (laser scanning microscopy) studies (Fig. 9.6a) confirms that the environment supports cell survival and proliferation. The assay used fluorescent markers to differentiate live and dead cells, showing significantly higher numbers of viable cells.

Green fluorescence (FDA—fluorescein diacetate) penetrates live cells where it is enzymatically converted to a green fluorescent compound, confirming metabolic activity and cell viability. Red fluorescence (PI—propidium iodide) is a membrane-impermeable dye that only penetrates dead cells with damaged membranes, staining them red. LSM images clearly show the dominance of green (live) cells over red (dead) cells, indicating a healthy and supportive cellular environment. This suggests that the material or scaffold used in the study promotes cell attachment, differentiation, and proliferation, which has important implications for tissue engineering and regenerative medicine.

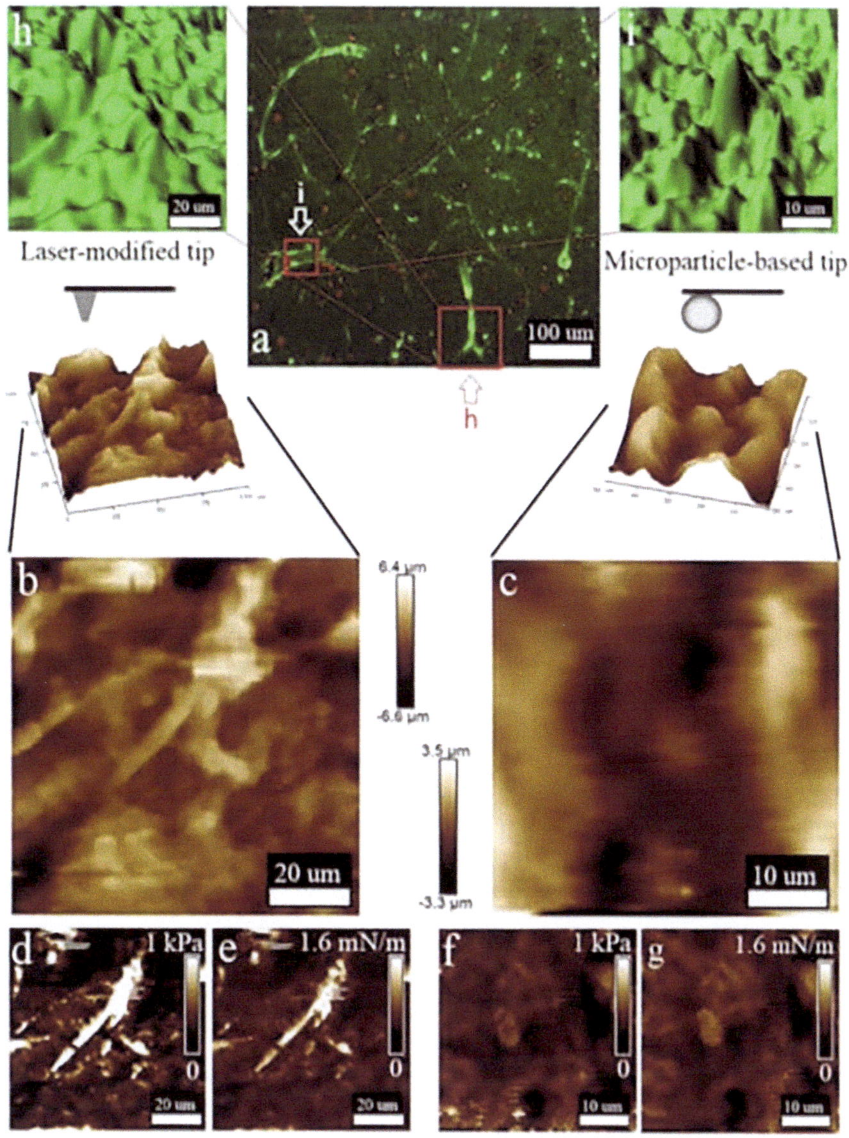

Fig. 9.6 Surface properties of mouse brain section examined with ScanAsyst Fluid High-Resolution laser-shaped probe (**b**, **d**, **e**), AFM (NP-10 microsphere cantilever microscopes (**c**, **f**, **g**), and LSM (**a**, **h**, **i**). **a** Histogram of the ratio of live (green) to dead (red) cells in the large-scale LSM. **h**, **i**. The two sections provide 3D visualization of smaller map regions selected for further AFM examination using two different cantilever types. The surface morphology of selected section is obtained using laser (**b**) and microsphere (**c**) cantilevers. The insets in both images show the 3D reconstructions of these locations. (**d**, **e**) AFM stiffness and elasticity maps generated with ScanAsyst Fluid High-Resolution laser-shaped probe. (**f**, **g**) AFM stiffness and elasticity maps using microsphere-based control panels. NP-10 [4] [Open access]

By ensuring optimal cell viability, this approach increases the potential for successful tissue regeneration, making it a promising strategy for biomedical applications such as the creation of artificial tissues, wound healing, and organ repair. Figure 9.6 illustrates the quantitative assessment of tissue section viability. The resulting image (Fig. 9.6a) clearly shows the cell bodies and neuronal processes. The morphological maps of rat brain section surface, reconstructed from the regions of interest in Z-stack images acquired by LSM in confocal mode are shown in insets (Fig. 9.6a). In addition, the labels of these regions are shown in Fig. 9.6a. There are rectangular tubular structures on the surface representing neuronal processes (neurons). In addition, both reconstruction regions are selected for stiffness analysis (Fig. 6e and g), nanomechanical mapping via elasticity (Fig. 9.6d and f) and shape mapping (Fig. 9.6b and c), all of which are performed using two different types of AFM arms. The position and shape of processes can be resolved thanks to the laser modulation arms (Fig. 9.6b–e) [4].

Finally, we offer some insight into the potential applications of nanotechnology in tissue engineering and regeneration. Specializing in both engineering and biological research, tissue engineering is a highly diverse field of research. This discipline relies heavily on mechanical support provided by porous 3D scaffolds, which create ideal conditions for development of cells in specific tissues and organs. When specialized cells are deposited on scaffold, the scaffold essentially functions as an artificial extracellular matrix for formation of new tissues. Cells attach to the scaffold and receive a variety of growth factors and biophysical stimuli that help them develop and differentiate into desired tissue type [7]. These scaffolds are either immediately implanted at wound sites or cultured in vitro to develop specific tissues that can later be applied to the injury. The combination of scaffolds, signals and cells are commonly known as tissue engineering triad as shown in Fig. 9.7 [5].

9.5 Requirements for 3D Biomaterials or Scaffolds

One of the main causes of morbidity and impairment in the elderly is bone abnormalities [4, 6]. Although bone tissue naturally can regenerate or self-heal, it has been observed that large bone lesions (critical size: 5 mm in diameter) cannot completely heal on their own without assistance [4]. One of the main goals of bone tissue engineering research is the repair and regeneration of bone defect tissue, but drug therapy for this purpose is still a challenge [8]. When human corneal epithelial (HCE) cells are seeded onto different films, morphological differences between the cells are observed (Fig. 9.8a–d). Cells grown on gelatin-lactose (GEL-LAC) film exhibit the characteristic cobblestone shape typical of corneal epithelial cells, similar to that seen in the control sample. On another hand, cells seeded on GEL-CA and SPI films have more rounded and elongated cell shape. In the same vein, it is less clear how the polygonal outline of cells is recorded on GEL-LAC and control films. Similar to 3T3 cells, in Calcein AM-EthD1 live/dead assays performed after 24, 48 and 72 h, very few dead cells are seen above the films (Fig. 9.8e–h) [5].

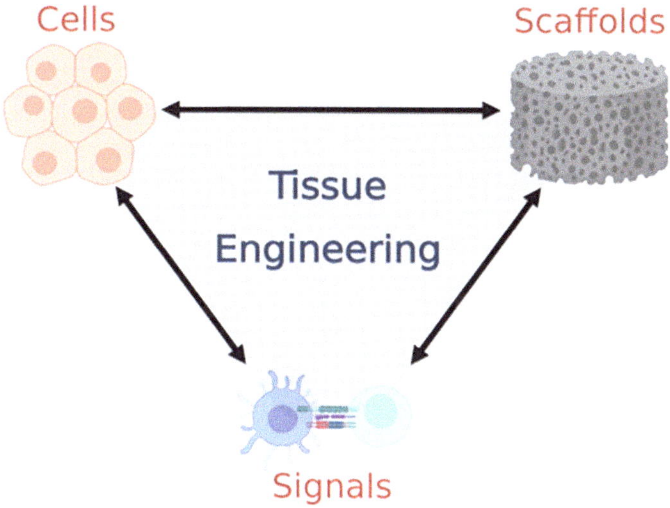

Fig. 9.7 The three components of tissue engineering triad are cells, scaffold (3D support) and external signals [5] [Open access]

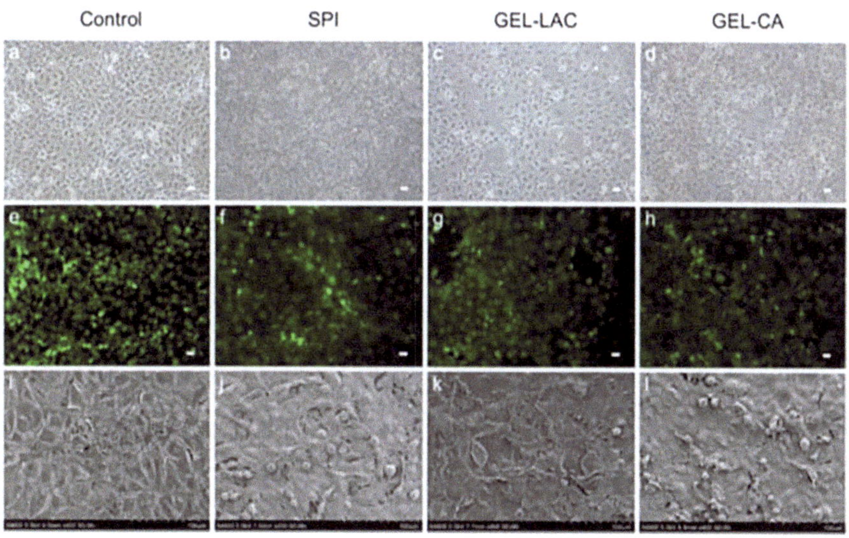

Fig. 9.8 HCE cells cultured on films. **a–d** Images of HCE cells on tissue culture wells (control group) and films 24 h after cell culture; **e–h** Images of representative CA-EthD1 analysis; **i–l** SEM images. **a–h** 10× magnification; 25 μm are shown by scale bars. **i–l** Enlarged images at 450× magnification [5] [Open access]

To regenerate, repair or replace critically sized bone lesions resulting from various diseases or injuries, advanced bone tissue engineering methodologies are universal therapeutic necessity [8]. For clinical use in patients, the development of methods and biomaterial systems capable of properly and permanently repairing/regenerating damaged bone tissue is critical [9]. Technologically engineered 3D biomaterial scaffolds have shown significant promise as a treatment for bone abnormalities throughout time [10]. However, the creation of 3D biomaterial scaffold systems with the required properties for bone tissue engineering applications is still highly needed.

In general, 3D biomaterial scaffold should be readily resorbed as new bone tissue grows and provide mechanical support comparable to that of natural tissue of interest such as bone [11]. Since they promote resorption and self-repair of native tissues in vivo in body, selecting the right quality of biomaterials is important for creating 3D biomaterial scaffolds [12]. Materials that interact with biological systems are called biomaterials [13]. In addition, when selecting biomaterials, it is necessary to take into account the binding force at scaffold interface [13]. The natural polymers most studied for use in bone tissue engineering are alginate, hyaluronic acid, chitosan, silk and collagen [14]. However, natural polymers have disadvantages when used exclusively in the creation of biomaterials for bone tissue engineering [15], including unsatisfactory mechanical properties and other problems [16]. When determining the compatibility, degradability and suitability of specific scaffolds used in tissue engineering, many important elements should be considered during development.

9.5.1 *Biocompatibility*

The ability of an implant material to function in vivo without causing harmful local or systemic reactions in the body is known as biocompatibility. Biomaterials are tested on tissues and animals before using human fracture fixation to ensure their safety and effectiveness. Inert materials are biomaterials that cause little to no reaction from the host such as metal alloys of cobalt and chromium. The effects of allergic reactions to stainless steel implants containing nickel and chromium highlight the importance of biocompatibility. When an implant is biologically rejected, immune cells cause an inflammatory response that may require the removal of implant. Various techniques have been used to improve the biocompatibility of titanium alloys such as sintering and laser micro texturing of the surface.

The use of resorbable magnesium is a relatively new idea in orthopedics and highly biocompatible. Much of research into new biomaterials is aimed at improving the biocompatibility of implants and preventing unnecessary problems [5]. Figure 9.9 illustrates the atomic field microscopy images of surface of Poly(3,4-ethylenedioxythiophene) (PEDOT). Nafion coating deposited by direct casting on a glass substrate. The conductive polymer coating in Fig. 9.9a and b is subjected to topographic images obtained at relatively large scanning sizes, 20×20 and 5×5 μm squared. The scanning has revealed a very rough and poorly ordered surface,

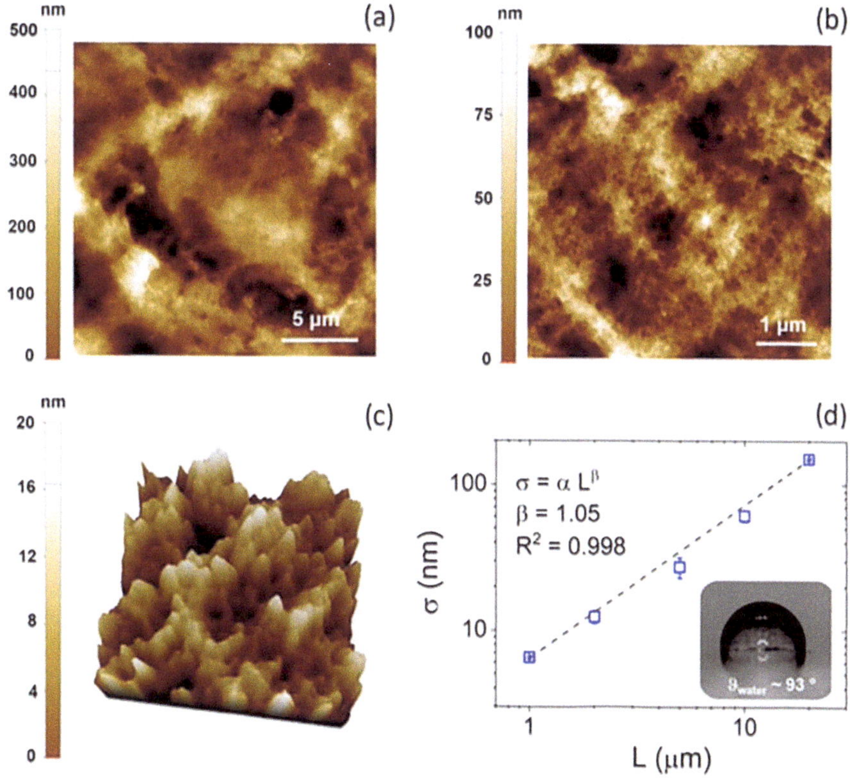

Fig. 9.9 Nafion coatings are characterized on surface. 2D and 3D topographic images (**a**, **b**) of PEDOT: Nafion coatings acquired at different length scales. One micrometer represents the side-scan size in (**c**). RMS (σ) versus side-scan size (L) is plotted in (**d**). The power law ($\sigma = \alpha\ L\beta$) indicates that dot size represents best fit. The inset image in (**d**) shows water droplet on PEDOT: Nafion coating [6] [Open access]

which is typical for polymer coatings formed using direct casting procedures [18, 19].

On another hand, as shown in Fig. 9.9c, PEDOT: Nafion coating has a nanostructured surface with nanoparticles of approximately 30–50 nm in size at smaller length scale. This surface is similar to the direct cast films made from the most common water-based PEDOT: PSS formulation (Clevios PH1000) [6].

9.5.2 Biodegradability

An important component of assessing the environmental risk of using biosurfactants is their biodegradability. Microorganisms that use surfactants as a source of carbon and energy usually mediate the biodegradation of surfactants [52]. Carbon chain scission and immediate loss of amphiphilicity are the first steps in the degradation process. The intermediate products of decomposition are converted mainly into carbon dioxide and water. Biosurfactants break down easily in environment due to their simple chemical structure, which prevents accumulation. The thickness of each fabricated membrane is about 200 μm (error < 3%). Figure 9.10 shows the images of poly(glycerol sebacate) (PGS) and ammonium persulfate (APS) membranes. FESEM and AFM are used to analyze the surface morphology and topography of each polymer membrane. Some patterns are visible on the surface of PGS membrane. PGS (RMS = 17.65 nm, Ra = 13.67 nm) > APS (RMS = 12.50 nm, Ra = 8.71 nm) > PDMS (RMS = 7.98 nm, Ra = 4.92 nm) is the order in which the surface roughness characteristics are found [7].

Additional studies have shown that biosurfactants are more biodegradable than their synthetic equivalents [5]. In addition, biosurfactants may be able to break down toxic and environmentally hazardous substances at the site of contamination without causing much harm to biological systems [4]. Minimal toxicity and high biodegradability indicate the environmental friendliness of biosurfactants, which is supported by the use of the words "green chemicals" [5] and "easily biodegradable" [6, 8].

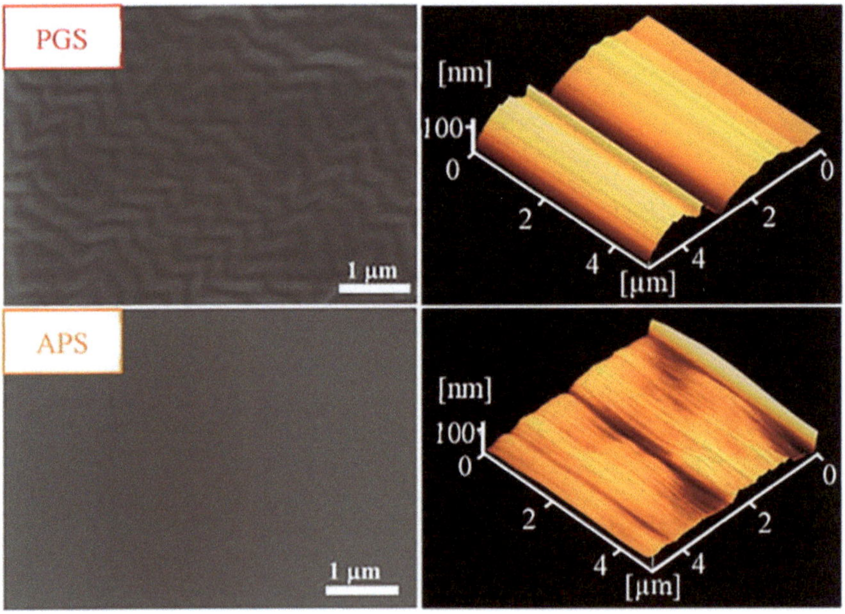

Fig. 9.10 AFM and FESEM images of two biodegradable membranes obtained for this study (× 20S000) [7] [Open access]

9.5.3 Mechanical Properties

Due to their anisotropic, heterogeneous structure and advanced mechanical properties associated with size effects during micromachining, nanocomposites are considered difficult materials to micromachine. Due to high cutting force, poor surface quality and high tool wear rate, it results in poor machinability. This two-part review study will be divided into two sections; the first will provide a thorough analysis of mechanical properties of various nanocomposites and second will focus on the micromachinability of materials [13]. The synthesis of extracellular polymeric substances (EPS) is expected to increase once Clostridium clostridium (C. clostridium) enters the stationary phase. At this stage, EPS isolation is not necessary since silica nanoparticles (SiO_2) are added directly to the medium and atomic force microscopy (AFM) is used to study the interaction between SiO_2 nanoparticles and EPS at the single-cell level.

When C. clostridium is treated with SiO_2 nanoparticles (10 μg/mL, day 15 of incubation), AFM imaging (Fig. 9.11) shows 40-long bacterial cell maintains its typical morphology with distinct cellular components. This indicates that SiO_2 exposure does not significantly alter the structural integrity of bacteria.

Understanding how SiO_2 nanoparticles interact with EPS is made possible by this study, which may impact surface adhesion characteristics and bacterial biofilm formation. High-resolution nanoscale studies are made possible by AFM, providing a better understanding of the effects of nanoparticles on microbial systems while preserving their native cellular environment [11]. SiO_2 nanoparticles are mainly associated with the surrounding EPS and on the cell surface as shown in AFM images (Fig. 9.11a and b). EPS near the cell formed a layer of several nanometers in height and fibers forming a network are visible away from the cell (Fig. 9.11b–e). Silica nanoparticles, individual particles, 10–15 nm in height are found in association with EPS. In contrast, no EPS-associated particles are detected in cultured C. clostridium cells not enriched with silica nanoparticles (Supplementary Material, Fig. 9.11 S) [8].

The anisotropic, heterogeneous structure and improved mechanical properties of nanocomposites, combined with size effects during micromachining make them difficult to micromachine. High cutting forces, poor surface quality and high tool wear are a result of poor machinability. A thorough analysis of mechanical properties of several nanocomposites will be presented in the first part of these two-part reviews, while second part will focus on the micromachinability of materials.

9.5.4 Scaffold Architecture

Among several tissue engineering (TE) techniques, the scaffold is a critical to provide a temporary scaffold for generating new tissue, either through in vitro culture of scaffold cells before implantation or by directing scaffold cell growth in vivo. High cell

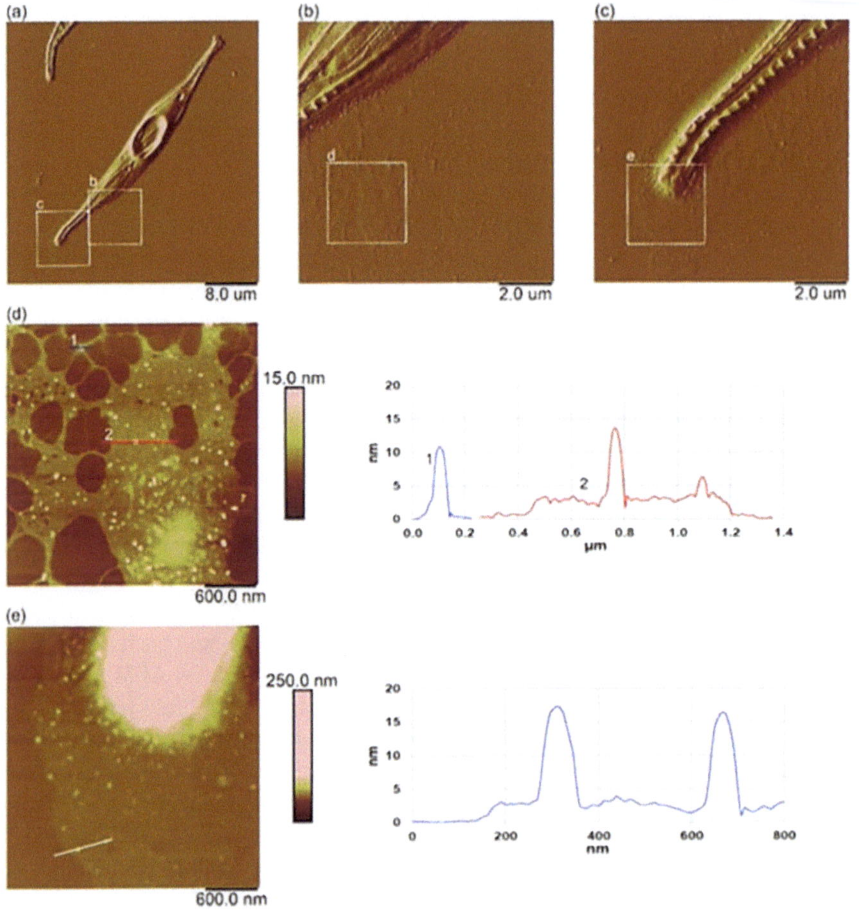

Fig. 9.11 AFM images showing Cylindrotheca closterium cells and EPS released with embedded SiO_2 nanoparticles, supplemented with SiO_2 (10 μg mL^{-1}). Contact mode is used to capture the images, which are displayed as height (**d, e**) and amplitude (**a, c**) data. EPS and heights of embedded nanoparticles are displayed as vertical profiles along indicated lines for following scan sizes: 40 μm × 40 μm (**a**); 10 μm × 10 μm (**b, c**); and 3 μm × 3 μm (**d, e**). The same color and number are used to represent line crossing EPS, nanoparticles, and vertical profile that belongs to indicated line in (**d**) [8] [Open access]

densities often occur before tissue formation during in vivo developmental processes. When designing 3D scaffolds for TE applications, this important biological factor must be taken into account. From prefabricated scaffolds to patient-specific implants, various processing techniques have been developed over the past few decades for the design and fabrication of 3D scaffolds. These scaffolds, manufactured using different technologies have different physical architectures such as porous, fibrous, irregular and homogeneous.

It is well known that scaffold design is important in controlling the spatial distribution of cells, which in turn influences the expression of cell signals and ultimately tissue development [2–4]. Cell migration and tissue ingrowth in scaffolds treated with standard techniques (porous leaching, gas foaming solvent flooding) are often limited to periphery, resulting in the creation of localized heterogeneous tissue [2] On the other hand, rapid prototyping (RP) scaffolds have low resolution, resulting in huge pore sizes and highly geometric designs that create an open pore architecture with 100% interconnectivity, which influences tissue growth [5]. Polylactic acid (LA) and polycaprolactone (PCL) are widely used in the scaffolding industry. PCL is the polymer of choice for RP procedures, but its inherent hydrophobicity and lack of functional groups are disadvantages due to poor cell attachment [6] Several surface modification strategies have been reported to improve the bioactive properties of PCL quite popular [5].

The diversity of functional groups and variable nature of induced surface properties are two major disadvantages of plasma technologies [18]. The use of plasma-treated surfaces as interfacial tie layers for subsequent immobilization of various chemical species or molecules has become the subject of recent research. In this way, more controllable surface chemistry can be achieved that is designed to trigger specific biological reactions [18, 20]. One of the most popular biomaterials for using biomedical applications is collagen (COL) [21]. AFM is performed on the top, middle and lateral regions of structure to gain insight into how wrinkle properties varied at different points within the 3D scaffold structure (Figs. 9.12 and 9.13). It is found that silk fibroin (SF) concentration affected the wrinkle wavelength (Figs. 9.12b) and amplitude (Figs. 9.12c) at each location within the scaffold (Figs. 9.12 and 9.13) [9].

However, there is a trend towards increasing surface area-modification of implants using COL or gelatin due to certain disadvantages [8, 22]. Fibronectin (Fn), is a glycoprotein, but it must remain immobilized in a biologically active form. This promotes early cell attachment. Fn has been reported to have a preference for gelatin [23]. Previous studies [9] have shown that PCL films and scaffolds can be successfully replaced using a multistep method that includes plasma pretreatment, UV-induced polymerization of 2-aminoethyl methacrylate (AEMA) and gelatin grafting immobilization and physisorption.

9.6 Biomaterials

Biomedical engineering is a discipline that largely involves the science of biomaterials. Biomedical engineering and biomaterials have adopted biology as the fundamental science upon which they are built, while engineering and materials science have consequently drawn their foundation from mathematics, physics and chemistry. This expands the reach of parent disciplines in a way that is currently not achieved by any other engineering field, at least not to the same extent as biomedical engineering or biomaterials community. Biological sciences adhere to strict schedule of important advances. As twenty-first century enters the biological age, biomedical

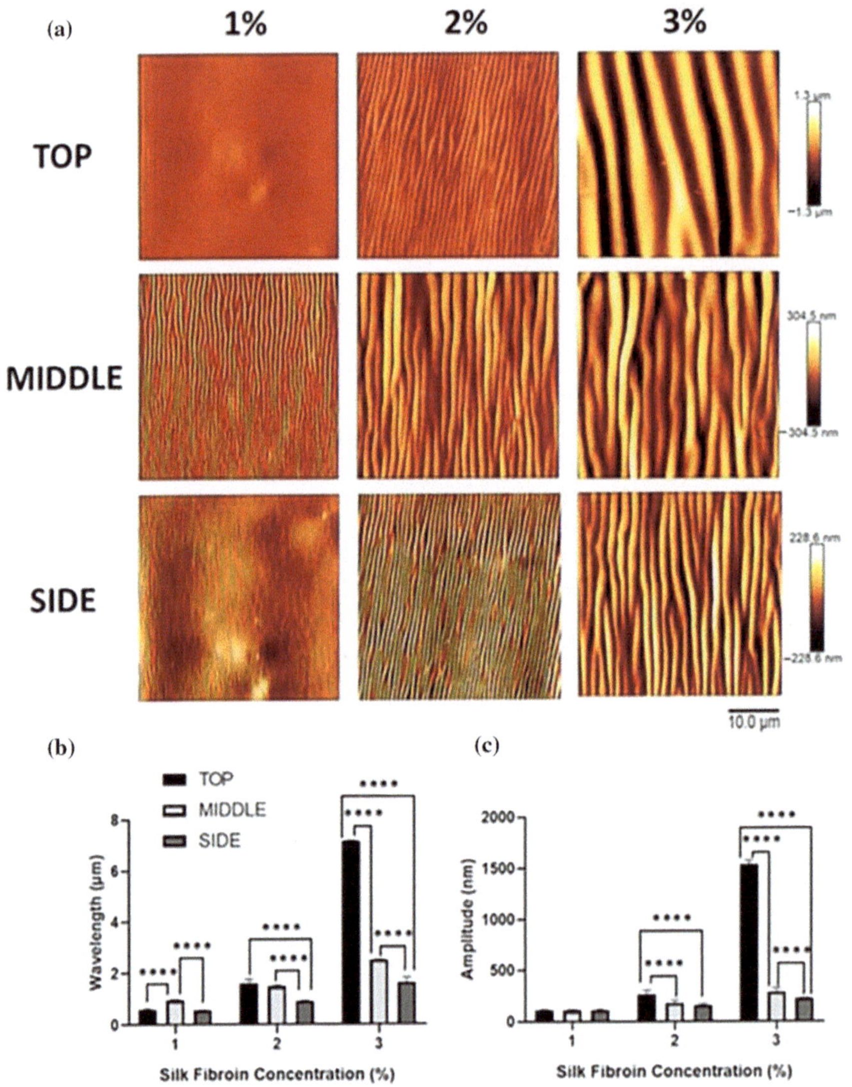

Fig. 9.12 At different points in the scaffold structure, wavelength and amplitude of silk wrinkles increase in tandem with increasing silk fibroin concentration and, therefore, film thickness. **a** AFM images show the wrinkled topography of silk recovered after heat-shrinking shape memory polymers (SMPs) scaffold at different silk concentrations (scale bar = 10 μm). Note that the pseudocolor range in each row is scaled to fit the data in that row. As silk concentration increases, wrinkle amplitude (**c**) and wavelength (**b**) increase at top, middle and side of scaffold. (n = 3, one-way ANOVA, $p < 0.0001$), and groups are then compared using Holm-Sidak multiple comparisons test [9] [Open access]

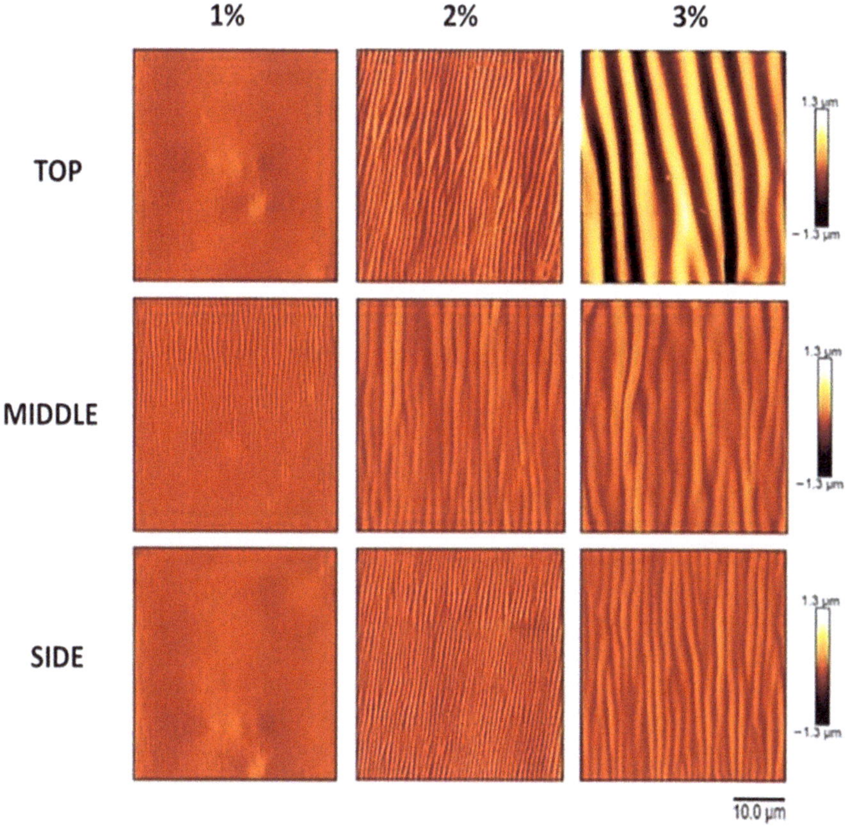

Fig. 9.13 The amplitude of wrinkles increases with increasing silk fibroin concentration (resulting in increased film thickness) and varies at top, middle and side of scaffold which appears as single pseudocolor band. SMP scaffold is heat-shrinked and treated with different silk concentrations as shown in AFM images of wrinkle topography (scale bar = 10 μm). Each image contains a single pseudocolor band as you can see [9] [Open access]

engineering and biomaterials advance this vital science and benefit society. Biomaterials and biomedical engineering are growing together as a global discipline. All over the world, departments dedicated to biomedical engineering are constantly being created and full-fledged undergraduate programs are being launched in this field [15].

Synthetic polymers such as poly(L-lactic acid), polyglycolic acid (PGA) and poly(lactic-glycolic) acid (PLGA) are used to create three-dimensional scaffolds for regeneration of damaged tissues [11]. Because these materials can be designed and their degradation properties can be controlled by changing the mixture of individual polymers or polymer itself, they have demonstrated remarkable success. One of its disadvantages is offering fewer opportunities for failure. The hydrolysis of PLLA and PLGA produces dangerous gases such as carbon dioxide, which lowers the pH of environment and causes cell necrosis [12].

3D scaffolds are created using biological materials or natural polymers as an artificial extracellular matrix (ECM) for tissue regeneration. Biological components and scaffolds are fabricated using materials such as collagen, silk fibroin chitosan and alginate-based substrates for using in various tissue engineering applications. Unlike scaffolds based on synthetic polymers, scaffolds based on natural polymers are physiologically active and significantly improve cell adhesion and proliferation. They are biodegradable, allowing host cells to gradually replace the degraded scaffold with their extracellular matrix (ECM).

9.7 Fabrication of 3D Scaffolds for Tissue Engineering

One of the simple and accessible methods used in 3D printing of polymers for bone tissue engineering is fused deposition modeling (FDM) [2]. Various thermoplastic polymers are used in 3D printing, including polycaprolactone (PCL) [3], poly(l-lactic acid) (PLLA) [4] and polyvinyl alcohol (PVA) [55]. Due to its FDA (Food and Drug Administration) approval, biocompatibility and biodegradability, PCL has been one of those frequently used to create bone scaffolds [56]. PCL has a low melting point, 60 °C for 3D printing . In addition, PCL has excellent mechanical properties including high flexibility and significant elongation at break [58], making it a promising scaffold material for the treatment of cranial deformities and cranio-facial bone loss [9]. However, optimization of PCL scaffold is achieved using 3D printing by incorporating additional characteristics including mechanical properties and antibacterial activity without compromising biocompatibility.

Infection complicates 2–5% of orthopedic implants [14] and in open fractures, this number can reach 30%. Implant-related infections are associated with serious illness and sometimes even death. These problems lead to implant removal, surgical debridement of affected tissue and ongoing antibiotic therapy [7]. One of the problems in imparting antibacterial properties to scaffolds is loading them with antibiotics or metal components. For example, created 3D-printed PCL scaffolds coated with an antibiotic (Cefazolin and macro- and micropores (produced using Porogen). Staphylococcus aureus (S. aureus) has effectively inhibited by these scaffolds [4]. Although this concept has great potential, the alarming rate of bacterial resistance to antibiotics suggests the need to develop alternatives to this strategy.

It is well known that the metal elements zinc, copper and silver have antibacterial properties [5]. In addition to widespread use as broad-spectrum antibacterial agent in research and clinical settings, silver nanoparticle-polymer composites have found applications as catalytic agents [16], supercapacitors [15], electromagnetic interference (EMI) shielding [18] and fuel cells [9]. Uncertainty surrounds the mechanism of action of silver nanoparticles (AgNps) against bacteria. However, recent research has suggested several plausible answers, such as: (i) Bacterial cell death occurs when AgNps attach to the bacterial membrane, causing membrane rupture and loss of cellular contents; (ii) Small nanoparticles penetrate the membrane and interact with biomolecules, destroying the bacterial cell; (iii) AgNps likely catalyze the reduction of O_2 to reactive oxygen species (ROS), which accumulate and cause DNA damage, apoptosis and suppression of antioxidant enzyme expression [17].

9.7.1 Techniques for Synthesising Nanofibers

The most significant external properties are possessed by nanofibers, which are solid fibers with two external dimensions in nanometers; diameter less than 100 nm and fiber length greater than the diameter. The length of nanofibers varies from a few nanometers to several micrometers and their diameter varies depending on the polymer and production method. Nanotubes, nanorods and conductive nanowires are terms for hollow, rigid and conductive nanofibers, respectively. In nanotechnology, polymer-based nanofibers are of primary interest, although many different types of nanofibers exist. Due to their high surface-to-volume ratio, nanofibers are an excellent alternative to conventional fibers. This characteristic improves their functionality making the materials stronger and more flexible. Consequently, this expands the potential applications of nanofibers for specific applications such as within larger structures such as scaffolds and nanomats.

Another important characteristic of nanofibers is their porosity, which affects things like water sorption, controlled chemical release, efficient exchange of biological signals and fluid transport. The porosity of the sample can be characterized as internal, bulk or surface property. The term "3D hollow gaps" describes the spaces created between nanofiber elements that are confined by fiber molecules. The hydraulic connection of pores creates continuous channels through which chemicals and liquids can pass through the nanofibers. The stiffness of nanofibers is often higher than that of traditional fibers. The mechanical properties of individual nanofibers can be improved compared to bulk or thin materials.

Improved alignment of polymer backbones along the length of nanofiber as well as intensification. The main factor that improves the mechanical properties of materials in the form of filaments is their strength when the fiber diameter is reduced. Orienting the molecular chains along the longitudinal axis will improve the tensile strength of the nanofiber. Surface cracks and nanofiber cracks damaged the fiber structure and facilitated mechanical damage [5].

9.7.2 Electrospinning: A Method for Producing Nanofibers

Chitosan nanofibers are produced by electrospinning that is affected by numerous factors including viscosity, operating voltage, temperature, pressure and flow rate. Since viscosity depends on how entangled the polymer molecule chain is in solution, viscosity is an important parameter for electrospinning. The electrospun chitosan nonwoven fabric is cleverly designed. The effect of chitosan concentration and electrospinning solvent on morphology of final nonwoven fabric is investigated. Trifluoroacetic acid (TFA) is found to be the most effective solvent because amino groups of chitosan form salts with TFA, which can effectively break the rigid bonds between chitosan molecules. With increasing chitosan concentration, the deposition shape of complex changed from spherical spheres to entangled fibrous networks.

Using dichloromethane and TFA as solvents to create electrospun chitosan nanofibers. In addition, they have used pure trichloromethane as a solvent to create porous electrospun chitosan nanofibers. FESEM is used to study the morphology and dispersion of nanoparticles on the surface of nanofibers (Fig. 9.14). Bead-free, bundled and randomly oriented nanofibers with an average diameter, 425 nm are present in all formulations. The surface of AP/PAN (Fig. 9.14b) and S-AP/PAN (Fig. 9.14c) nanofibers is no longer smooth compared to electrospun Na_2HPO_4/PAN nanofibers (Fig. 9.14a) since secondary nanoparticles are present uniformly attached to the surface of PAN nanofibers with some agglomerations. A high-magnification image of S-AP/PAN illustrating the fiber/particle interaction is shown in Fig. 9.14d. All the nanoparticles are firmly attached to nanofibers, which prevented them from falling off under the working conditions. In addition, the cross-sectional FESEM image is used to measure the thickness of the nanofiber mat (Fig. 9.14e). Based on the appearance, the estimated thickness of the S-AP/PAN nanofiber mat is 38 μm. The color change of Na_2HPO_4/PAN nanofibers from white to yellow, indicating the development of Ag_3PO_4 nanoparticles on the surface of PAN nanofibers, is depicted in the appendices (Fig. 9.14a–c). In addition, FESEM-EDS is used to investigate the elemental composition of composite nanofibers. AP/PAN spectra shown in Fig. 9.14f to confirm the composition of sample from Ag_3PO_4 and PAN, showing a significant amount of C, O, P and Ag. C (carbon), O (oxygen), P (phosphorus), S (sulfur) and Ag (silver) that present in considerable levels in S-AP/PAN FESEM-EDS (Field Emission Scanning Electron Microscopy—Energy Dispersive X-ray Spectroscopy) spectra (Fig. 9.14c). With no sign of further impurity components, this implies the presence of SO_4^{-2} (sulfate) ions in Ag_3PO_4 (silver phosphate) nanoparticles. The geographical distribution of these elements is further demonstrated by the elemental mapping of S-AP/PAN (Fig. 9.15). In particular, the consistent distribution of O, S, P and Ag across the nanofibers indicates that Ag_3PO_4 nanoparticles contain SO_4^{2-}, verifying their presence and even spread on the nanofiber surface [34].

A disadvantage of electrospray is that long-chain natural polymers cannot be used to separate the spray into nanoparticles; however, this same property can be used to create nanofibers using electrospinning.

9.7.3 The Electrospinning Process's Parameters

Numerous elements influence the electrospinning procedure. These elements fall into three categories: environmental, solution and electrospinning parameters. Needle diameter, flow rate, needle-to-collector distance and applied electric field are

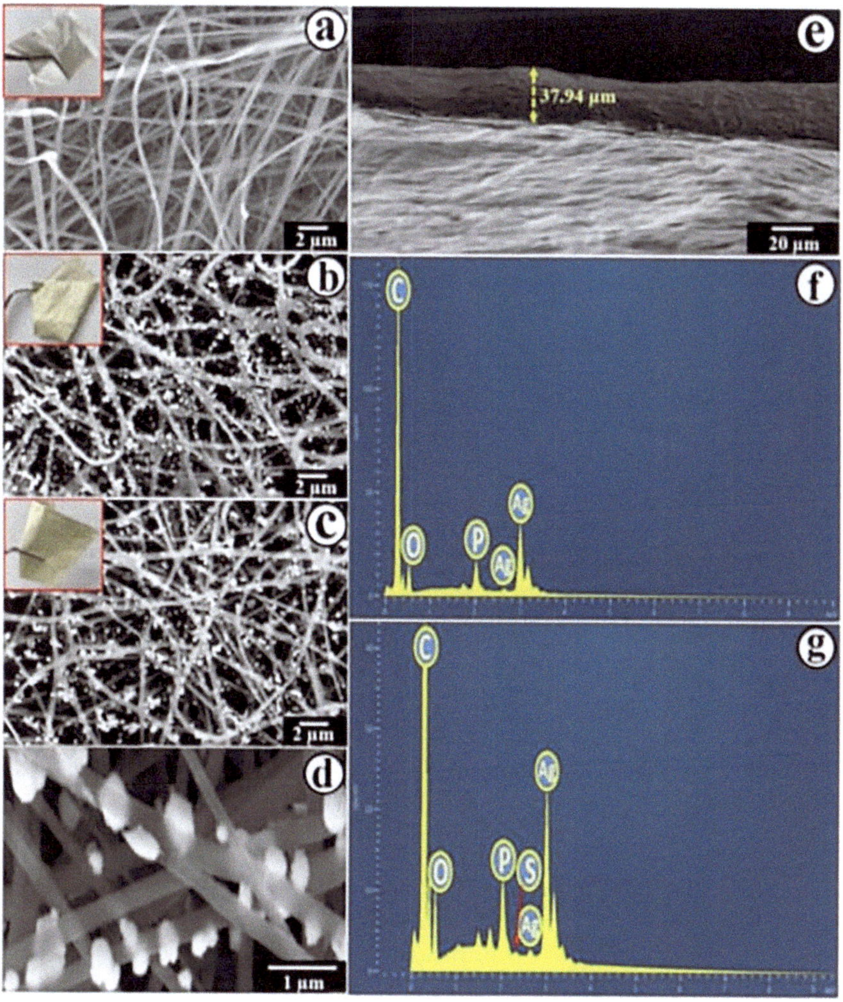

Fig. 9.14 FESEM images of **a** Na_2HPO_4/PAN nanofibers, **b** AP/PAN, **c** S-AP/PAN, **d** a high-magnification image of S-AP/PAN, and **e** a cross-sectional image of S-AP/PAN showing thickness of nanofiber mat. The corresponding EDS of (**b**) and (**c**) panels (**f**) and (**g**), respectively. A digital image of the comparable nanofibers is included in the appendices [34] [Open access]

among the electrospinning parameters. Solution parameters include solvent, polymer concentration, viscosity and conductivity of the solution. Temperature and relative humidity are some of the environmental parameters. Each of these variables directly impacts the production of electrospun fibers that are smooth and bead-free. Therefore, it is critical to fully understand the significance of each of these regulatory parameters to better understand the electrospinning technique and the production of polymer nanofibers.

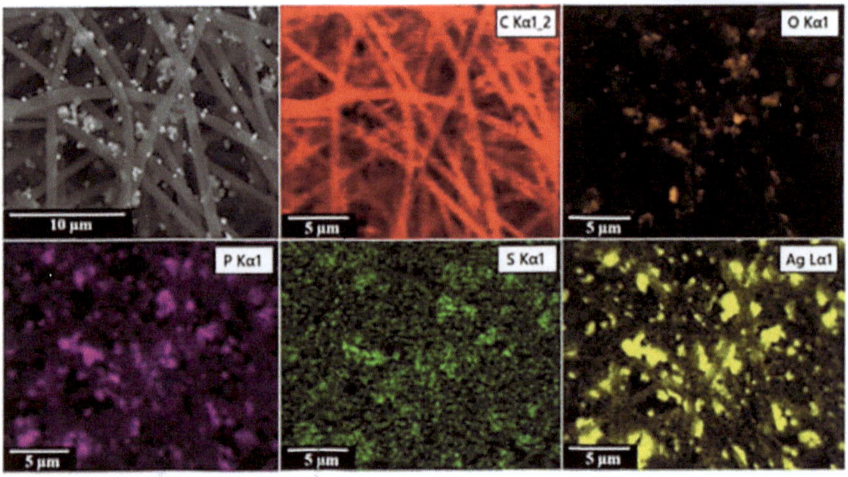

Fig. 9.15 S-AP/PAN elemental mapping [34] [Open access]

System and process parameters are two main sets of parameters that have a significant impact on electrospinning processes:

- The degradation rate of electrospun nanofibers and other characteristics of polymer solutions such as conductivity, viscosity and surface tension, which contribute to the reduction of bead formation are determined by system factors such as molecular weight of polymers and their distribution.
- The diameter of nanofibers is influenced by process variables such as needle tip diameter, pump flow rate and voltage applied to solution. The distance between the tip of needle and metal collector on which the nanofibers are deposited determines the amount of solvent evaporation from the electrospun nanofibers [8].

9.8 Polymeric Materials Using Electrostatic Process to Manufacture Nanofibers

9.8.1 Organic Polymeric Substances for Nanofibers

A relatively new but well-established scientific field is nanotechnology. It is common knowledge that most materials behave differently at molecular and nanoscale sizes than in bulk materials. It has been proposed that scientists use these qualities in a variety of fields [22]. Through the application of nanotechnology-based structures, biomedical researchers have been able to take advantage of several opportunities, especially in drug delivery, by exploiting the physical, chemical and biological features of these nanoscale delivery systems [4]. The special properties of nanoparticles such as controlled release of their contents, biocompatibility, biodegradability and improved cellular uptake have generated great interest in the development of innovative drug delivery systems (NDDS). Interaction with biological environment and drug penetration into the cell are two important factors to consider when creating NDDS. In this context, many nanosystems have been reported using different types of nanocomposites such as nanofibers [4, 6–8].

Polymer-based nanofibers, a broad class of materials comprising natural, synthetic and/or hybrid polymer structures with nanoscale fiber-like shapes, represent significant and promising benefits in medical research. Their exceptional properties include large surface area, high porosity, high encapsulation efficiency, ease of functionalization and manipulation and enhanced thermal and chemical stability [5, 7]. The extent to which materials can move within these nanosystems is greatly enhanced by the larger surface area of polymer nanofibers [32]. Polymer nanofibers are created using a variety of approaches and performance of final products is greatly influenced by the method used. Among the most widely used methods are drawing processes, phase separation procedures, self-assembly, template synthesis, lyophilization synthesis, interfacial polymerization and electrospinning of nanofibers and spinneret-based tunable engineering parameter (STEP).

9.8.2 Artificial Polymeric Substances for Nanofibers

In the research and industrial sectors, polymer-based nanofibers have received much attention as an important class of materials. Polymer nanofibers with submicrometer diameters, <1 μm, have special characteristics such as high surface area per mass, which makes it easier to add functional properties to the surface for specific applications. The most common methods used to create polymer nanofibers are electrospinning, drawing techniques, phase separation/inversion, solution blowing (air jet spinning), power spinning (rotational spinning), electrospinning, template synthesis and interfacial polymerization of nanofibers. Due to its mass production capabilities, affordability and ability to create continuous fibers from a variety of polymers, electrospinning is the most widely used method. However, the pros and cons of synthesis methods will also be discussed. In addition, the methods for characterizing polymer nanofibers and effective influencing factors on nanofibers will be proposed.

Polymers, both synthetic and natural, can be used to create polymer nanofibers. Polymer blends are used to enhance conductivity in sensors and brain tissue engineering, as well as biodegradability for medical applications. This chapter will review recent advances in the application of polymer nanofibers as scaffolds for tissue engineering, wound dressings, drug delivery, filters, protective clothing, sensors and reinforcement of composite materials [14].

9.9 Applications: Tissue Engineering Using Electrospun Nanofibers

Tissue healing based on transplantation is one of the most promising methods of tissue regeneration. In a transplant, a tissue or an organ is removed from one person (donor) and surgically transplanted into another (recipient) or may be moved within the same person from one location to another. Allogeneic and autologous grafts are two categories of grafts. The organs or tissues of transplant recipient are called autologous grafts. Its most common uses include the treatment of blood malignancies such as leukemia and lymphoma. Donor site morbidity and limited availability are some of disadvantages of autologous procedures. An alternative to autografts is allograft transplantation, which involves transferring tissue or an organ from one person to another. Allografts are used to promote healing and restoration of damage to skin, cornea, heart valves, tendons and limbs.

However, they can cause an immunological reaction and pose a risk of spreading the disease. Consequently, tissue engineering (TE) has become an excellent method to replace or repair damaged tissues and can overcome all the disadvantages associated with allogeneic and autologous tissue repair [1]. TE is an emerging interdisciplinary field that uses concepts from biology and engineering to achieve tissue regeneration [2]. TE techniques use biomaterials, cells, or a combination of both to preserve, restore or improve tissue function. For example, thermally sensitive polymer brush coatings (TRPBCs) are among advanced biomedical applications that act as a mixture of biomaterial and cells for treating TE [5].

9.9.1 Bone Tissue Engineering Using Nanofibers

This focuses on the use of nanofiber composites for bone tissue engineering. In particular, the first section provides an overview of the subject as well as definitions of the most commonly used terminology. The biomimetic approach underlies the development of nanofiber composites for bone tissue engineering, the choice of organic phase as the filler, new trends in electrospinning process and combination of the electrospinning method with other scaffold fabrication methods [4, 8]. Nonwoven PCL nanofiber scaffolds for bone tissue engineering are fabricated by electrospinning [9]. To understand mesenchymal stem cells (MSCs) derived from the bone marrow of newborn rats, they are seeded on nanofibrous scaffolds to study their behavior. The results showed that MSCs proliferated on the nanofibrous scaffold and caused scaffolds to produce an abundant extracellular matrix. The synthesis and mineralization of collagen are investigated in combination with these PCL nanofibers.

These results effectively illustrated the potential applications of PCL-based nanofiber scaffolds in hard tissue engineering [10]. HA, and β-tricalcium phosphate (β-TCP) are mixed to form a porous scaffold that is biodegradable and appropriate for the regeneration of injured bone using a combination of gel casting techniques

[11]. The polymer sponge method has been effectively used to create porous β-TCP/HA scaffolds, which consist of HA nanofibers integrated into a β-TCP matrix. These in vitro studies demonstrated that mechanical strength of porous composite scaffolds is effectively increased by adding HA nanofibers as a second component in β-TCP. This is demonstrated the feasibility of using nanocomposite hydroxyapatite nanofibers as a scaffold material to improve the mechanical performance of load-bearing tissues such as bone.

9.9.2 Cartilage Tissue Engineering Using Nanofibers

Complete regeneration or restoration of damaged cartilage tissue requires a lot of effort. Cartilage tissues lack the regenerative properties of other tissues because they are not vascularized and because chondrocytes are confined to lacunae, which limits their ability to migrate (skin and bones). The purpose of this chapter is to describe the many critical factors that influence the creation of nanofiber composites including selection of materials, fabrication methods, surface treatments, cell sources, growth factors and so on. Two different fibrous scaffolds are fabricated for potential using in cartilage tissue regeneration. Fmoc-valyl acrylamide (FVC) fibers are used as templates to create both scaffolds, which are subsequently attached to ECM mimics. It is requested to compare the effect of varying scaffold composition on their ability to adhere to chondrocytes and promote their proliferation. Figure 9.16a shows the SEM image of FVC template. Individual fibers with a diameter of 200–500 nm define the FVC template. These nanofibers exhibit strong aggregation and appear to create matrix-like structures by stacking on each other. The long hexadecyl carbon chain of FVC, which allows self-assembly due to its strong hydrophobic properties, likely gives it its fibrous texture. In addition, π–π stacking interactions mediated by aromatic groups from the Fmoc moiety promote fiber formation. Figure 9.16b–e shows further addition of each layer to prepare Construct I [33].

Scaffold-based tissue engineering has emerged as a crucial method to repair or regenerate damaged or wounded cartilage tissues to overcome this inherent failure. Under this regime, there has been great interest in creating complex biomimetic scaffolds to replace natural cartilage tissue. Nanofiber composites are a popular scaffold material because they closely match the extracellular matrix found in nature and can support cell proliferation and direct differentiation during new tissue formation. The fibrous element improves mechanical and surface properties, while other components such as polymers, ceramics, metals and quantum dots provide improved functionality and biocompatibility. However, by making further chemical or physical changes, these composite scaffolds can be tailored to a specific tissue. The biological significance of these factors for applications in cartilage tissue engineering is being explored. For the convenience of readers, an overview of composition and function of articular cartilage is given [9]. In another work, the same cells were cultured in a monolayer on a polystyrene surface and then a serum-free medium was created, which is a more sulfated cartilage matrix rich in proteoglycans. These

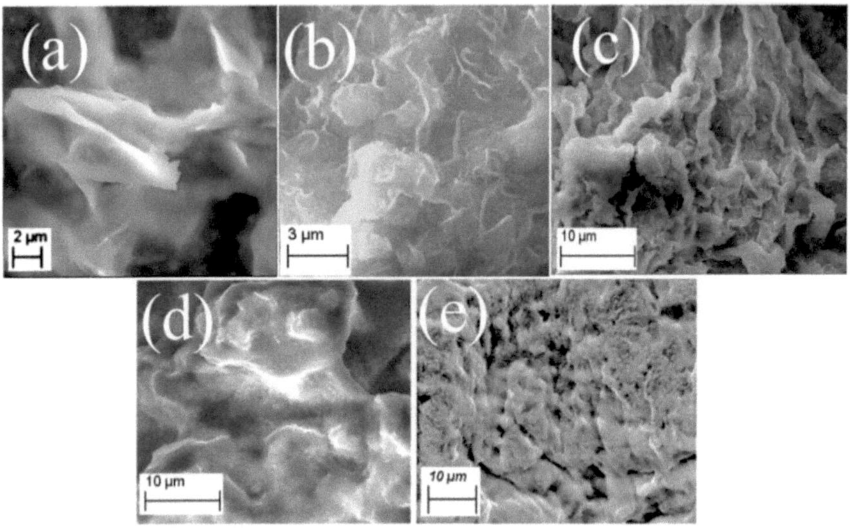

Fig. 9.16 SEM of layer-by-layer architecture of Construct I. FVC fibrous assemblies (**a**); FVC fibers coated with DSp peptide sequence (**b**); collagen type I (Col-I) incorporated into FVC-DSp (**c**); chitosan (Ch) incorporated into FVC-DSp-Col-I (**d**); and the final Construct I scaffolds after attachment of chondroitin sulfate to FVC-DSp-Col-I-Ch (**e**). [33] [Open access]

results indicate that the scaffold morphology and culture medium composition affect the organic molecularity of FBCs. For this reason, there is a trend to further explore Polycaprolactone (PCL) fibers as versatile scaffolds for tissue engineering [10].

To enhance the regenerative effect of cartilage, modified 3DS-2 scaffold cross-linked with hyaluronic acid is successfully created. For the regeneration of cartilage tissue, porous 3D scaffold 3DS-1 based on electrospun gelatin/PLA nanofibers is prepared. Additionally, in vivo cartilage regeneration study in rabbits using an articular cartilage injury model has shown that 3DS-2 could improve cartilage repair. 3D wireframes are used in this study. The results showed that 3DS-1 and 3DS-2 scaffolds have excellent cytocompatibility and superabsorbent properties. The current 3D nanofiber scaffold 3DS-2 may be promising in the field of cartilage tissue engineering [11].

9.9.3 Ligament Tissue Engineering Using Nanofiber Technology

Over the past few decades, the tissue engineering and regenerative medicine have shown great promise in treating tissue damage and degeneration that can cause serious health problems. These approaches are intended to improve clinical outcomes in the treatment of these conditions. This is especially important because human tissue does not regenerate very well [12]. The goal of emerging discipline of regenerative engineering is to improve and regenerate damaged tissue while restoring its functionality. The fabrication of artificial scaffolds has been revolutionized by significant advances in materials science, tissue engineering and regenerative medicine. These matrices are designed to harness electrochemical signals, promote cell adhesion and promote biosorption [13].

An important factor in the development of nano- or microstructured materials, known as scaffolds, which are critical for controlling cell behavior and promoting repair of damaged tissues [14]. Because of they can replicate the natural extracellular matrix (ECM), materials made from electrospun fibers with dimensions ranging from nanometers to micrometers are especially desirable [15]. High surface area to volume ratio, good porosity, fiber uniformity, compositional diversity, flexibility and ease of functionalization with bioactive compounds are just some of the unique qualities that electrospun fiber scaffolds can offer. These scaffolds are excellent candidates for the delivery of proteins, growth hormones and antibiotics because they can effectively regulate the rate of release of bioactive compounds [16]. There are several ways to incorporate these bioactive chemicals into the electrospinning process; blending, emulsion, coaxial electrospinning or even surface functionalization during subsequent electrospinning [17].

As a result, electrospinning is now recognized by scientists as a useful additive manufacturing technology. This allows precise control of fiber deposition, facilitating the creation of nano- or microstructured materials known as scaffolds. Scaffolds are essential to control cell behavior and promote tissue regeneration [26]. Ability to control properties such as hydrophilicity, stimulus–response, and nano- or micro structuring, such as microwires [24], nanotubes [25], nanobelts [26], nanofibers [26], core–shell [27], multilayering [28] or two- or three-dimensional assembly from one-dimensional structures allow the electrospinning of nanofibers to have noticeable advantages. This allows for complex geometries and structures [24].

9.9.4 Nanofibers for the Engineering of Skeletal Muscle Tissue

Traumatic injuries, aggressive tumor ablation and chronic denervation are common clinical circumstances that often result in significant muscle tissue loss and require surgical restoration by injecting healthy donor tissue. However, this strategy can result in significant donor site morbidity. For example, musculoskeletal diseases such as Duchenne muscular dystrophy and spinal muscular atrophy could be treated using tissue engineering procedures that produce skeletal muscle tissue. The resulting tissue is usually of poor mechanical quality because damaged soft tissues, including skeletal muscle, often do not regenerate effectively. 3D skeletal muscle structures represent a viable way to address this issue by restoring body shape and soft tissue function.

In many cases, many more cells are required to compensate for significant tissue defects. It is often ineffective to directly inject muscle progenitor cells into the defective area due to their poor integration and survival rates. Even if the passage number is increased, the proliferation and differentiation capacity of these cells can be increased by co-culture with multipotent mesenchymal stem cells and providing them with sufficient growth factors. A core–shell nanofiber composite structure is used to create an aligned scaffold that more closely resembles native skeletal muscle. This scaffold is made of core nanofiber yarn (NFY) encapsulated in a photocurable hydrogel shell using a simple and efficient electrospinning technique. Similar regenerative approaches are being explored in cardiac tissue engineering in addition to skeletal muscle repair. When cardiac muscle is damaged, fibroblasts, endothelial cells and macrophages are attracted to the area by inflammatory signals, causing tissue scarring. The mechanical function of the heart is impaired by this scar tissue, reducing the heart's ability to pump blood effectively and ultimately leading to heart failure. To combat this, scientists are exploring the use of cardiac patches that inhibit scar tissue formation while promoting cardiac regeneration. The mechanical and electrical properties of graft or patch are critical in this situation. To achieve successful signal propagation and functional recovery, the synthetic tissue must cooperate with native myocardium by matching its mechanical and electrical resistance (Fig. 9.17) [30].

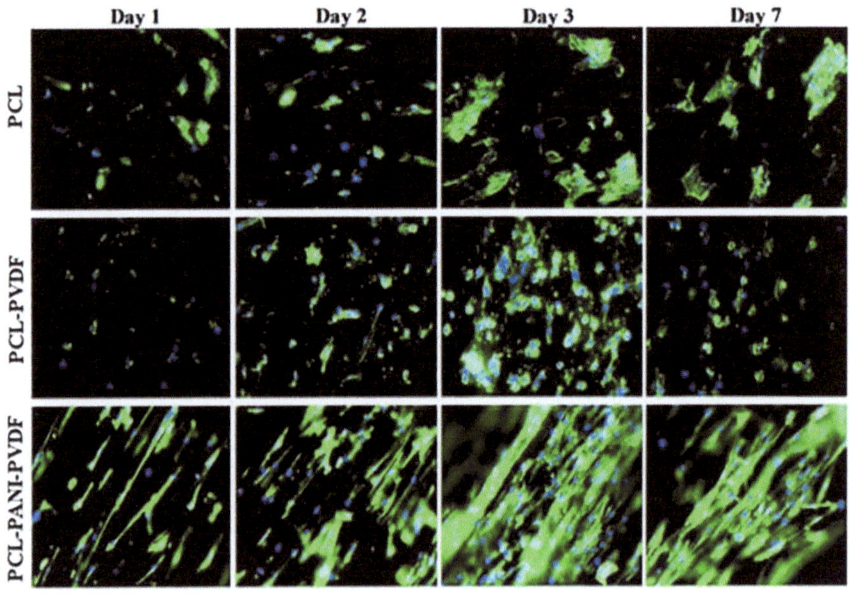

Fig. 9.17 rat cardiomyoblasts (H9c2) on PCL, PCL-PANI and PCL-PANI-PVDF scaffolds on different days are compared. green: Alexa fluor 488 Phalloidin for F-actin filaments; blue: DAPI nuclear staining [30] [Open access]

However, the generation of functional muscle tissue constructs requires in vitro culture of muscle progenitor cells on an appropriate matrix for tissue engineering of the three-dimensional skeletal muscle suitable for therapeutic purposes. Such a matrix can be created by electrospinning nanofibers, which expand the possibilities of biomaterials research. Regarding this procedure, several polymers can be electrospun to produce synthetic, composite, natural nanofibers or a mixture of two types of nanofibers [30].

In addition, topographical or mechanical modification of the nanofibers is possible. The ability of conductive polymers to stimulate myogenic differentiation has made them very popular recently. A scaffold promotes fiber alignment, which is a very promising method for skeletal muscle regeneration [16]. For the first time, it is reported that electrospun PAN-PANi nanofiber membrane could be used as an effective scaffold for muscle regeneration. Unlike electrospun PAN nanofibers alone, PAN-PANi composite nanofiber membrane controls a higher rate of cell proliferation and differentiation. According to these findings, conductivity, mechanical properties and surface modification of electrospun PAN-PANi nanofibers play a crucial role. They offer a fresh perspective on the most biocompatible scaffolds for muscle tissue engineering [17].

9.9.5 Nanofibers for Engineering Skin Tissue

To repair damaged tissues, polymer nanofibrous scaffolds can interact and control certain regenerative processes at the molecular level. Numerous techniques can be used to create nanofibers, including electrospinning, phase separation and self-assembly. The most popular method for creating nanofibers for using in tissue engineering is electrospinning. Due to their structural similarity to the natural extracellular matrix, electrospun ultrafine fibers can be modified to exhibit appropriate pore distribution, high surface area to volume ratio, cell adhesion and proliferation. Since they can stop the loss of fluid and protein from the wound area, promote exudate removal, inhibit microbial infection, have excellent anti-adhesive properties and promote endogenous cell proliferation and remodeling, electrospun polymer nanofibers have several advantages such as skin substitutes. Currently, nanofibrous scaffolds are produced in combination with cells, growth factors and/or both to promote carefully orchestrated wound healing. This details the application of electrospun nanofiber scaffolds for skin tissue engineering. In addition, new developments in the creation of biocomposite nanofiber scaffolds electrospun-coupled cells and electrosprayed stem cells for skin tissue engineering are also described [17–20]. First, the raw AFM images in different regions of the nanofiber mat used are shown in Fig. 9.18. The top surface of nanofiber mat is captured in Fig. 9.18a, its bottom surface (during electrical connection to the polypropylene substrate) is captured in Fig. 9.18b, its outer boundary is captured in Fig. 9.18c, and its outer edge is captured in Fig. 9.18d. Using the sample as an example, Fig. 9.19 shows the results of the post-processing operations. When comparing these images with the original image

(Fig. 9.18a), it becomes clear that the polynomial background subtraction, the first step, significantly changes the luminance distribution. As a result, it is reasonable to expect that it will change the grayscale-resolved Hurst exponent distributions. Finally, the sharpening step can be expected to have some effect as well as shown in Fig. 9.19c and d [35].

Human dermal keratinocytes are shown to proliferate and infiltrate the electrospun scaffold [19–22]. Furthermore, a novel approach is developed to fabricate gelatin-GAG hybrid nanofibrous mats using response surface methodology to optimize the electrospinning process. The potential application of the gelatin-GAG hybrid nanofibrous mat in skin tissue engineering is demonstrated by culturing high-density human dermal fibroblasts [23–25]. In the fifth step, the primary gland is used to extract the silk sericin protein. This protein is combined with polycaprolactone (PCL) to fabricate the nanofibrous scaffold using the emulsion electrospinning method. The effect of silk sericin content in the hybrid nanofibrous scaffold on its morphology and proliferation of primary human dermal fibroblasts (FEK4) for 1–5 days is investigated. The results are shown in a rectangular cell shape and revealed an increased proliferation of FEK4 fibroblasts with a sufficient proportion of electrospun PCL/SS nanofibers [27].

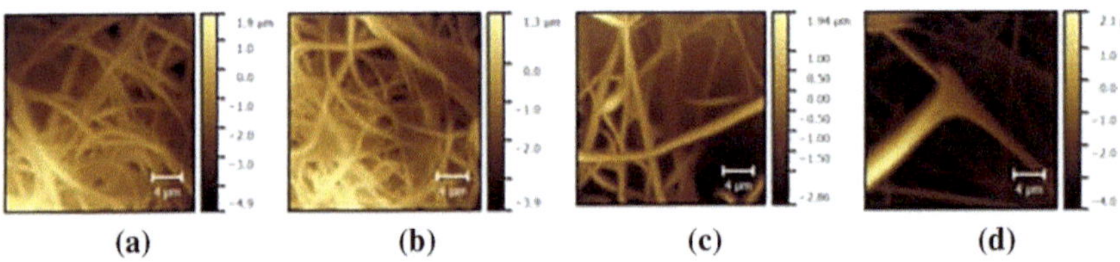

Fig. 9.18 AFM images (**a–d**), described in detail in text are obtained at different locations on electrospun nanofiber mat [35] [Open access]

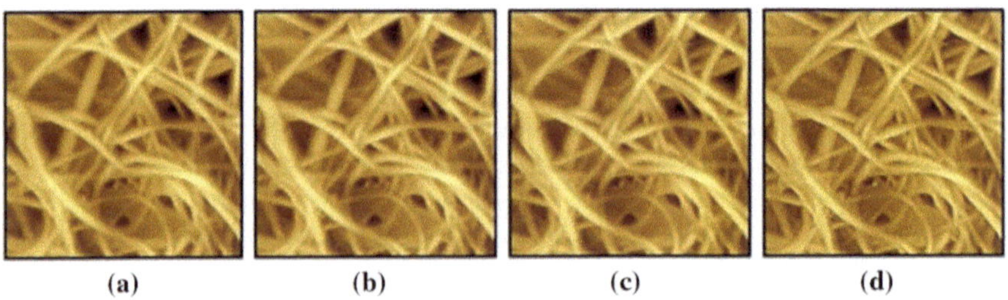

Fig. 9.19 Sample A underwent the following post-processing procedures: (**a**) polynomial background subtraction; (**b**) line alignment; (**c**) horizontal error removal and (**d**) one sharpening [35] [Open access]

9.9.6 Nanofibers for Engineering Vascular Tissue

Blood arteries are a special type of conduit that help remove ARETE products from tissues and transfer nutrients and oxygen from one tissue to another throughout the body. These channels resist high pressure and shear stresses, which control blood flow and permeability, prevent thrombosis under basal conditions and essential for immunological reactions. One of the main vascular diseases, atherosclerosis, causes tissue ischemia by blocking blood arteries. To restore blood flow in this scenario, the blocked blood artery must be replaced, and angiogenesis is required to remodel and regenerate the damaged tissue at microcirculatory level.

An ideal tissue-engineered vascular graft (TEVG) of moderate diameter would resemble the mechanical and biological characteristics of autologous vasculature and be capable of operating in the hemodynamic environment of small-diameter vessels, which is characterized by increased resistance and low flow [28]. Developing a functional TEVG will require satisfying several design requirements. For TEVG to be associated with fully grown and functional endothelium, it must be biocompatible in the sense that it is non-immunogenic and non-thrombogenic. To withstand hemodynamic stress both immediately after implantation and over time, it must have mechanical properties similar to those of the native vessel. It has to be resistant to long-term deformation, which could lead to an aneurysm. Successful tissue engineering and regenerative medicine (TEV) also requires a good healing response to reduce inflammation and prevent intimal hyperplasia and fibrosis.

To promote remodeling and allow endothelial development and vascular smooth muscle cell (SMC) infiltration for efficient integration into the vasculature, physiologically active TEVG would be beneficial. Finally, from a surgical perspective, TEVG must be strong enough to hold sutures for implantation and flexible enough to be manipulated [29–31]. They have concluded that by varying the spacing between the patterned scaffold structures, such 3D porous structures can effectively constrain the mechanical properties of the electrospun-aligned nanofibrous scaffold for blood vessel regeneration in both longitudinal and transverse directions. In addition, they have suggested that electrospun nanofiber scaffolds having heterogeneously and anisotropically aligned structures can effectively constrain the mechanical performance and biological activity of the scaffold [33].

9.9.7 Nanofibers for Engineering Neural Tissue

Catastrophic conditions include lesions of the peripheral or central nervous system (CNS) in severely ill patients. The researchers have found that the substrate structure affected cell morphology. Cells grown on larger diameter fibers have exhibited pseudopodia rich in microfilaments attached to individual fibers (Fig. 9.20a–c), whereas cells grown on finer fibers are more dispersed and have a smooth muscle-like morphology (Fig. 9.20d), comparable to cells grown on glass substrates (Fig. 9.20e)

[26]. This may be explained by the fact that the fiber substrates are rougher than smooth glass and tissue culture-treated polystyrene substrates, which are generally considered the gold standard and thus provide a larger surface area for cell adhesion [32].

By limiting nerve regeneration in the spinal cord, resulting in the formation of dense scar tissue at the site of injury, damage to the central nervous system results in irreversible loss of neurological function. The original tissue architecture required for signal transmission is disrupted as a result of scar tissue inhibiting axonal growth

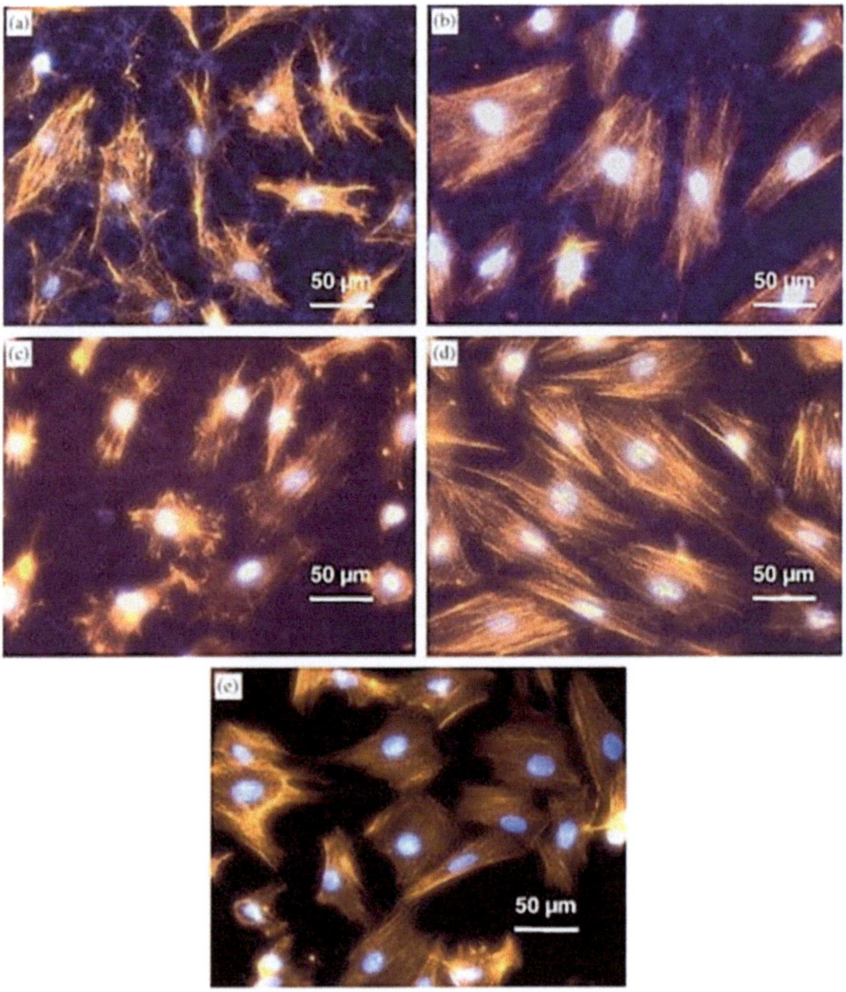

Fig. 9.20 Characteristics of muscle cells 20 h after seeding on the following surfaces: **a** glass scaffolds; **b** 15:85 gelatin-anisole nanofibers; **c** 30:70 gelatin-anisole nanofibers; **d** 45:55 gelatin-anisole nanofibers; and **e** gelatin nanofiber mats. Actin cytoskeleton (phalloidin) and nuclear (bisbenzimide) staining [32] [Open access]

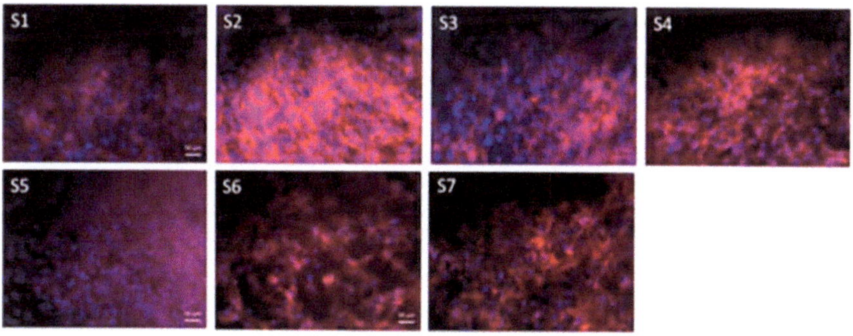

Fig. 9.21 fluorescence images of ST-2 cells after 7 days of incubation on self-woven nanofiber mats with different PCL/PGS ratios S1–S7 [32] [Open access]

through the gap. To restore the ability of the axon to grow at the site of injury, bridges, such as pipelines, from natural or synthetic materials [32–34] are required. All the nanofiber composite mats have shown the release of bioglass particles, indicating their potential utility for the release of therapeutic ions. However, PGS is found to reduce cell viability (samples S1-S4 in Fig. 9.21) when a lineage of stromal cells is seeded on these substrates compared to pure PCL, with and without bioglass (samples S5-S7 in Fig. 9.21). This is attributed to the degradation of PGS from nanofiber surface, which in turn reduces the adhesion ability of the cells [32].

The potential of electrospun nanofibers as fibrous 3D scaffolds that can serve as a synthetic extracellular matrix for various cell types to repair lost or damaged tissue is enormous. Synthetic ECM is created using various methods and synthesized nanofibers. Three key methods for creating nanofibers are electrospraying, electrospinning and phase separation or self-assembly. Of these three methods, electrospinning has been studied the most and produces the most promising results. Several biomaterials such as synthetic and natural polymers are used to create electrospun nanofibers. Consequently, these electrospun nanofibrous scaffolds are currently being explored for applications in musculoskeletal tissue engineering, including skin, vascular and neural tissue engineering as well as for the controlled delivery of anticancer drugs and biomolecules (proteins, DNA and RNA). In summary, these results demonstrate the superior properties of electrospun nanofiber scaffolds, including their high surface area and porosity for enhancing cell adhesion and migration, leading to the generation of novel tissues or organs for a range of tissue engineering applications [31].

With their unique properties, nanofibers, which range in diameter from 1 to 1000 nm, are revolutionizing tissue engineering. Their roughness and large surface area mimic real tissue very well, making them ideal for stimulating cell proliferation and creating artificial tissues. Cell development-promoting nanofibers should provide a stable and favorable environment that stimulates cell differentiation and proliferation for successful tissue engineering. The development of nanofibers that resemble the microstructure of natural tissues can improve the biological interactions and capabilities of artificial tissues. Enhanced bioconductivity nanofibers improve arterial and nutrient transmission, maintaining the health and viability of transplanted tissues. The use of nanofibers in the production of replacement tissues such as skin, muscle and bone confirm their potential to advance regenerative medicine. Nanofibers are a significant advancement in tissue engineering due to their special qualities and capabilities that promise better treatment outcomes and faster tissue regeneration.

Problems

1. What are the specifications of 3D biomaterials or scaffolds, mentioning the composition of the material?
2. List some important applications: For the use of electrospun nanofibers in tissue engineering
3. How will the characterization and nanofiber techniques have used in vascular tissue engineering work?
4. How many factors determine the use of diversified fiber technology in ligament tissue engineering?
5. Describe the process of interaction of different nuclear forces of nanofibers used in structural ligament tissue engineering with each other.

References

1. Mozaffari, A., & Parvinzadeh Gashti, M. (2022). Air plasma functionalization of electrospun nanofibers for skin tissue engineering. *Biomedicines, 10*(3), 617.
2. Sambani, K., Kontomaris, S. V., & Yova, D. (2023). Atomic force microscopy imaging of elastin nanofibers self-assembly. *Materials, 16*(12), 4313.
3. Kwon, O., Lee, J., Son, H., & Park, J. (2024). Quantitative understanding of ionic channel network variation in nafion with hydration using current sensing atomic force microscopy. *Polymers, 16*(5), 604.
4. Farniev, V. M., Shmelev, M. E., Shved, N. A., Gulaia, V. S., Biktimirov, A. R., Zhizhchenko, A. Y., & Kumeiko, V. V. et al. (2022). Nanomechanical and morphological AFM mapping of normal tissues and tumors on live brain slices using specially designed embedding matrix and laser-shaped cantilevers. *Biomedicines, 10*(7), 1742.
5. Romo-Valera, C., Guerrero, P., Arluzea, J., Etxebarria, J., de la Caba, K., & Andollo, N. (2021). Cytocompatibility and suitability of protein-based biomaterials as potential candidates for corneal tissue engineering. *International Journal of Molecular Sciences, 22*(7), 3648.
6. Guzzo, S., Carli, S., Pavan, B., Lunghi, A., Murgia, M., & Bianchi, M. (2021). Evaluation of the in vitro biocompatibility of PEDOT: Nafion coatings. *Nanomaterials, 11*(8), 2022.
7. Chang, P. Y., Wang, J., Li, S. Y., & Suen, S. Y. (2021). Biodegradable polymeric membranes for organic solvent/water pervaporation applications. *Membranes, 11*(12), 970.
8. Vukosav, P., Pašalić, L., Bakarić, D., Domazet Jurašin, D., & Mišić Radić, T. (2023). Interaction of silica nanoparticles with microalgal extracellular polymers. *Water, 15*(3), 519.

9. Oguntade, E., Fougnier, D., Meyer, S., O'Grady, K., Kudlack, A., & Henderson, J. H. (2024). Tuning the topography of dynamic 3d scaffolds through functional protein wrinkled coatings. *Polymers, 16*(5), 609.
10. Cheng, Y., Koh, L.-D., Li, D., Ji, B., Han, M.-Y., & Zhang, Y.-W. (2014). On the strength of β-sheet crystallites of Bombyx mori silk fibroin. *Journal of the Royal Society, Interface, 11*, 20140305.
11. Chu, H., Yang, W., Sun, L., Cai, S., Yang, R., Liang, W., Yu, H., & Liu, L. (2020). 4D printing: A review on recent progresses. *Micromachines, 11*, 796.
12. Shie, M. Y., Shen, Y. F., Astuti, S. D., Lee, A. K. X., Lin, S. H., Dwijaksara, N. L. B., & Chen, Y. W. (2019). Review of polymeric materials in 4D printing biomedical applications. *Polymers, 11*, 1864.
13. Kumar, R., Kumar, M., & Chohan, J. S. (2021). The role of additive manufacturing for biomedical applications: A critical review. *Journal of Manufacturing Processes, 64*, 828–850.
14. Siwal, S. S., Mishra, K., Saini, A. K., Alsanie, W. F., Kovalcik, A., & Thakur, V. K. (2022). Additive manufacturing of bio-based hydrogel composites: Recent advances. *Journal of Polymers and the Environment, 30*, 4501–4516.
15. Sinha, R., Cámara-Torres, M., Scopece, P., Verga Falzacappa, E., Patelli, A., Moroni, L., & Mota, C. (2021). A hybrid additive manufacturing platform to create bulk and surface composition gradients on scaffolds for tissue regeneration. *Nature Communications, 12*, 500.
16. Aryal, M. R., & Pun, S. (2022). Additive manufacturing of prosthetic hands: A brief overview. *International Journal on Interactive Design and Manufacturing, 16*, 1099–1112.
17. Schadte, P., Rademacher, F., Andresen, G., Hellfritzsch, M., Qiu, H., Maschkowitz, G., Gläser, R., Heinemann, N., Drücke, D., Fickenscher, H., et al. (2023). 3D-printed wound dressing platform for protein administration based on alginate and zinc oxide tetrapods. *Nano Converg, 10*, 53.
18. Van Manen, T., Janbaz, S., Jansen, K. M. B., & Zadpoor, A. A. (2021). 4D printing of reconfigurable metamaterials and devices. *Commun Mater, 2*, 56.
19. Chang, Y.-Y., Jiang, B.-C., Chen, P.-Y., & Chiang, Y.-Y. (2021). An affordable and tunable continuous wrinkle micropattern for cell physical guidance study. *Journal of the Taiwan Institute of Chemical Engineers, 126*, 288–296.
20. Ojstršek, A., Jug, L., & Plohl, O. (2022). A review of electro conductive textiles utilizing the dip-coating technique: Their functionality, durability and sustainability. *Polymers, 14*, 4713.
21. Mashhadi Keshtiban, M., Taghvaei, H., Noroozi, R., Eskandari, V., Arif, Z. U., Bodaghi, M., Bardania, H., & Hadi, A. (2024). Biological and mechanical response of graphene oxide surface-treated polylactic acid 3D-printed bone scaffolds: Experimental and numerical approaches. *Advanced Engineering Materials, 26*, 2301260.
22. Shi, H., Wu, X., Sun, S., Wang, C., Vangelatos, Z., Ash-Shakoor, A., Grigoropoulos, C. P., Mather, P. T., Henderson, J. H., & Ma, Z. (2022). Profiling the responsiveness of focal adhesions of human cardiomyocytes to extracellular dynamic nano-topography. *Bioact Mater, 10*, 367–377.
23. Jung, H., Choi, H., Lee, J. H., & Kang, J. W. (2002). Sonolytic degradation of methyl tert-butyl ether: The role of coupled Fenton process and persulphate ion. *Water Research, 36*, 4699–4708.
24. Monteagudo, J. M., Duran, A., Martin, I. S., & Garcia, S. (2014). Ultrasound-assisted homogeneous photo-catalytic degradation of Reactive Blue 4 in aqueous solution. *Applied Catalysis, B: Environmental, 152–153*, 59–67.
25. Ziemba, A. M., Fink, T. D., Crochiere, M. C., Puhl, D. L., Sapkota, S., Gilbert, R. J., & Zha, R. H. (2020). Coating topologically complex electrospun fibers with nanothin silk fibroin enhances neurite outgrowth in vitro. *ACS Biomaterials Science & Engineering, 6*, 1321–1332.
26. Blachowicz, T., Domino, K., Koruszowic, M., Grzybowski, J., Böhm, T., & Ehrmann, A. (2021). Statistical analysis of nanofiber mat AFM images by Gray-scale-resolved Hurst exponent distributions. *Applied Sciences, 11*(5), 2436.
27. Sun, W., Gregory, D. A., Tomeh, M. A., & Zhao, X. (2021). Silk fibroin as a functional biomaterial for tissue engineering. *International Journal of Molecular Sciences, 22*, 1499.

28. Kundu, B., Kurland, N. E., Bano, S., Patra, C., Engel, F. B., Yadavalli, V. K., & Kundu, S. C. (2014). Silk proteins for biomedical applications: bioengineering perspectives. *Progress in Polymer Science, 39*, 251–267.
29. Kochhar, D., DeBari, M. K., & Abbott, R. D. (2021). The materiobiology of silk: exploring the biophysical influence of silk biomaterials on directing cellular behaviors. *Front Bioeng Biotechnol, 9*, Article 697981.
30. Oguntade, E., Wigham, C., Owuor, L., O'Grady, K., Acierto, A., Zha, R.H., & Henderson, H.J. (2023). Dry and wet wrinkling of a silk fibroin biopolymer by a shape-memory material with insight into mechanical effects on secondary structures in silk network. *J Mater Chem B, submitted*.
31. Dolgin, J., Hanumantharao, S. N., Farias, S., Simon, C. G., Jr., & Rao, S. (2023). Mechanical properties and morphological alterations in fiber-based scaffolds affecting tissue engineering outcomes. *Fibers, 11*(5), 39.
32. Sukumar, U. K., & Packirisamy, G. (2015). Bioactive core–shell nanofiber hybrid scaffold for efficient suicide gene transfection and subsequent time resolved delivery of prodrug for anticancer therapy. *ACS Applied Materials & Interfaces, 7*(33), 18717–18731.
33. Tanzli, E., & Ehrmann, A. (2021). Electrospun nanofibrous membranes for tissue engineering and cell growth. *Applied Sciences, 11*(15), 6929.
34. Romanelli, S. M., Knoll, G. A., Santora, A. M., Brown, A. M., & Banerjee, I. A. (2015). Comparison of engineered peptide-glycosaminoglycan microfibrous hybrid scaffolds for potential applications in cartilage tissue regeneration. *Fibers, 3*(3), 265–295.
35. Panthi, G., Hassan, M. M., Kuk, Y. S., Kim, J. Y., Chung, H. J., Hong, S. T., & Park, M. (2020). Enhanced antibacterial property of sulfate-doped Ag3PO4 nanoparticles supported on PAN electrospun nanofibers. *Molecules, 25*(6), 1411.

Chapter 10
Applications of Nanocomposites

Composite materials, which are used to improve the performance of existing materials in several non-industrial materials are distinguished from each other by the fact that they are made up of nanomaterials embedded in a base matrix such as metal, ceramics, or polymers. Their unique qualities allow for improved strength and durability, which increases the efficiency or longevity of the materials. In addition, composites have better electrical and thermal properties, making them ideal for additional applications. They have improved resistance to wear and corrosion, which prevents wear. In electronics, they are used to make sensors and sockets work better, and to create institutions for financing solar energy and more energy-efficient batteries. They help to increase the caliber of building materials in the construction sector, including the help to improve the quality of building materials such as implants and insulation and they can be used in the medical field to create new medical devices such as implants and artificial resuscitation. The distinction between composites has begun to generate revenue and numerous growth prospects in several industries.

10.1 Introduction

One of the most pressing problems in the modern world is environmental pollution, which can be both direct and indirect and poses a serious threat to living beings. Many industries contribute to environmental degradation by discharging both organic and inorganic pollutants into nearby rivers. However, with the rapid advancement of science, technology, engineering, special materials and methods have been developed to reduce this problem. Adsorption, oxidation, photocatalytic degradation, enzymatic and microbiological degradation are common phases in the wastewater treatment process. Due to its simplicity, ease of use and operational efficiency, adsorption stands out among these methods as one of the most successful technologies. A common and very effective adsorbent are activated carbon. However, its prohibitive cost is still a significant drawback. When choosing, the main factor in choosing adsorbents

R. A. Al-Samarai and Y. Al-Douri, *Engineering of Nanomaterials*,
https://doi.org/10.1007/978-981-95-5499-7_10

is their cost effectiveness, with preference given to widely available materials, that require little processing or are by-products of industrial operations. Even if these methods work well, they do not always work perfectly. Organic pollutants that are partially oxidized can sometimes produce secondary pollutants are potentially even more dangerous than original toxins.

An example from the study is shown in Fig. 10.1, where two poly(acrylonitrile) (PAN) nanomats are electrospun (Fig. 10.1a) and hot pressed at 180 °C (Fig. 10.1b), and then subjected to identical conditions for atomic force microscopy (AFM) measurements. As expected, the topographic maps show that heat pressing results in the thickening of fibers. However, the fibers are unrealistically depicted in Fig. 10.1a as looking more like impermeable walls. This artifact is not visible in the image of untreated nanofibrous mat (Fig. 10.1c), which displays separate phases between the individual fiber surfaces. This result is explained by small movements of nanofibers during scanning, which creates the false impression that they are wider in topographic image. On another hand, the phase map of hot-pressed sample in Fig. 10.1d shows dark lines. These lines have nothing to do with the physical differences described above, and they are weakly visible in topography (Fig. 10.1b). Rather, they indicate topographic limitations that occur during heating, a feature is observed in thermal stability of PAN nanofiber mats [1].

Due to the use of oxygen as an oxidizing agent and ability to completely mineralize organic materials at low temperatures and concentrations, this approach is superior to others. As a result, high-energy UV radiation is required to excite electrons from the valence to conduction bands in some semiconducting metal oxides such as ZnO and TiO_2. These oxides have a large band gap. To harness solar energy, a large amount of research has been done to modify the band gap of semiconducting metal oxides or develop substitutes. These procedures involve (a) doping with additional components, (b) using metals or other semiconductors to create the heterojunction structure and (c) using clay or graphene/graphene oxide as a substrate to increase sensitivity to sunlight [9]. Peak Force QNM is used to study bovine cortical bone immersed in water [27]. As shown in Fig. 10.2 [1], they have observed changes in the elastic moduli of osteons, interstitial bone, cement lines and various sub-lamellar lamellae. They are able to see soft mineralized collagen fibers within the stiffer matrix. The Deryagin-Muller-Toporov (DMT) model with Hertzian contact profile, which is used to fit the load-strain curves of limb shrinkage process, formed the basis for modeling and measuring the elastic moduli (Fig. 10.2c and d). We used DMT modeling to map the elastic moduli of phase-separated polyurethanes [1].

The adsorption method faces problems in determining the adsorption process and treatment of wastewater containing insoluble dyes [4]. The main disadvantages of these methods are the need to dispose of contaminated activated sludge and problems associated with organizing the correct reaction conditions and release of by-products [5, 6]. The ability to create nanomaterials that can convert toxic pollutants into less harmful ones has led to many advances in nanoscience and nanotechnology that have had a major impact. Heterogeneous photocatalysis is considered a potential method to solve environmental problems because it is environmentally friendly and extremely efficient and there are effective challenges ahead [7, 8]. Since clay minerals have

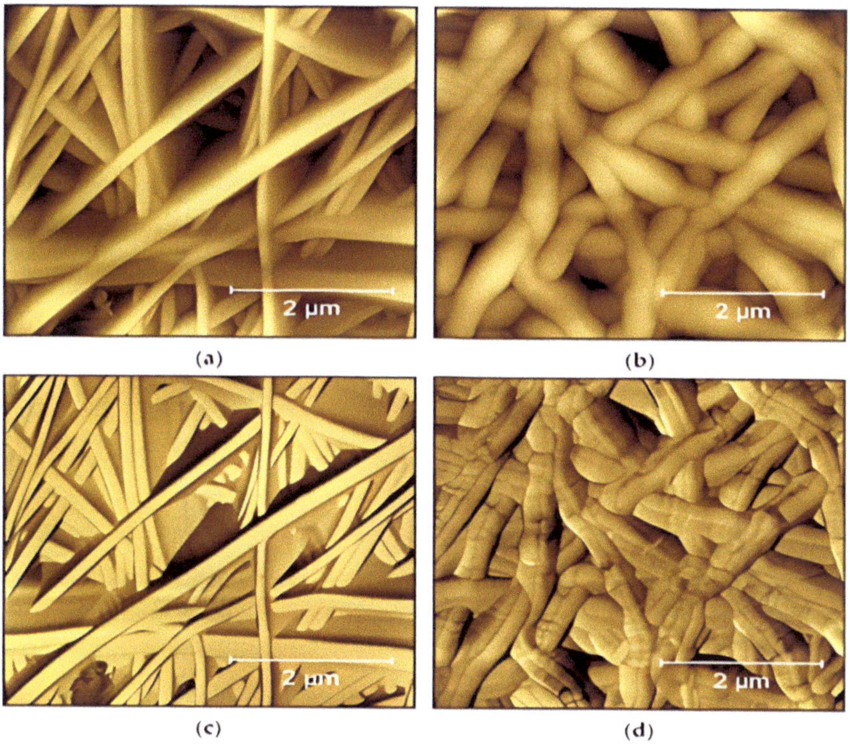

Fig. 10.1 AFM images of polyacrylonitrile nanofiber (PAN) mats (**a**–**d**), and **e**—untreated mat, phase image, untreated mat, topographic image, heat-pressed mat, phase image. Images are captured using following scanning parameters on a Nanosurf FlexAFM: 512 lines of 512 points, 1.4 s/line, p gain 550, i gain 1000, d gain 100, and free oscillation amplitude 6 V. Setpoint 55% [1] [Open access]

high adsorption–desorption capacity, clay or graphene/graphene oxide nanocomposites have been reported to exhibit different degradation behavior. They convert UV photocatalysis to visible photocatalysis. In addition, they are responsible for the separation of electron–hole pairs, which increases the photocatalytic potential of nanomaterials.

10.1.1 Advanced Methods of Oxidation

The use of oxidizing agents to oxidize pollutants is part of advanced oxidation processes. The most powerful oxidative radicals used to remove organic pollutants from water are the sulfate radical (SO_4) and the hydroxyl radical (OH). Compared to other traditional purification technologies, advanced oxidation processes have several

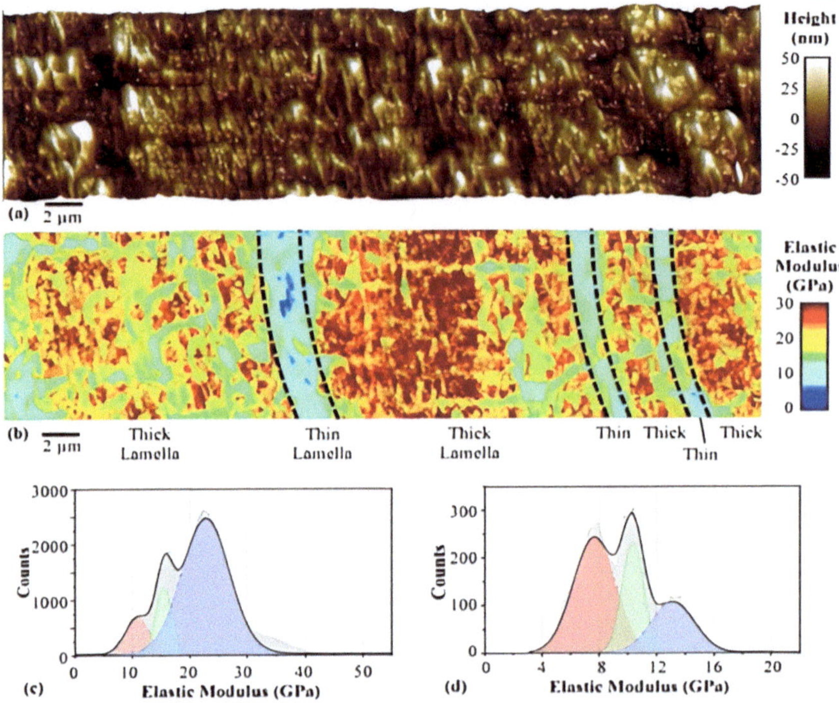

Fig. 10.2 **a** Atomic-wave microscopy of bovine cortical bone. Includes **b** topographic measurements, elastic modulus map, elastic modulus plots for **c** thick sub lamellae and **d** thin sub lamellae, and a stitched scan showing thick and thin sub lamellae in osteons [1] [Open access]

advantages, including faster reaction rates, resulting in shorter retention times and ability to handle required system flow rates without taking up enormous space.

Advanced oxidation processes are often divided into advanced electrochemical oxidation procedures, ultraviolet treatment, ozone treatment, photo-enhanced oxidation processes and catalytic advanced oxidation processes.Advanced oxidation procedures are environmentally friendly chemical methods that can convert organic contaminants into harmless byproducts that do not move contaminants from one phase to another or create large volumes of sludge.

The cross-linking of $C_{32}H_{66}$ ($C_{32}H_{66}$ is a long-chain hydrocarbon that can be used to create molecular monolayers, which are used in applications such as preparing or modifying surfaces) SAMs (a monolayer of $C_{32}H_{66}$ molecules) by HHIC (intermolecular impacts) shows the collision-induced C–H bond cleavage, a process that ultimately leads to C–C cross-linking as shown in Fig. 10.3 [21–23]. Small-scale AFM images of 1 μm × 1 μm are shown in the accompanying images [2]. Figure 10.3b depicts how washing the sample with hexane completely removed the $C_{32}H_{66}$ SAMs layer from the silicon substrate as shown in Fig. 10.3a.

This indicates that the molecules are dissolved in hexane as a result of alkane SAMs simply physically adsorbing onto the substrate. The topography of SAMs

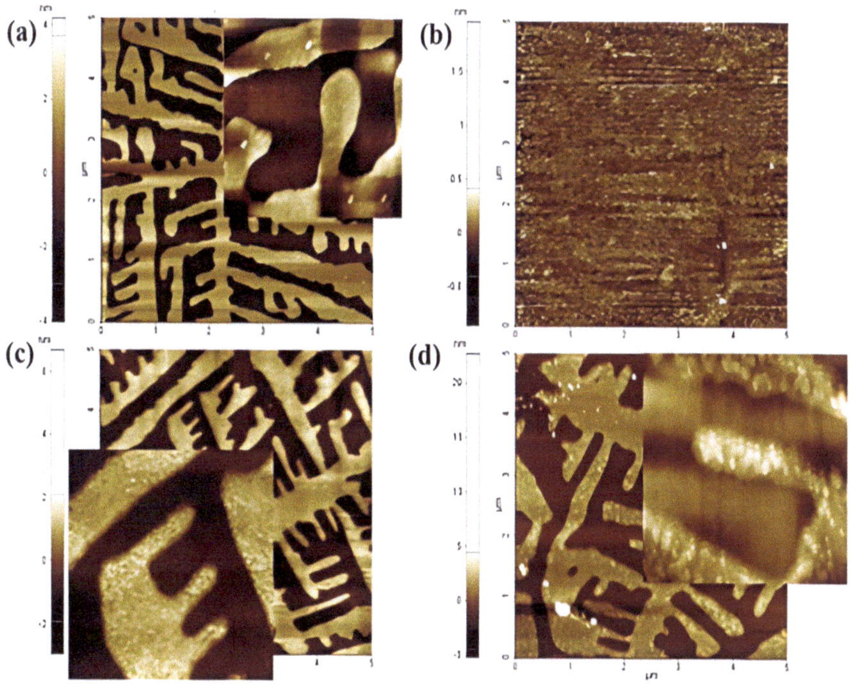

Fig. 10.3 AFM of $C_{32}H_{66}$ thin films prepared at a concentration of 0.13% (w/v): **a** after precipitation, **b** after precipitation and hexane washing, **c** after cross-linking by hyperthermic hydrogen bombardment, **d** after hexane washing and hyperthermic hydrogen bombardment (inserts: higher resolution images and scan scale 1 μm × 1 μm) [2] [Open access]

remained essentially the same after the alkane molecules are cross-linked as shown in Fig. 10.3c. The cross-linked SAMs are able to survive hexane washing as shown in Fig. 10.3d. In this case, although SAMs themselves are no longer soluble in hexane, no interactions are observed between them and the substrate [2].

However, using advanced oxidation processes (AOP) to destroy all pollutants contained in wastewater may not be sustainable. Combining them with heterogeneous photocatalysis is a viable alternative. AOPs have been shown to significantly increase the efficiency of wastewater degradation, which in turn promotes the breakdown of both organics and persistent chemicals. To accelerate the oxidation and destruction of various organic pollutants in wastewater, modern oxidation processes have recently become widespread, resulting in the in-situ formation of extremely powerful chemical oxidants such as hydroxyl radical (OHl) [11]. When used correctly, AOPs offer a good chance of reducing contaminant concentrations from several hundred parts per million to less than five parts per billion. For this reason, they are called "21st Century Treatment Processes." The production of reactive free radicals, in most situations hydroxyl radicals, as well as the chemical and physical characteristics of pollutants influence the efficiency of AOP degradation in refractory wastewater compounds [12]. The decomposition process of pollutants themselves is driven by the oxidation reaction that occurs between these radicals and the pollutants. Many methods can be used to create these technologies such as oncolysis [13], ozone-based procedures [14], Fenton reactions [15], heterogeneous photocatalysis [16] and other combinations of these technologies [17, 18]. Reactive agents, each of which can be described in terms of its unique approach, have some advantages.

- It is well known that there are significant deficiencies in traditional wastewater treatment systems that can be addressed by AOP. This method processes all distillery wastewater, so sample filtration is not required.
- AOP's ability to convert organic molecules into simpler, more stable inorganic compounds such as salts, carbon dioxide, and water, while producing virtually no precipitate and eliminating the need for additional processing steps, is one of its major advantages.
- AOP can be used as a post-treatment before release into the environment or as a pre-treatment to modify persistent pollutants so that they can be subjected to biological treatment.
- The use of HO makes it possible to achieve a rapid AOP reaction when treating wastewater from distilleries.
- Despite their advantages, AOPs have a number of disadvantages and limitations. AOP typically undergoes acidic conditions, especially Fenton oxidation, which results in significant consumption of acid and base.
- An AOP system using H2O2 may be hazardous to humans.
- In this dose-dependent mechanism, the right number of H_2O molecules are generated to provide the required degree of therapy.
- The main disadvantage of AOPs is the relatively expensive method of their production due to the need to use expensive chemicals, higher energy consumption, and the formation of unidentified, persistent by-products that in certain situations can be even more nephrotoxic than the parent compounds, the reason is still unknown.
- All AOPs are used to purify non-target chemicals from group radicals; however, they are not suitable for some classes of hazardous compounds that are resistant to assault carried out by •HO.

10.2 Preparation of Nanocomposite

When synthesizing nanomaterials, two approaches are used; "top-down" and "bottom-up". Massive bulk materials are crushed to the smallest particles using a top-down method. This method works best with solid and distributed materials. One atom or molecule at a time, nanoparticles are formed using the bottom-up method. Gas or liquid phases are mainly used. In most cases, bottom-up approaches produce purer nanoparticles with better control over surface chemistry and particle size. Mammals contain the highest protein, type collagen, which exists primarily in the form of fibrils and fibrils. The main task of collagen is to give things shape and mechanical strength. During supramolecular assembly, the latter occurs due to intermolecular cross-linking of collagen molecules [1]. The basic triple helix found in all collagen proteins is derived from multiple repeats of the basic tripeptide Gly–X–Y, where X and Y are typically occupied by proline and hydroxyproline, respectively [1–3]. At the nanoscale, triple helices self-assemble to form fibers [4]. Large fibrous structures can be formed by microfibrils that form as the nanofibers continue to intertwine linearly [5–8].

Nanocomposite membranes prepared from PSF and GO are commercially available. The morphology of the membranes is investigated using microscopic techniques. Due to the hydrophilic nature of GO and the hydrogen interaction between the matrix and nanofiller, the water flux of nanocomposite membranes is improved [18]. Polysulfone/graphene oxide (PSF/GO) membranes are prepared [19] using a solution casting technique. The membranes contained up to 2 wt% GO. The morphology of the membranes is investigated using atomic force microscopy (AFM) (Fig. 10.4) [3].

Self-assembled collagen fibrils can be observed under certain conditions both in vivo and in vitro [9]. Collagen-based hydrogels are widely used in biomedical applications due to their excellent biocompatibility, tunable biodegradability, minimal immunogenicity, and ability to stimulate cell proliferation [3]. However,

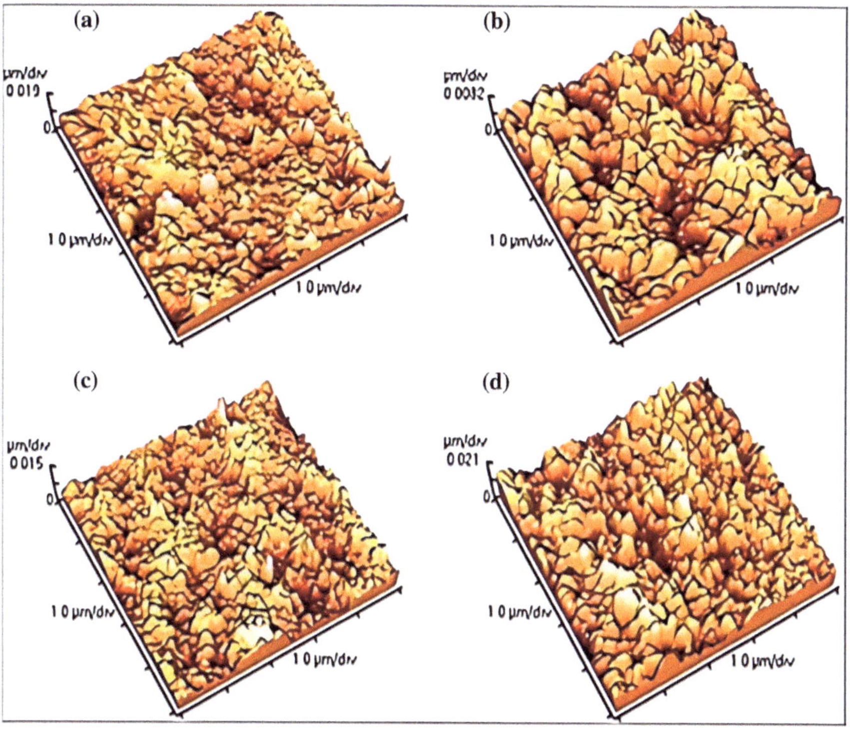

Fig. 10.4 3D AFM surface morphology of obtained membranes is shown in **a** for pure PSF, **b** for PSF/GO 0.5, **c** for PSF/GO 1, and **d** for PSF/GO 2 membranes. AFM; polysulfone (PSF); and polysulfone/graphene oxide (PSF/GO) [3] [Open access]

collagen-based hydrogels often have lower mechanical properties than synthetic polymers, limiting their use as tissue engineering scaffolds and wound dressings [10]. Collagen is usually chemically cross-linked using glutaraldehydes and carbodiimides or by enzymatic catalysis to improve the mechanical properties and reduce the biodegradation rate of collagen-based hydrogels [11]. On another hand, the use of standard chemical crosslinkers may result in tissue calcification or cytotoxicity. Natural macromolecules demonstrate excellent biocompatibility with collagen and low toxicity compared to synthetic ligands. As a result, an increasing number of studies have led to the creation of composite hydrogels with the addition of biopolymers such as cellulose and chitosan [10–14].

10.2.1 Precipitation

Filtration or centrifugation is used in sedimentation to remove the solvent. This process produces insoluble compounds by combining or reacting molecules. Insoluble molecules nucleate and grow to form molecules. In visible light, silver-containing compounds work well as photocatalysts. One of these plasmonic photocatalysts and photosensitizers is AgBr, but in a single-phase form, it is optically unstable. Thus, AgBr can be combined with other materials to create highly efficient photocatalysts that are sensitive to visible light. Authors in [19] attempted to synthesize AgBr@Bi_2O_3 composite via a facile deposition method using innovative hierarchical α-Bi_2O_3 rods with a special shape suitable for photocatalysis. They ultrasonically dispersed 1 g of Bi_2O_3 in 50 ml of distilled water, followed by dissolving 0.62 g of cetyltrimethylammonium bromide (CTAB) with stirring. Next, a certain amount of $AgNO_3$ dissolved in 2 ml of NH_4OH is added to the above mixture. The reaction mixture is then stirred for 24 h at room temperature. The AgBr@BiO_3 composite is then prepared by following centrifugation and drying procedure at 60 °C. The feasibility of fabricating Ag/AgBr-Ag_2CO_3 hybrids using the co-precipitation method is reported by Zhang et al. In addition, they have presented a completely new photocatalytic method for elemental mercury (Hg0) removal in the presence of light. They prepared Ag/AgBr-Ag_2CO_3 by dissolving 0.7 g KBr and 2.77 g Na_2CO_3 in 150 ml distilled water, adding them to a solution of 0.2 $MgNO_3$, and stirring for 20 min. They used KBr and Na_2CO_3 as precipitants.

10.2.2 Intercalation

The process of inserting molecules, ions or atoms between layers of a base material–usually a crystalline structure–is known as intercalation. In materials science, this technique has been widely studied and applied, in particular, to improve the performance of materials used in a variety of applications, including electronics, energy storage, and catalysis. Intercalation is a reversible and controllable process in which the introduced species populate the spaces between layers rather than disrupting the lattice structure of the base material.

Lithium ions intercalate into the graphite anode during discharge and are released during charging in lithium-ion batteries, making this one of the most common causes of intercalation. This allows for efficient energy storage and release. Additionally, intercalation has been investigated in many other materials that can contain a wide range of intercalated molecules such as carbon-based materials, clay minerals and transition metal dichalcogenides.

Intercalation has various advantages including (1) appropriate nano-sized photocatalyst is dispersed throughout solid supports to create unique reaction sites; (2) solid support enhances adsorption, which increases the rate of photocatalysis; and (3) less photocatalyst is required for the degradation of 250 ml dried zinc acetate in a solution of dimethylformamide with purified bentonite clay. The mixture is then treated with ultrasound for three hours. Finally, ZnO-bentonite nanocomposite is formed by calcination at 200 °C for 2–3 h. ZnO-bentonite is used to digest phenol in a continuously stirred tank reactor. Using a montmorillonite clay surface authors in [21] are able to stop the aggregation of guest molecules. Using a simple ion exchange procedure, they have incorporated zinc oxide into silicate layers of montmorillonite clay.

They have replaced zinc ions with sodium ions in montmorillonite (semiconductor) clay ZnO. Improvement in emission intensity is observed in ZnO hybrids due to increased destruction of oxygen vacancies occurring in the environment surrounding montmorillonite. BiO_3 is intercalated into montmorillonite [22] to improve the photocatalytic activity of Bi_2O_3. According to their research, sunlight and bentonite Bi_2O_3 can be used to treat textile wastewater on a large scale.

The ability of intercalation to modify the properties of materials has been proven to be a powerful research tool, especially for creating novel materials with specialized optical, electrical, and mechanical properties. Intercalation is the most widely used technique for forming nanocomposites in clay materials. To change their structure, clay layers are mixed with the original material. For example, a temperature-dependent shift in the slope of the b line will cause the CO_2 region to shrink or expand at pH values lower (or higher) than 0. As shown in the context of Current–Voltage (I–V) scenario, this behavior can be distinguished when examining the T function of the blistering chemical reactions, which can only occur when the pH is close to zero.

In general, temperature changes have little effect on the upper limit of water stability, usually greatly reduced by temperature changes. This is accomplished by preparing a fresh sample and immersing it in an electrolyte solution heated to 343 K. A single CV (cyclic voltammetry) scan ($v0 = 600$ mV/s) is performed on HOPG (highly oriented thermal graphite) and surface morphology acquired in situ throughout the procedure is shown in Fig. 10.5a [4].

The experimental setup used in the study, the effects of pH on the material properties, and detailed interfacing process are included in this discussion.

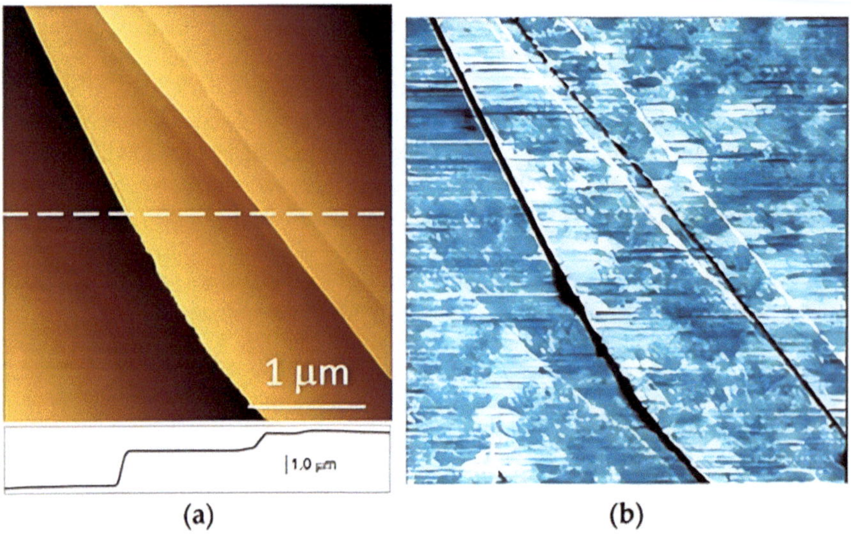

Fig. 10.5 **a** In-situ AFM image of HOPG surface morphology after cyclic voltammetry (CV) in hot (343 K) electrolyte at v0 = 600 mV/s. **b** Graphite terrace in phase contrast. Adsorbates are visible in darker areas and are most likely located in electrode surface defects [4] [Open access]

10.2.3 Method of Ion Exchange

The advantage of using nanoparticles over bulk materials is that their size and shape can be used to control their chemical and physical properties. Numerous applications including electrocatalysts [7], solar cells [8–11], batteries [12–15], sensors [19], biomedical [20–22], and electrical devices [23–25] benefit greatly from their special properties. Consequently, a large amount of research has been conducted on how to control the phase, physical size and nanoscale shapes of materials such as nanospheres, nanorods (NRs), and nanoplates [26–28]. The surface morphology of AEM can affect the functions of the membrane. In particular, some studies have observed that ionic conductivity of the membrane is enhanced by the creation of ion channels caused by the separation of hydrophilic and hydrophobic nanophases [28]. Figure 10.6 shows AFM images of surface topography of fabricated quaternary ammonium-poly(arylene ether sulfone)(Q-PAES). There is no noticeable phase separation in Q-PAES-10%, Q-PAES-20%, or Q-PAES-30%. However, Q-PAES-40% and Q-PAES-50% show a clear nanophase separation, which allowed the light and dark regions to be separated. Since the quaternary ammonium groups used as functional groups absorb water, the hydrophilic regions are visible in the dark regions and vice versa. The image illustrates how the substitution of quaternary ammonium occurs as the degree of branching (DB) value increases. This likely refers to a chemical process in which a quaternary ammonium group is replaced or modified in a material or system and the extent of this substitution is dependent on the increase in the DB value. DB value may refer to a specific parameter such as degree of

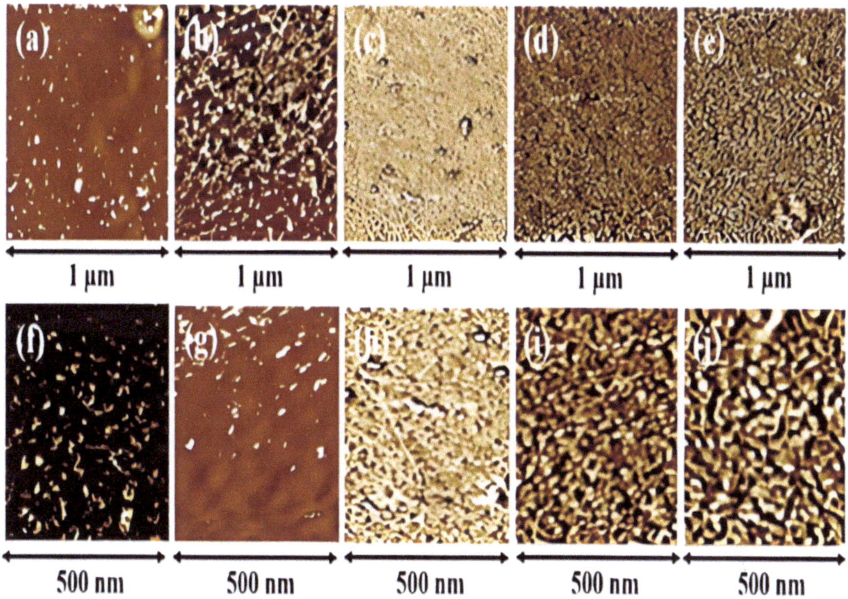

Fig. 10.6 **a**, **f** AFM phase images of Q-PAES-10%, **b**, **g** Q-PAES-20%, **c**, **h** Q-PAES-30%, **d**, **i** Q-PAES-40%, **e**, **j** Q-PAES-50% [5] [Open access]

substitution, reactivity or concentration that affects the rate or extent of substitution process. It is obvious that nanophase separation exists with the increase of salt level. As shown in Fig. 10.6e and j, hydrophilic domains in Q-PAES-50% membrane can have an average size of 100 nm [5].

Solution-based colloidal synthesis has been widely studied among various chemical and physical approaches as it can easily control the size and shape of nanoparticles (NPs) and produce uniform NPs. This bottom-up methodology creates NPs at molecular or atomic level as constituent parts. Colloidal synthesis is often carried out at elevated temperatures using ligands and inorganic or organometallic precursors. The synthesis procedure follows the LaMer paradigm, where nanoparticles (NPs) grow to a certain size after a nucleation step [21]. The hot injection method, which involves rapid addition of precursors to hot mixtures, triggers the nucleation of colloidal NPs [22].

10.2.4 Synthesis of Hydrothermal Process

A hydrothermal autoclave reactor is a specialized apparatus required for the synthesis of nanoparticles using a hydrothermal process. The autoclave reactor is made of cylindrical cylinders with thick steel walls and is hermetically sealed. It allows to withstand high pressure and temperature regularly and safely for long periods. In addition, the autoclave fabric must be solvent-resistant. However, "closure" is perhaps the most important component of a hydrothermal autoclave reactor. In addition, seals are an additional important component of the autoclave. SEM is used to analyze WO_3 coating formed on the glass substrate. The micrographs (Fig. 10.7) show that the oxide film consists of many microstructural units that are uniformly distributed over the sample surface. Firstly, spherical shapes can be identified that have a diameter, 400–550 nm and are composed of smaller nanoparticles ranging in size from 15 to 55 nm (Fig. 10.7). Secondly, there are distinct elongated ridges with width, 100–400 nm and a length that varies, most of which are less than 1 μm. Two or more units that are once submicron spheres combine to form some of these ridges. Finally, nanoparticles with a rounded or somewhat elongated shape can be distinguished; the rounded ones typically have a diameter, 50–55 nm, while the more elongated ones have a diameter of 80–140 nm [6].

To further study the distribution of silver particles on the surface of ceramic substrates, it used a scanning diffusion impedance microscope, which creates a potential difference between the material under study and the probe and measures the resulting current. In addition, the microstructural characteristics and local electrophysical properties of the obtained oxide films are investigated using AFM techniques. Figure 10.6 shows AFM analysis results for the WO_3 film formed on the surface of the glass substrate. Three different types of formations constituting the film microstructure are visible in the topographic scans; ridges, individual formations of 350–450 nm in a size that resembles slightly distorted submicron spheres and spherical nanoparticles between them. The latter, which are mainly 60 ± 15 nm

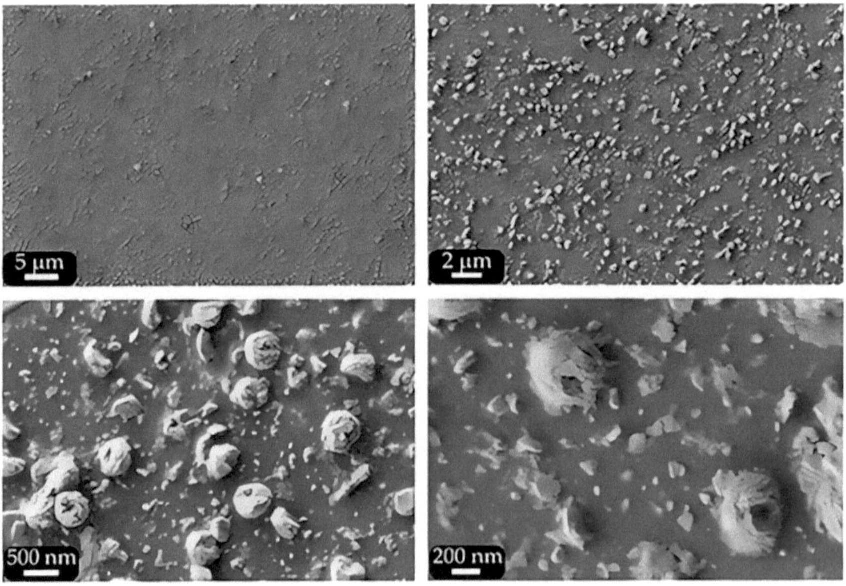

Fig. 10.7 WO_3 film microstructure (SEM) produced on a glass substrate [6] [Open access]

in size, while larger particles up to 160 nm in size may exist, uniformly cover the space between the ridges and individual submicron formations, which is especially noticeable in Fig. 10.8c, e and f. The ridges detected by AFM are 2–5 μm long, and they are widened in the center, consisting of elongated rods located next to each other and have a maximum width of more than 1 μm (Fig. 10.8c and e). The surface potential distribution map shows that ridges and nanoparticles between them have a certain potential concentration, while at the edges of ridges, potential is slightly lower.

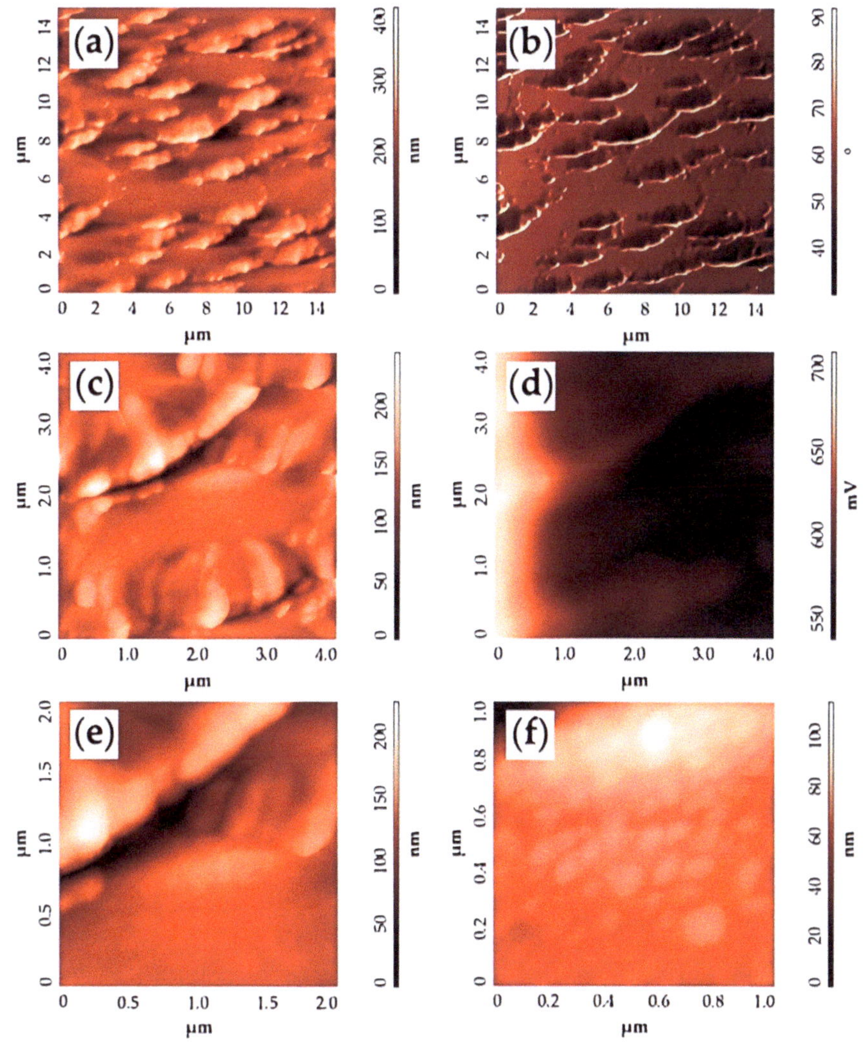

Fig. 10.8 Topography (**a**, **c**, **e**, **f**), phase contrast mode (**b**), and surface potential distribution map (**d**) are AFM results of WO_3 film produced on a glass substrate [6] [Open access]

This method allows us to obtain high-contrast maps of mutual distribution of aluminum oxide and silver, despite the large differences in their electrophysical properties. This is because many hydrothermal processes require the use of fluids that are corrosive to the internal material of the autoclave. In addition, it is added with a unique protective coatings to prevent corrosion, so to create a design that blends seamlessly with the interior of the autoclave and can cover the whole or just part of it. The process of creating a single crystal by hydrothermal synthesis depends on the solubility of minerals in hot water under high pressure. The hydrothermal approach has an advantage over other methods that can produce crystalline phases that are unstable at melting point. This method is well suited for the production of massive, high-quality crystals with controlled composition.

Solvents in a sealed vessel (bomb, autoclave) are heated to temperatures well above their boiling point by increasing autogenous pressure. Solvent thermal treatment or hydrothermal treatment when water is used as a solvent, is the process of carrying out a chemical reaction under such circumstances. The disadvantages of the method include the need to use expensive autoclaves and difficulty of monitoring crystal growth. A new AgBr/ZnO/RGO photocatalyst is developed [23] using a simple hydrothermal impregnation method. Under visible light, the described composite photocatalyst is found to have high degradability in removing methyl orange. Their procedure is to add the ZnO/RGO composite first to 0.04 M KBr solution, stir it with ultrasonic vibration for 10 min while maintaining the pH at 6, and then add $AgNO_3$. Finally, the mixture is stirred in a water bath for two hours at 55 °C.

10.2.5 Method of Sol–Gel

At a relatively low production cost, the sol–gel technology makes it possible to synthesize various materials with unique, predetermined characteristics. In the sol–gel process, the liquid (sol) is first chemically converted into gel state and on further post-processing transitions into an oxide solid. Its advantages are homogeneous nanostructures created by this method with excellent purity that can be achieved at low temperatures. Figure 10.9 illustrates the steps of the sol–gel process. Currently, nanoparticles are produced and synthesized by various methods, including sol, vapor pressure method, mechanical alloying or high-energy bead collision method, plasma method, and electrochemical methods. Although all of the above methods can produce large quantities of nanomaterials, sol–gel method is more widely used and has greater industrial application than other methods [1–5]. Due to its special qualities and features, this process allows the production of high-quality nanoparticles of uniform size on an industrial scale [6–8].

By combining two or more metal (or metal oxide) precursors in specific ratios, this method can create two or more types of nanoparticles simultaneously allowing the creation of alloy products in a single step [9–11]. Other methods such as plasma method and electrochemical methods allow the synthesis of alloy products in a

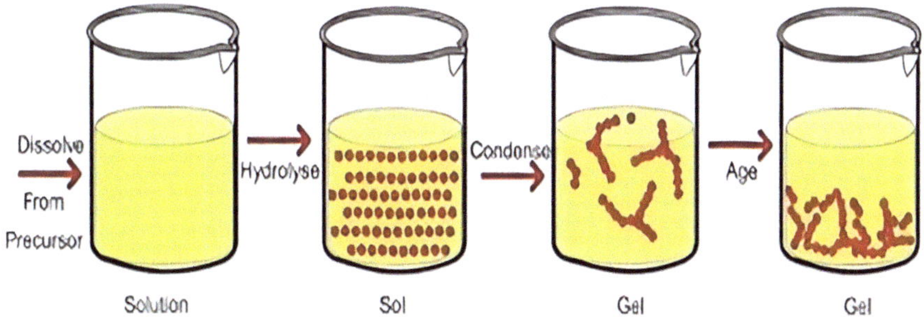

Fig. 10.9 The steps in sol–gel technique [7] [Open access]

single step; however, the fundamental difference between these methods and sol–gel approach lies in the industrial scale of the latter [12–15]. In addition, the sol–gel method can be used to obtain highly homogeneous compounds of extremely high purity (99.99% purity) [16–20]. The low process temperature of this method makes it possible to obtain nanometallic and ceramic materials in the temperature range of 70–320 °C, which is another advantage over traditional methods [21–24]. The morphological differences between the base sol–gel and FAS sol–gel films are shown in Fig. 10.10. According to atomic energy microscopy (AEM) both coatings are uniform. However, the shape of modified coating is significantly changed by adding fluorinated FAS precursor. FAS stands for fluoroalkyl silane, a precursor that modifies the surface properties of coating and sol–gel refers to the coating deposition method. As shown for FAS-modified sol–gel [13], this modification has resulted in a change from the previously observed smooth surface of the base sol–gel to a rough textured surface characterized by spherical protrusions. These changes may be a factor in the increased hydrophobicity of coating. The Rz values of FAS sol–gel and base coating are different; the latter is 10.4 nm while the former is 4.8 nm [7].

The temperature range in which other specific methods make it possible to obtain nanomaterials is 1400–3600 °C [25–28]. One method of bottom-up synthesis is the sol–gel procedure. During this procedure, several irreversible chemical processes are carried out to create the final compounds [28]. The original homogeneous molecules (sols) undergo these reactions to form a three-dimensional heavy infinite molecule known as gel [22–25]. Wet gel is produced by converting tetra-n-butyl orthosilicate (TBOS) into a gel using a procedure known as "consolidation process" [26]. There are several stages of the sol-gelation process, from feedstock to aerogel. A mixture of salts of different chemical compositions is used to create binary or tertiary hybrid systems [23]. Each elemental salt has a different reaction rate [22–25]. The reaction rate of salts is influenced by many variables, including pH, concentration, type of solvent, and temperature [26, 27]. The gel, consisting of polymers, is created by cross-linking cavities and is the 3D structure resulting from density of cells [28]. After the gel dries, a strong and rigid structure is formed as a result of volumetric shrinkage [24]. It should be noted that nanoporosity can be achieved by adjusting the drying conditions of the gel [22, 23]. Compared to typical porosity, nanoporosity has

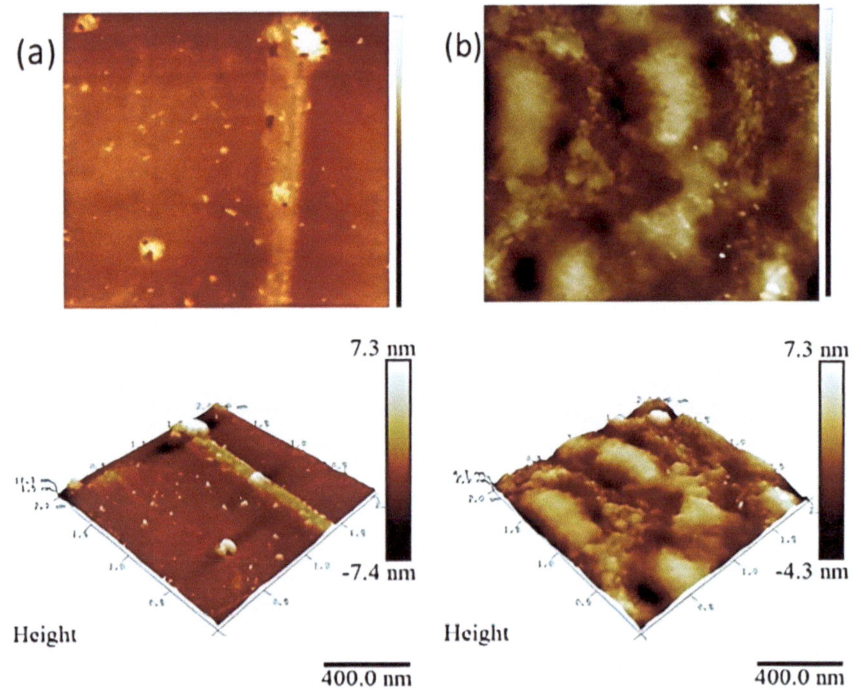

Fig. 10.10 AFM images of FAS Sol–Gel and Sol–Gel base, respectively [7] [Open access]

a relatively large specific surface area, which is an advantage [24–26]. For example, hydrogen is captured and stored in the porosity of carbon materials using nanoporous carbon materials [27].

However, the oxide solution and mixture are stirred for one hour before one milliliter of water is added for hydrolysis. After another 4 h of mixing, the liquid is allowed to stand for 24 h to allow the gel to fully form. Ultimately, evaporation is used to extract the solvent, resulting in a mixed photocatalyst. The catalyst thus created is used to remove methylene blue. A titanium-zirconium nanocomposite is created using a sol–gel process for the treatment of wastewater from the textile sector. For their synthesis, 50 ml of ethanol and 5 ml of titanium isopropoxide solution are used. After vigorously stirring the titanium oxychloride octahydrate solution dropwise, pH of solution is adjusted to 2. Finally, the resulting nanocomposite is dried at 120 °C.

10.2.6 Synthesis of Sonochemical

To initiate chemical processes, sonochemical synthesis depends on ultrasound-induced cavitation, that is, the formation, expansion, and collapse of bubbles [1, 2]. Electromagnetic energy comes in the form of ultrasound (US), frequency which

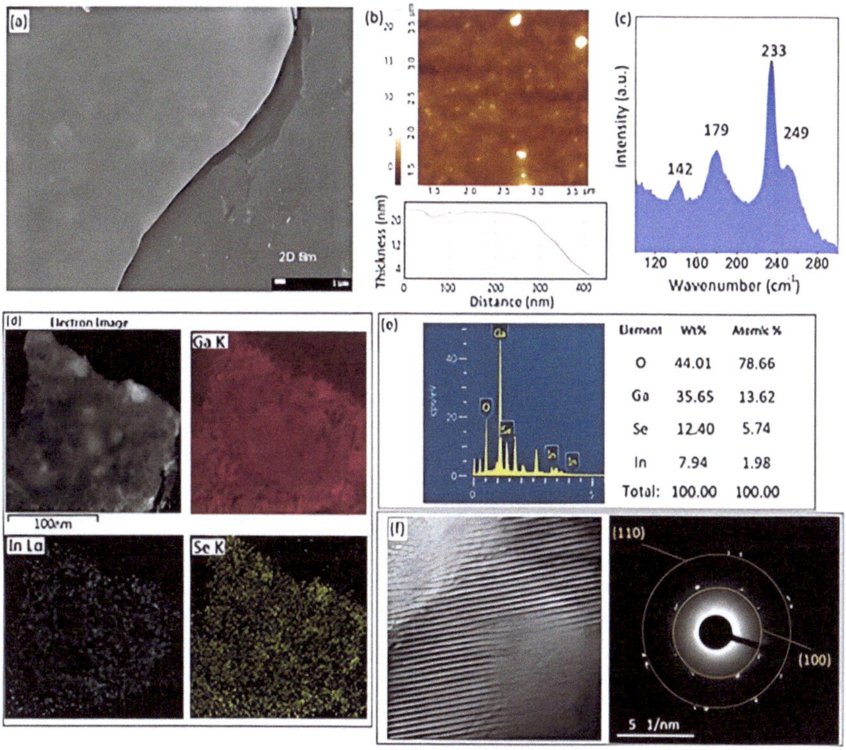

Fig. 10.11 2D GaxSey structures are as follows: **a** SEM image; **b** AEM surface morphology and thickness profile; **c** Raman spectra of 2D GaxSey structures; **d** EDS pattern of 2D membrane surface; **e** detailed results of EDS elemental composition of 2D membranes; **f** high-resolution TEM image and corresponding SAED patterns of 2D membranes [8] [Open access]

ranges from 20 kHz to 1 MHz [2]. Gases and particles in solution alternately compress and expand when ultrasonic waves, which are essentially pressure waves are applied to the reaction mixture. Cavitation bubbles can occur during expansion as a result of suppression of intermolecular forces in the liquid. When the bubbles reach a certain size, they expand and store energy. At this point, the bubbles burst, releasing the energy accumulated over a short period. The intense localized temperature and pressure or "hot spots" that have been caused by this phenomenon (known as cavitation) can reach ~5300 °C and ~ 1000 atmospheres, in that order [3]. It is important to note that the energy required for synthesis comes from ultrasound-induced cavitation and there is no direct molecular interaction between the ultrasonic waves and chemical species [4]. It is known that the primary by-products of sonochemical reactions that occur in the early stages of ultrasonic process are crystalline nanodomains. The formation of crystalline selenium nanodomains is initiated by 2D surface oxide layers of Galinstan particles (Fig. 10.11). The formation of 2D nanostructures coincided with subsequent increase in selenium content to 0.5 μmol L^{-1} in anhydrous ethyl alcohol during the ultrasonic procedure for 1 h. It is important to remember that the shape and properties of 2D nanostructures vary. In particular, ultrasound is applied to Galinstan in selenium-containing solutions (0.5 μmol L^{-1}) for 1 h in anhydrous ethyl alcohol to create 2D nanostructures rich in gallium and selenide (Fig. 10.11a). The lateral dimensions of these nanostructures are found to vary from a few millimeters to several hundred micrometers. As a result, these structures are easily detected (Fig. 10.11a). The thickness of these nanostructures is found to be between 20 and 100 nm. Figure 10.11b shows the surface morphology of 2D nanostructure and its corresponding thickness profile [8].

Ionic diffusion and sonochemistry promoted the increase of reactive wettability of the interface, which accompanied the decrease of ionic selenium on the surface of Galinstan nanoparticles. The coalescence process progressed actively with selenium consumption due to increased interface reactions at heterogeneous interfaces with increased selenium concentration and reaction time. The hybrid multilayer structures with nanoscale dimensions are formed by the long-term ultrasonic procedure (Fig. 10.12a). Transmission electron microscope (TEM) image of the crystal plane of layered structures is shown in Fig. 10.12b. The interlayer distance of 0.29 nm is determined at different points of the crystal structure. The EDS map of these nanostructures shows oxygen as the main component along with gallium, enzyme, and selenium (Fig. 10.12c). A typical EDS spot analysis based on the surface of multilayer nanostructures is shown in Fig. 10.12d. The main components of these layered structures are gallium and oxygen. In contrast, the atomic ratio of selenium to the enzyme is very close (Fig. 10.12d) [8].

10.3 Properties of Nanocomposites

Due to their anisotropic, heterogeneous structure, and advanced mechanical properties associated with size effects during micromachining, nanocomposites are considered difficult materials to micromachine. This results in increased cutting force, poor surface quality, and high tool wear, resulting in poor machinability. Understanding and analyzing various aspects, characteristics, and morphologies of nanocomposites requires the characterization of nanocomposite materials. Some surface modification effects include (a) filler dispersion and composite properties, (b) effects of doping/modification of operating range (UV or visible) of nanocomposites for photocatalysis, (c) influence of process parameters on the final morphology and properties; and (d) extensive property analysis to identify potential uses of nanocomposites.

Various techniques have been used to characterize the nanocomposites including Raman spectroscopy, XRD, SEM, EDS, TEM, AFM, and IR. When atoms in a crystal are deflected by radiation beams, the atomic and molecular structure of the crystal can be determined using X-ray crystallography (XRD). The use of SEM is the most common method for determining the size of the substance. Elemental analysis is carried out using an analytical method known as EDS. This depends on how the sample interacts with the X-ray excitation source.

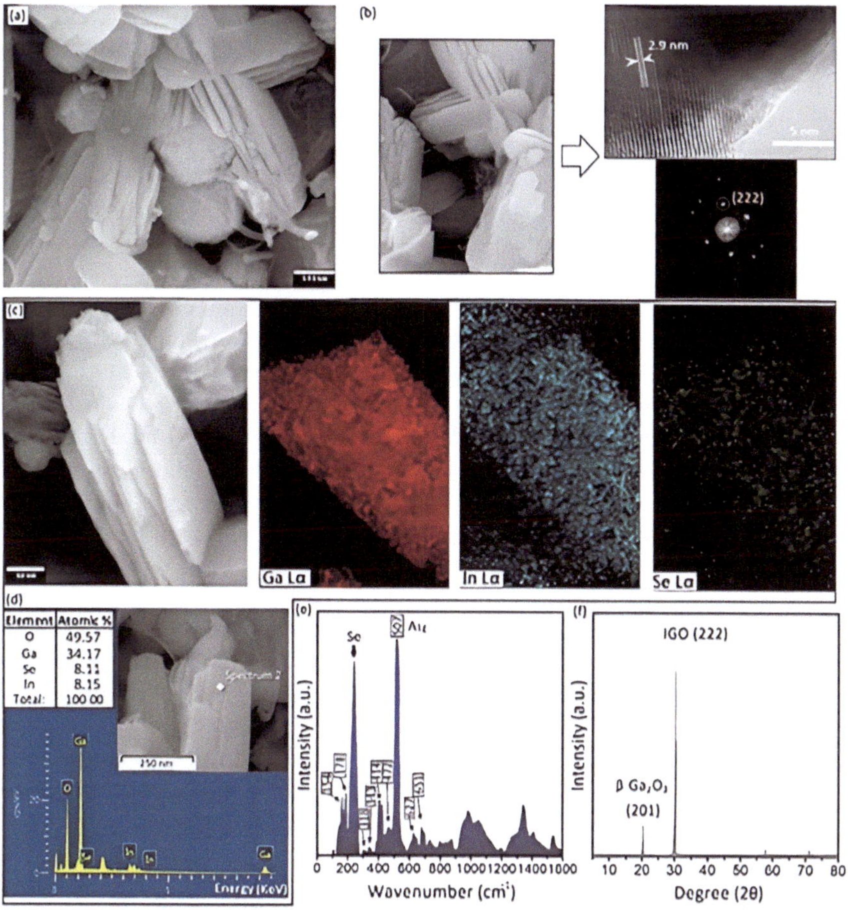

Fig. 10.12 SEM image of the InxGayO/Se layered structure, corresponding transmission electron microscope image, the fast Fourier transform patterns, the EDS pattern of nanostructure surface, detailed elemental composition of structure, Raman spectrum, and XRD pattern of the nanostructure are shown [8] [Open access]

10.4 Destruction of Pollutants Using Nanocomposites

Industrialization and urbanization in recent decades have damaged the environment, which has a direct impact on human life. Toxic organic dyes are waste from the food, textile, paper, plastic, rubber, leather, and cosmetics industries [1–3]. Industrial wastewater contains both organic and inorganic contaminants, including heavy metals [4, 5]. Industrial waste can cause harm to the environment and public health [6–9]. Manufacturers are required to convert contaminating dyes to acceptable results before disposing of them to minimize environmental waste [9]. The use of water for waste disposal must be sustainable and cost-effective [10].

Chemical precipitation, photocatalysis, carbon modification, nanofiltration, reverse osmosis, ion exchange, electrolysis, adsorption, and microbial control are examples of common commercial water removal technologies [3, 11–13]. Catalysis is widely used due to its availability, energy efficiency, and environmental friendliness [14]. Semiconductors are "green technologies" that can purify wastewater through photocatalysis [15, 16]. In addition to the normal toxicity of these compounds, textile wastewater can disrupt photosynthetic activity, gas solubility, and oxygenation in ecosystems by reducing water clarity and sunlight penetration. Initially, these formations appeared as seemingly random formations, spaced further apart than before, as a result of decay and accumulation and eventually acquired a tower-like shape connected at the upper layers. Occasionally, a taller building would emerge from the cluster and extend upward like an antenna. SEM images in Fig. 10.13 illustrate the differences in the morphology and patterns of ammonium magnesium potassium (AMK) nanostructures. The size and dispersion of nanospikes are measured using SEM images (Fig. 10.13a). The assembled nanostructures have dimensions of 340 ± 174 nm for the spike height, 83 ± 32 nm for average spike diameter and 544 ± 150 nm for the space between connected "peaks" ($n = 20$). In addition, the surface density of AMK nanostructures is 8 ± 2 spikes/μm^2. AFM images are obtained for comparison with SEM results (as shown in Fig. 10.13c and d). Based on AFM images of AR-Ti and AMK surfaces, the root mean square deviation (RMS) and average roughness (Ra) are calculated (Fig. 10.13e). For the AR-Ti surface, the average roughness (Ra) is 3.2 ± 0.7 nm and the root mean square deviation (RMS) is 4.1 ± 0.8 nm [9].

Features of this type of wastewater include increased chemical oxygen demand (COD), complex composition, high emissions, widespread distribution, and complex degradation [28]. Unwanted compounds can have catastrophic consequences and seriously threaten immediate recipients when they are detected in industrial effluents. Discharged wastewater containing dyes is toxic to humans, aquatic organisms, and microorganisms. These concerns have led to serious environmental problems that have negative impacts on the well-being, health, and standard of living of the people on earth. Therefore, to reduce the environmental impact of wastewater, it becomes mandatory to remove these dyes from wastewater before discharging. Several solutions have been proposed to solve this problem. These wastewater treatment processes use numerous traditional techniques including adsorption on activated carbon and agricultural waste, ultrafiltration, reverse osmosis, membrane separation, chemical coagulation, ion exchange resins, distillation, and AOP. Compared to other wastewater treatment methods, AOPs are a potentially effective way to reduce wastewater loading and discoloration. AOP uses a variety of techniques, including heterogeneous photocatalysis, ionizing radiation, photo-Fenton-like reactions, sono-Fenton reactions, conventional Fenton reactions and electrochemical oxidation. Heterogeneous semiconductor photocatalysis is a new technology among AOPs used in wastewater treatment, which has the advantage of operating under environmental conditions [4]. Supported, unsupported, bound, or doped semiconducting metal oxides are examples of heterogeneous photocatalysis.

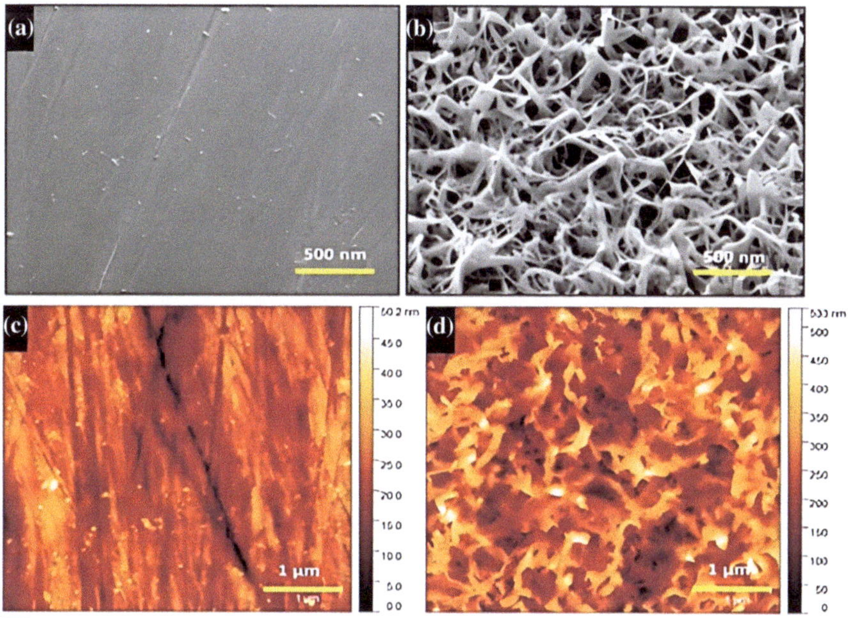

(e)

	RMS (nm)	Ra (nm)
AR-Ti	4.1 ± 0.8	3.2 ± 0.7
AMK	73.0 ± 6.0	58.2 ± 4.8

Fig. 10.13 SEM images of Ti6Al4V alloys, revealing the morphology and distribution of the resulting nanostructure: nanostructured AR-Ti (**a**) and AMK **b** surfaces. **c** AFM image of AR-Ti; **d** AMK surfaces. **e** Roughness values (RMS and Ra) with mean ± SD and n = 5 for a 1.5 × 1.5 µm scan [9] [Open access]

10.4.1 Clay Supported Nanocomposites

In this chapter, it is discussed how to create typical "micro composites" by combining clays (kaolin, talc) and other fine materials (calcium carbonate, carbon black, fiberglass, silica) with various polymers. As name suggests, a significant portion (often more than 25 weight percent) of the system consists of micron-sized aggregates, or "tactoids," which are clay components of these materials. In other words, although hardness and strength are often increased, the clay component in this case mainly acts as a filler or filler to reduce the cost of the finished product (Teng, 1970a). In contrast, the integrated clay particles in "nanocomposites" have at least one size in the nanometer range (<100 nm), which provides a large surface area for interaction with the polymer stage.

Composite materials can be treated in the same way as pure or pure polymer since the clay mineral component content rarely exceeds 5% by weight of the system. More importantly, with a small increase in cost, more or less uniform dispersion of nano-sized clay particles in a polymer matrix leads to a marked improvement in mechanical and thermal properties. Polymer clay nanocomposites exhibit marked improvements in elongation at break, storage stability, durability, flame retardancy, solvent resistance and gas barrier properties compared to corresponding parent polymers as well as superior tensile, flexural, impact strengths and thermal deformation temperatures. However, since the thickness of individual silicate layers is much smaller than wavelength of visible light, the optical properties of polymer are not significantly changed. The term "polymer–clay nanocomposites" is preferred to "clay-polymer nanocomposites" because polymer is the main component of the system.

The latter phrase refers to materials and products that are produced by overall interaction of clays and polymers and is synonymous with "clay-polymer complexes." It is also important to note that the word "clay" in polymer clay nanocomposites refers to a specific type of clay mineral that is often associated with organically altered montmorillonite. Clay nanocomposites are the subject of several research projects aimed at reducing harm caused by industrial effluents. Makarchuk et al. [2] have created magnetic nanocomposites in an attempt to extract colors from aqueous solutions. The pores of saponite clay are treated with magnetite (Fe_3O_4) of varying concentrations to create mesoporous and microporous structures. The dyes malachite green, Congo red, and indigo carmine are adsorptively removed using the formed magnetic nanocomposite. In terms of removal of malachite green, Congo red, and indigo carmine, the sorption capacity of the nanocomposite exceeded that of saponite clay by 3, 6, and 2.5 times, respectively.

In the presence of sonic cavitation [28], TiO_2-nano clay nanocomposite is used as a photocatalyst for phenol degradation. TiO_2-nano clay nanocomposite is formed in benzyl alcohol. Nanoparticles are developed in the spaces between the nanoclay platelets. By adding CTAB, cationic surfactant, bentonite clay, nanoclay is produced. TiO_2-nano clay nanocomposite is found to have higher pollutant removal percentage (59%) compared to 47% nanoclay.

Poly(acrylic acid)-bentonite-FeCo (PAA-B-FeCo) hydrogel nanocomposite is synthesized [26] using in situ emulsion polymerization under the influence of ultrasound. The exfoliated bentonite clay and Fe-Co platelets increased the strength and stability of the hydrogel and promoted the adsorption of organic pollutants. Crystal violet (CV) is used to test the performance of nanocomposite hydrogel under different cavitations, pH, and temperature conditions. The ideal temperature is found to be 35 °C along with a basic pH of 11, which is the reason for higher dye adsorption due to dissociation of COO– ions at higher pH levels. Bimetallic elements occur as hydrogel metal ions, which exhibit repulsion at low pH and reduced sensitivity. The hydrogel alone removed 87% of CV in 15 h and when combined with ultrasonic removal, this number drops to 97% in just 5 h. The reason for this improvement may be mechanical excitation caused by physical forces resulting from sound cavitation [26] that are synthesized polyacrylic acid-nano clay nanocomposite using ultrasound. The putative nanocomposite is used in experiments on the sonosorption of the malachite green dye.

10.4.2 Graphene and Graphene Oxide Supported Nanocomposites

Graphene is 2D sheet of atomically thin sp^2 carbon atoms arranged in a honeycomb pattern. Numerous positive characteristics have been demonstrated, including high mechanical strength [1], electrical conductivity [2], molecular barrier ability [3] and other noteworthy properties. For these reasons, the creation of polymer-based nanocomposites by incorporating graphene into polymers has been the goal of countless research projects [4–7]. However, due to the complexity of bottom-up production [8], poor solubility [9], and aggregation in solution due to van der Waals interactions [2], the use of pure graphene has proven difficult. Alternatively, a top-down strategy can be used to synthesize molecules with structures comparable to graphene from graphite or other carbon sources in an attempt to realize some of benefits of pure graphene while adding functionalized oxygen groups to surface. Graphite oxide, consisting of several stacked layers of graphene oxide (GO) is formed by the oxidation of graphite in protonated solvents.

Besides hydroxyl (OH single bond), alkoxy (C single O bond, C single bond), carbonyl (C double O bond), carboxylic acid (COOH single bond), and other oxygen-based functional groups, graphenes have a similar hexagonal carbon structure with GO [10]. In addition to being easier to synthesize, these oxygen-containing groups have other advantages over graphene, such as increased solubility [2] and surface functionalization potential, which have opened up many applications for nanocomposite materials. Additionally, GO can be processed in various ways to create reduced graphene oxide (rGO), attempting to reduce the number of oxygen groups and achieve characteristics that are more similar to those of pure graphene [11]. Graphene quantum dots (GQDs), which are essentially sheets of graphene with a lateral dimension of less than 100 nm are a new type of graphene derivative that have recently emerged. They use special edge effects of graphene.

Due to their ideal material characteristics and dispersability in polymer matrices, graphene derivatives (GO, rGO, and GQDs) have proven to be useful fillers in polymer nanocomposite materials, leading to several applications [12, 13]. The close arrangement of sp^2 carbon atoms has been demonstrated to provide a near-perfect barrier to gas molecules [3, 14], indicating its applications in packaging materials [12], shielding delicate electronic equipment [14] even in corrosion-resistant materials [15, 16]. For comparable reasons, with improvements in membrane technologies, the selectivity of molecules of certain size can be modified by changing the concentration of the filler in nanocomposites [17, 18]. There is no single process that can precisely bond each carbon atom to each hydrogen atom. This is because a graphene layer can be bonded with hydrogen atoms from the top or bottom. As shown in Fig. 10.14, there are three particularly well-known and stable graphene configurations: chair, passenger, and boat [12].

MO nanocomposite has shown a maximum photocatalytic degradation efficiency of 99%, while the bare Cu_2MoS_4 photocatalyst shows 57% efficiency. The improved photocatalytic performance of Cu_2MoS_4/rGO is primarily attributed to enhanced dye

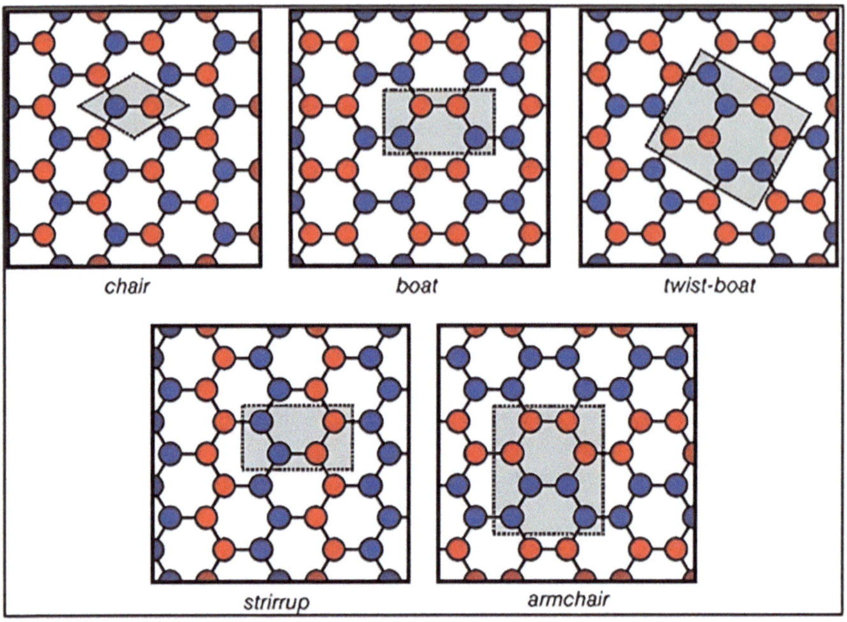

Fig. 10.14 Five different graphene structures with an equivalent carbon atom. Here, hydrogen absorption above and below the graphene layer represented by red and blue dots [12] [Open access]

adsorption and superior electrical performance of rGO sheets. In addition, the edges and active sites of rGO help to disperse the Cu_2MoS_4 hollow nanotubes uniformly and without much aggregation throughout the substrate. In nanocomposites, graphene can serve as an adsorbent, conductive carrier, photosensitizer, photo stabilizer, photocatalyst, and cocatalyst. To optimize the effect of graphene on photocatalytic performance of semiconductor photocatalysts, all variables affecting the photocatalytic performance of composites need to be thoroughly studied and improved. In addition, graphene can be used to create composite photocatalysts and their use in various photocatalysis. For example, for semiconductor manufacturing, customizable 2D nano platform graphene can serve as a novel template, carbon source and agent that guides and controls morphology and structure [21].

10.4.3 Metal Oxides, Binary and Ternary, and Heterostructures

Andres Manuel Del Rio identified vanadium (V) as a chemical element in 1801, and it is known as vanadinite ($Pb_5(VO_4)3Cl$) [1, 2]. At that time, vanadium is given name "panchrome" due to the wide range of colors that its various salts displayed. This chemical element was (re)discovered in 1831 by Nils Gabriel Sefstrom in the form of oxide [3]. In honor of the Scandinavian goddess Vanadis, he has named this chemical element "vanadium". Vanadium has the electronic configuration [Ar]$3d^3 4s^2$ [4], which makes it an early transition metal of the first row (group 5; Z = 23). Vanadium ranks fifth (0.019%) among all transition metals and 22nd among all elements found in the earth's crust [5, 6] or an average, 159 g per ton [7].

Effect on dye removal: The ZnO-AC/A + P system is very effective in eliminating COD and causes degradation of malachite green and Congo red by 99% and 92%, respectively. Hydroxyl radicals are mainly responsible for oxidative degradation. The resulting nanocomposites are more processable. A comparative analysis of Table 10.1 shows that, compared to alternative approaches, Rh B removal using Bi_2O_3-bentonite is an effective method in terms of varying removal efficiencies, operating times and initial dye concentrations as described in related studies in the literature [28].

A typical soil contains more than 100 mg kg^{-1} of vanadium, but this amount is highly dependent on both human activity and geology [2, 7]. Vanadium can be found in a variety of minerals, including vanadinite, carnotite, roscoelite, patronite, bravoite, and davidite, as well as in silicates, sulfides, phosphates, and oxides [2, 8]. Regarding the physical and chemical characteristics of this transition metal, vanadium has a melting point, 1910 °C and can form an oxide with a range of crystal structures and oxidation states (from +II to +V) [9]. Vanadium is a transition metal with significant biological and environmental value. Vanadium is widely present in soil, water, air, and living things. Vanadium, the second most abundant transition metal (TM) in seawater, primarily exists as hydrogen vanadate ($H_2VO_4^-$) with an average concentration of 2 $\mu g \cdot L^{-1}$ [2, 10]. In contrast, its levels in freshwater sources such as rivers, lakes and groundwater (GW) can be significantly higher, reaching up to 30 $\mu g \cdot L^{-1}$ [2, 11], depending on environmental and geochemical conditions. Vanadium is present in the environment as a result of both human and natural activities. Vanadium is commonly found in nature in the form of volcanic

Table 10.1 Rh B removal efficiency (%) using different methods [28] [Open access]

Nanocomposite	Removal efficiency (%)	Operation time (min)
$BiVO_4$/reduced graphene oxide	98.5	600
BiOBr/montmorillonite	98.9	120
BiOBr-graphene oxide	95	45
BiOBr-graphene	87.4	135
Bi_2O_3-bentonite	98.5	80
C60 or CdS/TiO2	33.1	120
Nb or Mn doped Bi2O3	77	90
Ag/$KNbO_3$	95.7	90
TiO_2@WO_3/Au	94	60
ZnO-Ag	100	20
TiO_2–xNx/BN	97.8	40
$MnFe_2O_4$/PANI	100	60

emissions, dust from soil and rock weathering, and marine aerosols [12–15]. Vanadium loading from anthropogenic activities is mainly caused by the extraction and combustion of crude oil, which pollutes the atmosphere. Vanadium contamination caused by human activities has been found in water resources such as rivers, lakes, and seas. It is estimated that 60,000 tons or more of vanadium are released into the atmosphere each year, mainly from the combustion of fossil fuels [16]. In addition, V can remain in the ground, water, and air for a long time [17]. Using a simple coprecipitation method, Raizada et al. [26] have synthesized nano-ZnO supported on activated carbon particles or brick grains to prepare ZnO-AC and ZnO-BGP. When exposed to sunlight, ZnO-AC and ZnO-BGP exhibited exceptional adsorption and photocatalytic activity toward the degradation of malachite green and Congo red. The order of adsorption capacity is ZnO-AC > ZnO-BGP > ZnO. Their studies have shown that while adsorption followed by photocatalysis (AP) has a retarding effect on dye degradation, simultaneous adsorption and photocatalysis (A+P) has a promoting effect.

10.5 Decomposition Mechanism

Before moving on to its consideration, it seems appropriate to define the term "decomposition mechanism", which will be discussed below. This should largely be attributed to the methods used by researchers to decipher it and the goals pursued in its further implementation.

In addition to the generally accepted kinetic explanation of the process, based on the precise study of shape of evaporation curves, Arrhenius method is widely used to study the decomposition mechanism at the fundamental level. Processes in solids are associated with the rearrangement of crystal lattice. These studies are based on two main areas of solid-state physical chemistry; disorder theory and transport theory, which were created in the 1920s and 1930s, respectively by the famous German chemical physicists Wagner and Schottky, and Russian physicist Frenkel.

The research into the relationship between disorder and crystal reactivity has begun in several laboratories in the 1950s and 1960s. Boldyrev [1, 2], who has led these studies in Russia since 1950s, believes that it is protons, ions, and electron displacements in the lattice that cause the breaking or formation of chemical bonds and, ultimately, the decomposition of the reagent. Using 10% TiO_2, the morphology and characteristic features of the RGO/BiOCl/TiO_2 nanocomposites (as shown in Fig. 10.15a) are chemically analyzed by EDX. According to XPS results, the EDX pattern of the RGO/BiOCl/TiO_2 nanocomposites in Fig. 10.4b confirmed that the synthesized nanocomposites comprised C, O, Bi, Cl, and Ti elements. It is found that the atomic ratio of Ti (A% = 2.07%) and Bi (A% = 9.68%) is approximately 19:90, which is much larger than the theoretical atomic ratio of Ti and Bi. The uniform distribution of these elements as shown in the corresponding element map images in Fig. 10.15c, indicated that the RGO nanosheets, TiO_2 nanoparticles, and BiOCl are fully bonded and well assembled [11].

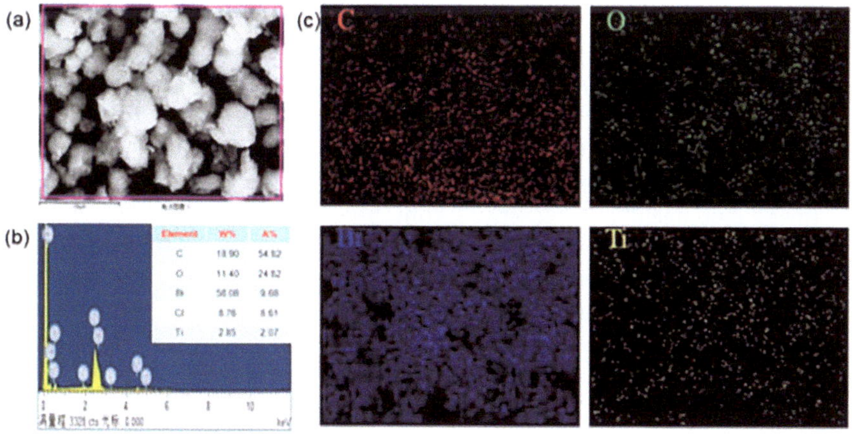

Fig. 10.15 SEM images, EDX spectra and elemental mapping of RGO/BiOCl/TiO_2 nanocomposites with 10% TiO_2 are shown in (**a**), (**b**), and (**c**), respectively [11] [Open access]

An additional obstacle is the wide range of systems used. AgBr can be photoexcited to form electron–hole pairs when exposed to visible light. The surface plasmon resonance of Ag nanoparticles in the visible light region simultaneously causes the generation of electron–hole pairs in metallic Ag nanoparticles. Photogenerated holes (h^+) in the highest occupied molecular orbital (HOMO) of silver (Ag) mix with photogenerated electrons (e^-) in the silver bromide (AgBr) composite (CP), facilitates charge transfer and influencing the material's photocatalytic properties. In this case, the photogenerated electrons in the lower unoccupied orbital of Ag can be efficiently injected into the ZnO CB [26] and then into RGO sheet. Here, Ag nanoparticles can act as a mediator for AgBr-Ag-ZnO-RGO vector electron transfer, effectively separating charges and enhancing photocatalytic activity. Unlike the heterojunction type AgBr/ZnO photocatalyst, ternary Z-scheme compound AgBr/ZnO/RGO has stronger redox ability and simultaneously higher charge separation efficiency. On another hand as shown in Fig. 10.4, AgBr is an excellent visible-light-driven photocatalyst due to its small size and the favorable interface between AgBrandZnO in the ternary composite for electron–hole separation.

10.6 Different Applications

In the food business, nanotechnology offers attractive prospects for improving food safety and quality control, producing new food additives and creating new flavors [1]. Nanotechnology can be used in the food business to produce packaging with improved mechanical, thermal, and/or security properties. Food packaging systems use nanosensors to notify consumers when food has expired. Foods prepared using nanotechnology may also be healthier [2]. Since nanostructures come in different shapes and have different qualities, they can be added to food or used to package food to enhance its nutritional value [3]. 55 of 633 available nanomaterials currently are used in agriculture and food science [4]. According to 2020s outcomes, food products using nanotechnology will account for almost half of all food products [5].

Without a doubt, the food business is being revolutionized by nanotechnology. Most of the documented applications of nanostructures in food products include (i) improving food quality, (ii) bioactive fortification, (iii) using nanocarrier encapsulation to control the release of bioactive compounds, (iv) modifying food textures and structures and (v) using intelligent packaging systems to detect and neutralize biochemical, microbiological, and chemical changes [6]. Numerous nanomaterials are widely used in food production to enhance their nutritional value. For example, protein solubility facilitates the assembly of protein nanoparticles with desired functional properties in food components, which is why protein nanoparticles are used in food production. So, two main areas are food nanosensing and nanostructured food components application of nanotechnology in the food industry.

Food formulations and food packaging are examples of nanostructured food ingredients. Nanodetection of food products is used to obtain improved quality and safety assessment results. Nanopolymers are replacing traditional materials in food packaging [8]. Modern nanotechnology capabilities allow the food sector to create new healthy food products. Another possible application of nanotechnology is the development of new nanomaterial-based engineering methods for the targeted distribution of nutrients and bioactive ingredients in functional foods [9]. Some food components have special properties because they improve texture, consistency, taste, and other aspects at the nanoscale; these components are nanostructures. Improving shelf life has been the subject of many forms of food nanotechnology [26]. Nanocarriers are now used in food production in the same way as delivery structures for various food flavors without changing the shape (Fig. 10.16) [10].

10.6.1 Various Aspects and Functions of Nanotechnology in the Food Sector

The food business is increasingly interested in nanotechnology. Numerous applications have been documented in various fields, including targeted delivery of nutrients and/or bioactive molecules through nanoencapsulation, the use of biosensors for the detection and quantification of pathogens and modification of food composition and the preservation of fruits and vegetables using edible films [11]. The food business is benefiting from improved production methods brought about by nanotechnology, which imparts new and improved characteristics to products. Nanotechnology can be used in production to promote plant development through technical innovations in precision agriculture, as well as for pest detection, pest resistance and allelopathy. Conventional insecticides, fertilizers and herbicides can be nanoencapsulated to: (i) provide slower and longer-lasting action than agrochemical products, (ii) allow for

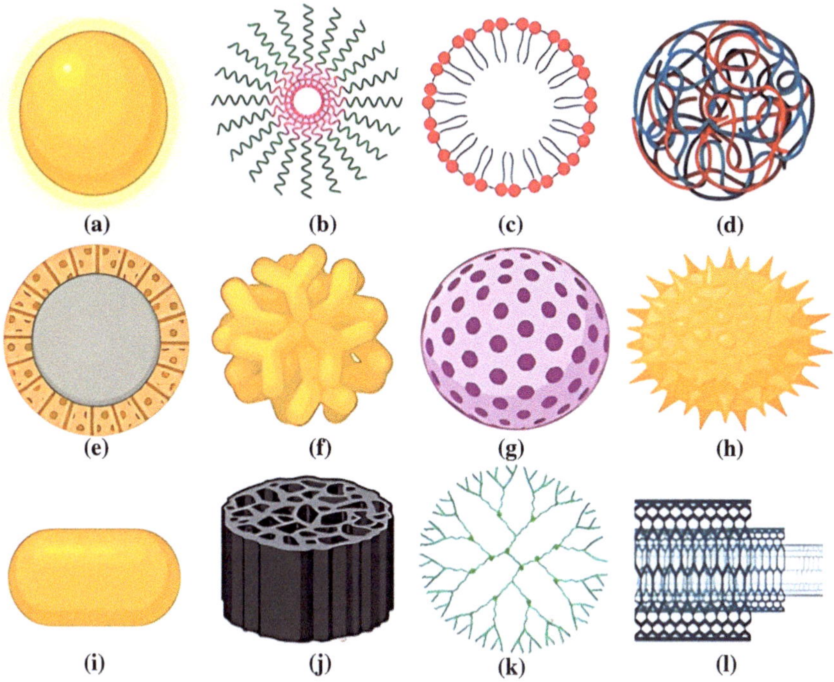

Fig. 10.16 Graphic representation of several types of nanomaterials used in the food sector. These nanoparticles are **a** metallic; **b** polymeric micelles; **c** liposomes; **d** polymeric; **e** mesoporous with a hard core; **f** branched gold; **g** mesoporous; **h** surface-functionalized; **i** nanorods; **j** porous silica nanoparticles; **k** dendrimers; and **l** carbon nanotubes [10] [Open access]

safer handling, more efficient use, and more precise dosing of these compounds with less environmental impact, which guarantees better environmental protection. In addition, there is a growing interest in the use of nanotechnology for rapid and early diagnosis of plant diseases. A detailed discussion of possible applications and benefits of nanotechnology in precision agriculture is elaborated in [27].

In addition, nanotechnology is used in food processing and preservation. This program provides protection that hides different aromas and tastes. In addition, controlled release and improved quality of food ingredients and water-insoluble food additives are possible through nanodispersibility technology [23]. Creating new tastes and sensations, encapsulating food additives, controlling the release of enhanced bioavailability flavors in food additives and creating the complete texture of food ingredients are just some of the benefits of using nanotechnology in the food industry. By creating new delivery routes for the synthesis of nano-agrochemicals, enhancing nutritional quality and creating new products through bioactive encapsulation, nanotechnology has completely transformed the food sector.

Nanoencapsulation increases the nutritional value of the product [28]. Processing food ingredients at the nanoscale not improves nutritional quality but also opens up new flavor and texture possibilities. Compared to traditional encapsulation methods, nanoparticles provide better product encapsulation properties and higher release efficiency. Nanomaterial encapsules protect foods from heat or moisture while controlling the active release of food ingredients and interaction with other components of the food matrix. For example, hydrophobic compounds are delivered via liposomes and oral bioavailability is enhanced by nanoemulsions [26]. Using nanocochleates, it is possible to obtain food with nutrients more efficiently and without compromising taste and color [24].

On another hand, expanded micelles of self-assembled nanoscale liquid structures less than 30 nm in size are used as carriers. They are used to distribute beverages and food additives. The study focuses on the potential use of lycopene, isoflavones, omega-3 fatty acids, and β-carotene. The American company Fresh Corporation has developed a new nanoceramic that allows nano-drying of food, reducing the amount of oil used due to its large surface area in fast food and snack bars [25]. As a result, nanoencapsulation allows a product to be precisely (bio)available at a given rate for a given period [27].

In many different industries, the incorporation of nanoparticles into composite materials known as nanocomposites, the results in significant improvements in quality and performance. These advanced materials offer several important benefits. Significant improvements in material quality are provided by nanocomposites, which consist of nanoparticles (NPs) embedded in a base matrix such as metals, ceramics or polymers. Nanocomposites increase the strength and durability by utilizing the special properties of nanoparticles, which increases the flexibility and elasticity of the base materials. In addition, NPs improve their electrical and thermal properties, allowing them to be used in increasingly complex and challenging applications. The service life of materials is further increased by their increased resistance to wear and corrosion.

Nanocomposites are essential for many different sectors. They help make solar panels, batteries, sensors, and connectors in electronics more efficient. They improve the quality of insulation and concrete in construction, which helps advance infrastructure and building technology. Nanocomposites are used in medical applications to improve the functionality and durability of implants and prostheses. Nanocomposites have enormous potential for technological development and innovation in various fields, contributing to breakthroughs in a wide range of industries due to their ability to significantly improve the properties of materials.

Problems

1. Explain advanced oxidation techniques in nanomaterials
2. How to change the surfaces of the parts without changing the composition of the material.
3. What are the different aspects and functions of nanotechnology in the food sector?

4. Describe the heterogeneous structures of Dascop, binary and ternary systems, and metal oxides.
5. When two metal elements are pressed together, why is there no obvious surface adhesion in nanomaterials?

References

1. Joshi, J., Homburg, S. V., & Ehrmann, A. (2022). Atomic force microscopy (AFM) on biopolymers and hydrogels for biotechnological applications—possibilities and limits. *Polymers, 14*(6), 1267.
2. Liu, J., Xu, R., Zhu, Y., Yang, D. Q., Nie, H. Y., & Lau, W. M. (2022). AFM/XPS analysis of the growth and architecture of oriented molecular monolayer by spin cast process and its cross-linking induced by hyperthermal hydrogen. *Applied Sciences, 12*(12), 6233.
3. Kausar, A., & Bocchetta, P. (2022). Polymer/graphene nanocomposite membranes: Status and emerging prospects. *Journal of Composites Science, 6*(3), 76.
4. Bussetti, G., Yivlialin, R., Goletti, C., Zani, M., & Duò, L. (2019). Temperature effects on the HOPG intercalation process. *Condensed Matter, 4*(1), 23.
5. Kim, J. H., Vinothkannan, M., Kim, A. R., & Yoo, D. J. (2020). Anion exchange membranes obtained from poly (arylene ether sulfone) block copolymers comprising hydrophilic and hydrophobic segments. *Polymers, 12*(2), 325.
6. Gorobtsov, P. Y., Grigoryeva, M. K., Simonenko, T. L., Simonenko, N. P., Simonenko, E. P., & Kuznetsov, N. T. (2023). Application of the [WO2 (C5H7O2) 2] complex in hydrothermal synthesis of WO3 film and study of its electrochromic properties. *Applied Sciences, 13*(15), 8729.
7. Mussa, M., Shtawa, A., & Takita, S. (2023). The influence of adding silica fluoroalkylsilane on the morphology, mechanical, and corrosion resistance properties of sol-gel derived coatings. *Engineering proceedings, 56*(1), 98.
8. Karbalaei Akbari, M., Siraj Lopa, N., & Zhuiykov, S. (2023). Sonochemistry of liquid-metal galinstan toward the synthesis of two-dimensional and multilayered gallium-based metal-oxide photonic semiconductors. *Micromachines, 14*(6), 1214.
9. Wood, J., Bright, R., Palms, D., Barker, D., & Vasilev, K. (2024). Damage behavior with atomic force microscopy on anti-bacterial nanostructure arrays. *Nanomaterials, 14*(3), 253.
10. Shafiq, M., Anjum, S., Hano, C., Anjum, I., & Abbasi, B. H. (2020). An overview of the applications of nanomaterials and nanodevices in the food industry. *Foods, 9*(2), 148.
11. Jing, Z., Dai, X., Xian, X., Zhang, Q., Zhong, H., & Li, Y. (2020). Novel Ternary heterogeneous reduction graphene oxide (RGO)/BiOCl/TiO2 nanocomposites for enhanced adsorption and visible-light induced photocatalytic activity toward organic contaminants. *Materials, 13*(11), 2529.
12. Kumar, S., Himanshi, Prakash, J., Verma, A., Suman, Jasrotia, R., & Ahmed, J. et al. (2023). A review on properties and environmental applications of graphene and its derivative-based composites. *Catalysts, 13*(1), 111.
13. Glaze, W. H., Kang, J. W., & Chapin, D. H. (1987). The chemistry of water treatment processes involving ozone, hydrogen peroxide, and ultraviolet radiation. *Ozone Science and Engineering, 9*, 335–342.
14. Chamarro, E., Marco, A., & Esplugas, S. (2001). Use of Fenton reagent to improve organic chemical bio-degradability. *Water Research, 35*, 1047–1051.
15. Fujishima, K. (1972). Honda, Electrochemical photolysis of water at a semiconductor electrode. *Nature, 238*, 37–38.

16. Spengler, C., Nolle, F., Mischo, J., Faidt, T., Grandthyll, S., Thewes, N., Koch, M., Müller, F., Bischoff, M., Klatt, M. A., et al. (2019). Strength of bacterial adhesion on nanostructured surfaces quantified by substrate morphometry. *Nanoscale, 11*, 19713–19722.
17. Bright, R., Hayles, A., Wood, J., Palms, D., Brown, T., Barker, D., & Vasilev, K. (2022). Surfaces containing sharp nanostructures enhance antibiotic efficacy. *Nano Letters, 22*, 6724–6731.
18. Del Valle, A., Torra, J., Bondia, P., Tone, C. M., Pedraz, P., Vadillo-Rodriguez, V., & Flors, C. (2020). Mechanically induced bacterial death imaged in real time: A simultaneous nanoindentation and fluorescence microscopy study. *ACS Applied Materials & Interfaces, 12*, 31235–31241.
19. Shi, Y., Cai, M., Zhou, L., & Wang, H. (2020). Measurement of mechanical properties of naked cell membranes using atomic force microscope puncture test. *Talanta, 210*, Article 120637.
20. Gulati, K., & Adachi, T. (2023). Profiling to Probing: Atomic force microscopy to characterize nano-engineered implants. *Acta Biomaterialia, 170*, 15–38.
21. Bright, R., Hayles, A., Wood, J., Ninan, N., Palms, D., Visalakshan, R. M., Burzava, A., Brown, T., Barker, D., & Vasilev, K. (2022). Bio-inspired nanostructured Ti-6Al-4V alloy: The role of two alkaline etchants and the hydrothermal processing duration on antibacterial activity. *Nanomaterials, 12*, 1140.
22. Bright, R., Fernandes, D., Wood, J., Palms, D., Burzava, A., Ninan, N., Brown, T., Barker, D., & Vasilev, K. (2022). Long-term antibacterial properties of a nanostructured titanium alloy surface: An in vitro study. *Mater Today Bio, 13*, Article 100176.
23. Bright, R., Hayles, A., Fernandes, D., Visalakshan, R. M., Ninan, N., Palms, D., Burzava, A., Barker, D., Brown, T., & Vasilev, K. (2021). In vitro bactericidal efficacy of nanostructured Ti6Al4V surfaces is bacterial load dependent. *ACS Applied Materials & Interfaces, 13*, 38007–38017.
24. Wood, J., Hayles, A., Bright, R., Palms, D., Vasilev, K., & Hasan, J. (2022). Nanomechanical tribological characterisation of nanostructured titanium alloy surfaces using AFM: A friction vs velocity study. *Colloids and Surfaces. B, Biointerfaces, 217*, Article 112600.
25. Karbalaei Akbari, M., Siraj Lopa, N., & Zhuiykov, S. (2023). Crystalline nanodomains at multifunctional two-dimensional liquid–metal hybrid interfaces. *Crystals, 13*, 604.
26. Chiew, C., Morris, M. J., & Malakooti, M. H. (2021). Functional liquid metal nanoparticles: Synthesis and applications. *Mater Adv, 2*, 7799–7819.
27. Kalantar-Zadeh, K., Tang, J., Daeneke, T., O'Mullane, A. P., Stewart, L. A., Liu, J., Majidi, C., Ruoff, R. S., Weiss, P. S., & Dickey, M. D. (2019). Emergence of liquid metals in nanotechnology. *ACS Nano, 13*, 7388–7395.
28. Rahman, M. M., Hosano, N., & Hosano, H. (2022). Recovering microalgal bioresources: A review of cell disruption methods and extraction technologies. *Molecules, 27*(9), 2786.

Index

R. A. Al-Samarai and Y. Al-Douri, *Engineering of Nanomaterials*,
https://doi.org/10.1007/978-981-95-5499-7

MIX
Papier aus verantwortungsvollen Quellen
Paper from responsible sources
FSC® C105338

If you have any concerns about our products,
you can contact us on
ProductSafety@springernature.com

In case Publisher is established outside the EU,
the EU authorized representative is:
Springer Nature Customer Service Center GmbH
Europaplatz 3, 69115 Heidelberg, Germany

Printed by Libri Plureos GmbH
in Hamburg, Germany